Lecture Notes of
the Unione Matematica Italiana

3

Editorial Board

The Editorial Policy can be found at the back of the volume.

Luc Tartar

An Introduction to Sobolev Spaces and Interpolation Spaces

 Springer

Author

Luc Tartar
Department of Mathematical Sciences
Carnegie Mellon University
Pittsburgh, PA 15213-3890
USA
e-mail: tartar@andrew.cmu.edu

Library of Congress Control Number: 2007925369

Mathematics Subject Classification (2000): 35-XX, 46-xx, 46B70, 46M35

ISSN print edition: 1862-9113
ISSN electronic edition: 1862-9121

DOI 10.1007/978-3-540-71483-5

In memory of Sergei SOBOLEV, 1908–1989

He pioneered the study of some functional spaces which are crucial in the study of the partial differential equations of continuum mechanics and physics, and the first part of these lecture notes is about these spaces, named after him.

In memory of Jacques-Louis LIONS, 1928–2001

He participated in the development of Sobolev spaces, in part with Enrico MAGENES, applying the general theory of interpolation spaces which he had developed with Jaak PEETRE, who further simplified the theory so that it became more easy to use, and the second part of these lecture notes is about these interpolation spaces.

To Lucia

To my children
Laure, Michaël, André, Marta

Preface

After publishing an introduction to the Navier[1]–Stokes[2,3] equation and oceanography [18], the revised version of my lecture notes for a graduate course that I had taught in the spring of 1999, I want to follow with another set of lecture notes for a graduate course that I had taught in the spring of 2000; that course was divided into two parts, the first part on Sobolev[4] spaces, and the second part on interpolation spaces. The first version had been available on the Internet, and after a few years, I find it useful to make the text available to a larger audience by publishing a revised version.

When I was a student at Ecole Polytechnique, which was still in Paris, France, on the "Montagne Sainte Geneviève",[5] I had the chance to have

---

[1] Claude Louis Marie Henri NAVIER, French mathematician, 1785–1836. He worked in Paris, France.

[2] Sir George Gabriel STOKES, Irish-born mathematician, 1819–1903. He worked in London, and in Cambridge, England, holding the Lucasian chair (1849–1903).

[3] Reverend Henry LUCAS, English clergyman and philanthropist, 1610–1663.

[4] Sergei L'vovich SOBOLEV, Russian mathematician, 1908–1989. He worked in Leningrad, in Moscow, and in Novosibirsk, Russia. I first met him when I was a student, in Paris in 1969, then at the International Congress of Mathematicians in Nice in 1970, and conversed with him in French, which he spoke perfectly (all educated Europeans did learn French in the beginning of the 20th century). I only met him once more, when I traveled with a French group from INRIA (Institut National de la Recherche en Informatique et Automatique) in 1976 to Akademgorodok near Novosibirsk, Russia, where he worked. There is now a Sobolev Institute of Mathematics of the Siberian branch of the Russian Academy of Sciences, Novosibirsk, Russia.

[5] Geneviève, patroness of Paris, c 419 or 422–512.

Laurent SCHWARTZ[6-8] as my main teacher in mathematics in the first year
(1965–1966), and the course contained an introduction[9] to his theory of distributions,[10] but I only heard about Sobolev spaces in my second year (1966–
1967), in a seminar organized by Jacques-Louis LIONS[11-13] for interested students, in addition to his course on numerical analysis. I learnt a little more in
his courses at the university in the following years, and I read a course [13]
that he had taught in 1962 in Montréal, Québec (Canada), and I also read a
book [1] by Shmuel AGMON,[14] corresponding to a course that he had taught
at Rice[15] University, Houston, TX.

[6] Laurent SCHWARTZ, French mathematician, 1915–2002. He received the Fields
Medal in 1950. He worked in Nancy, in Paris, France, at École Polytechnique,
which was first in Paris (when I had him as a teacher in 1965–1966), and then in
Palaiseau, France, and at Université Paris VII (Denis Diderot), Paris, France.

[7] John Charles FIELDS, Canadian mathematician, 1863–1932. He worked in
Meadville, PA, and in Toronto, Ontario (Canada).

[8] Denis DIDEROT, French philosopher and writer, 1713–1784. He worked in Paris,
France, and he was the editor-in-chief of the Encyclopédie. Université Paris 7,
Paris, France, is named after him.

[9] Which means that he only considered questions of convergence for sequences, and
he did not teach anything about the topologies of $\mathcal{D}$ or $\mathcal{D}'$, which I first learnt in
his book [15].

[10] Laurent SCHWARTZ has described something about his discovery of the concept
of distributions in his biography [16].

[11] Jacques-Louis LIONS, French mathematician, 1928–2001. He received the Japan
Prize in 1991. He worked in Nancy and in Paris, France, holding a chair (analyse
mathématique des systèmes et de leur contrôle, 1973–1998) at Collège de France,
Paris, France. I first had him as a teacher at Ecole Polytechnique in 1966–1967,
and I did research under his direction, until my thesis in 1971. The laboratory
dedicated to functional analysis and numerical analysis which he initiated, funded
by CNRS (Centre National de la Recherche Scientifique) and Université Paris VI
(Pierre et Marie Curie), is now named after him, the Laboratoire Jacques-Louis
Lions.

[12] Pierre CURIE, French physicist, 1859–1906, and his wife Marie SKŁODOWSKA-
CURIE, Polish-born physicist, 1867–1934, jointly received the Nobel Prize in
Physics in 1903, and she also received the Nobel Prize in Chemistry in 1911.
They worked in Paris, France. Université Paris 6, Paris, France, is named after
them.

[13] Alfred NOBEL, Swedish industrialist and philanthropist, 1833–1896. He created
a fund to be used as awards for people whose work most benefited humanity.

[14] Shmuel AGMON, Israeli mathematician, born in 1922. He worked at The Hebrew
University, Jerusalem, Israel.

[15] William Marsh RICE, American financier and philanthropist, 1816–1900.

I first read about interpolation spaces (in a Hilbert[16,17] setting) in a book that Jacques-Louis LIONS had written with Enrico MAGENES[18] [14], and then he gave me his article with Jaak PEETRE[19] to read for the theory in a Banach[20,21] setting, and later he asked me to solve some problems about interpolation for my thesis in 1971, and around that time I did read a few articles on interpolation, although I can hardly remember in which of the many articles of Jaak PEETRE I may have read about some of his results. For the purpose of this course, I also consulted a book by BERGH[22,23] & LÖFSTRÖM[24] [2].

I also learnt in other courses, by Jacques-Louis LIONS or others, in seminars, and the usual process went on, learning, forgetting, inventing a new proof or reinventing one, when asked a question by a fellow researcher or a student, so that for many results in this course I can hardly say if I have read them or filled the gaps in statements that I had heard, and my memory may be inaccurate on some of these details. Some of the results may have been obtained in my own research work, which is concerned with partial differential equations from continuum mechanics or physics, and my personal reason for being interested in the subject of this course is that some of the questions studied have appeared in a natural way in a few practical problems. Of course, although a few problems of continuum mechanics or physics have led to some of the mathematical questions described in this course, many have been added for the usual reason that mathematicians are supposed to discover general structures hidden behind particular results, and describe something

---

[16] David HILBERT, German mathematician, 1862–1943. He worked in Königsberg (then in Germany, now Kaliningrad, Russia) and in Göttingen, Germany. The term Hilbert space was coined by his student VON NEUMANN, when he worked on his mathematical foundation of quantum mechanics.

[17] János (John) VON NEUMANN, Hungarian-born mathematician, 1903–1957. He worked in Berlin, in Hamburg, Germany, and at IAS (Institute for Advanced Study), Princeton, NJ.

[18] Enrico MAGENES, Italian mathematician, born in 1923. He worked in Pavia, Italy.

[19] Jaak PEETRE, Estonian-born mathematician, born in 1935. He worked in Lund, Sweden.

[20] Stefan BANACH, Polish mathematician, 1892–1945. He worked in Lwów (then in Poland, now Lvov, Ukraine). There is a Stefan Banach International Mathematical Center in Warsaw, Poland. The term Banach space was introduced by FRÉCHET.

[21] Maurice René FRÉCHET, French mathematician, 1878–1973. He worked in Poitiers, in Strasbourg and in Paris, France. I do not know who introduced the term Fréchet space.

[22] Jöran BERGH, Swedish mathematician, born in 1941. He has worked in Lund, and at Chalmers University of Technology, Göteborg, Sweden.

[23] William CHALMERS Jr., Swedish merchant, 1748–1811.

[24] Jörgen LÖFSTRÖM, Swedish mathematician, born in 1937. He worked at Chalmers University of Technology, Göteborg, Sweden.

more general after having done a systematic study, akin to a cleaning process. For those who do not yet know much about continuum mechanics or physics, I recommend looking first at more classical descriptions of the problems, for example by consulting the books which have been prepared under the direction of Robert DAUTRAY[25] and Jacques-Louis LIONS [4–12]. For those who already know something about continuum mechanics or physics, I recommend looking at my other lecture notes for reading about the defects which I know about classical models, because other authors rarely mention these defects even when they have heard about them: I suppose that it is the result of having been raised as the son of a (Calvinist) Protestant minister that I learnt and practiced the point of view that one should not follow the path of the majority when reason clearly points to a different direction. However, although I advocate using reason for criticizing without concessions the points of view that are taught in order to find better "truths", one should observe that this approach is more suited to mathematicians than to engineers or physicists; actually, not all "mathematicians" have been trained well enough for following that path, and that might explain why some people initially trained as mathematicians write inexact statements, which they often do not change even after being told about their mistakes, which others repeat then without knowing that they propagate errors; if their goal had not been to mislead others, a better strategy would have been to point out that some statements were only conjectures.

I have decided to write my lecture notes with some information given in footnotes about the people who have participated in the creation of the knowledge related to the subject of the course, and I have mentioned in [18] a few reasons for doing that: I had great teachers[26] like Laurent SCHWARTZ and Jacques-Louis LIONS, and I have met many mathematicians, for whom I use their first names in the text, but I have tried to give some simple biographical data for all people quoted in the text in order to situate them both in time and in space, the famous ones as well as the almost unknown ones; I have seen so many ideas badly attributed and I have tried to learn more about the mathematicians who have introduced some of the ideas which I was taught when I was a student, and to be as accurate as possible concerning the work of all.[27] Another reason is that I enjoy searching for clues, even about questions that might be thought irrelevant for my goals; I might be stopped by a word,

---

[25] Ignace Robert DAUTRAY (KOUCHELEVITZ), French physicist, born in 1928.

[26] Although I immediately admired their qualities, like pedagogical skill, I later became aware of some of their defects, the discussion of which I shall postpone until I decide to publish all the letters that I wrote to them.

[27] Although I have never read much, it would be quite inefficient for me to change my method of work for the moment, because too many people have recently shown a tendency to badly quote their sources. In some cases, information that I had proven something in the 1970s has been ignored, for the apparent reason that I had told that to people who wanted to avoid mentioning my name, the strange thing being that instead of trying to find someone who would have done similar

wondering about its etymology, or by a new name, wondering about who this person was, or even by a name which has been attached to a well-known institution and I want to discover who was that forgotten person in honor of whom the institution is named; the Internet has given me the possibility to find such answers, sometimes as the result of many searches which had only given small hints, and I hope that I shall be told about all the inaccuracies that are found in my text.

I was glad to learn a few years ago the motto of Hugo of Saint Victor[28] "Learn everything, and you will see afterward that nothing is useless", and to compare it with what I had already understood in my quest about how creation of knowledge occurs. I have often heard people say about a famous physicist from the past, that luck played an important role in his discovery, but the truth must be that if he had not known beforehand all the aspects of his problem he would have missed the importance of the new hint that had occurred, and so this instance of "luck" reminds me of the saying "aide toi, le ciel t'aidera" (God helps those who help themselves). Those who present chance as an important factor in discovery probably wish that every esoteric subject that they like be considered important and funded, but that is not at all what the quoted motto is about. My reasons for publishing lecture notes is to tell the readers some of what I have understood; the technical mathematical aspects of the course are one thing, the scientific questions behind the theories are another, but there is more than that, a little difficult to express in words: I will have succeeded if many become aware, and go forward on the path of discovery, not mistaking research and development, knowing when and why they do one or the other, and keeping a higher goal in mind when for practical reasons they decide to obey the motto of the age for a while, "publish or perish".

When I was a graduate student in Paris, my advisor invited me a few times to join a dinner held for a visitor, who had usually talked in the Lions–Schwartz seminar, which met every Friday at IHP (Institut Henri Poincaré[29]). It was before Université de Paris split into many smaller universities, which happened in 1970 or 1971, and I had heard my advisor mention a special fund from ZAMANSKI,[30] the dean of "Faculté des Sciences". The buildings for sciences were then known as "Halle aux Vins", because they were being built on a place previously used for the wine market, and it was only after all the wine merchants had moved to Bercy, on the other bank of the river

---

work before me they sometimes preferred to quote one of their friends who had used the result in the 1990s, without any mention of an author for it.

[28] Hugo VON BLANKENBURG, German-born theologian, 1096–1141. He worked at the monastery of Saint Victor in Paris, France.

[29] Jules Henri POINCARÉ, French mathematician, 1854–1912. He worked in Paris, France. There is an Institut Henri Poincaré (IHP), dedicated to mathematics and theoretical physics, part of Université Paris VI (Pierre et Marie Curie), Paris, France.

[30] Marc ZAMANSKI, French mathematician, 1915–1996. He worked in Paris, France.

Seine, that the complex of buildings became known as "Jussieu",[31] and I wonder if part of the plan was not to support the local restaurants, who were losing a lot of their customers because of the transfer of the wine market to Bercy. It was at one of these dinners, which all took place in the restaurant "Chez Moissonnier", rue des fossés Saint Bernard,[32-34] that I first met Sergei SOBOLEV, probably in 1969, but I had not been aware that he had given a talk; my understanding of English was too poor at the time for conversing with visitors, but fortunately Sergei SOBOLEV spoke French, perfectly.[35]

I met Sergei SOBOLEV a second time, at the International Congress of Mathematicians in Nice in 1970, and as he was waiting in front of me in a line at a cafeteria, I took the occasion to ask him a question, about what he thought were interesting mathematical areas to study, and I must have mentioned applications, because his answer was that it was difficult to know what could become useful, as even questions in number theory had found applications.

I met Sergei SOBOLEV a third time in 1976, when I traveled to Novosibirsk with a group from INRIA (Institut National de Recherche en Informatique[36,37] et Automatique[38]), and he was working then on cubature formulas [17], i.e., quadrature formulas in three dimensions, a subject which I did not find interesting enough to enquire about it. Apart from the fact that he seemed eager not to be too involved with the political establishment, which may explain why he worked alone, the subject may have actually been more important in the Soviet Union than in the West, as I learnt during the same

---

[31] Antoine Laurent DE JUSSIEU, French botanist, 1748–1836. He worked in Paris, France.

[32] Bernard DE FONTAINES, French monk, 1090–1153. He founded the monastery of Clairvaux, France, and is known as Saint Bernard de Clairvaux. He was canonized in 1174 by ALEXANDER III, and PIUS VIII bestowed on him the title of Doctor of the Church in 1830.

[33] Orlando BANDINELLI, Italian Pope, 1105–1181. Elected Pope in 1159, he took the name ALEXANDER III.

[34] Francesco Xaverio CASTIGLIONE, Italian Pope, 1761–1830. Elected Pope in 1829, he took the name PIUS VIII.

[35] Until the beginning of the 20th century, every educated person in Europe learnt French. I was told that Sergei SOBOLEV was born into an aristocratic family, and that without the 1917 revolution in Russia he would have become a duke.

[36] Informatique is the French word for computer science, and ordinateur is the French word for computer, but these words were in use much before DE GAULLE created a special committee for coining French words that had to be used in replacement of the American words invented in technology.

[37] Charles DE GAULLE, French general and statesman, 1890–1970. Elected President of the Republic in 1959 (by the two legislative chambers), he had then a new constitution for France accepted (5th republic), and he was reelected by direct election in 1965; he resigned in 1969.

[38] Automatique is the French word for control theory.

trip, when my good friend Roland GLOWINSKI[39] told me about some discussions with Nicolai YANENKO,[40] who had said that a numerical scheme mentioned by the French team did not work; together with Jean CÉA,[41] he was trying to understand why the scheme did not work on Russian computers, while it worked well in France (although it was not a very efficient scheme), and after a few days, the explanation was found, which was that the Russian computers did not have double precision, a feature which had existed for some time on the American computers used in France, but such computers could not be imported into the Soviet Union, because of the embargo decided by United States, a consequence of the Cold War; it could have been precisely the limitations of the computers which made the study of good cubature formulas useful.

In his description of lives of great men, PLUTARCH[42] told a story about ARCHIMEDES[43,44] and Cicero.[45] At the time when Cicero became governor of Sicily, he wanted to visit the tomb of ARCHIMEDES, and the people of Syracuse had no idea where it was, but Cicero knew something about the tomb, which permitted him to discover it: ARCHIMEDES had wanted to have on his tomb a reminder of what he thought was his best result, that the surface of a sphere of radius R is equal to the lateral surface of a tangent cylinder of same height (i.e., with a circular base of radius R and height $2R$), which is then $4\pi R^2$, so Cicero's aides just had to go around the cemeteries of Syracuse and look for a tomb with a sphere and a cylinder on it.

Apparently, there was no mathematical result that Jacques-Louis LIONS was really proud of having proven, because after his death people who had been in contact with him insisted that what he had been most proud of was one of his successes in manipulating people. In 1984, in a discussion with Laurent SCHWARTZ, I had said that I was not good at following complicated proofs and he had said that one cannot follow another person's mind and that only Jacques-Louis LIONS was capable of following every proof. Roger

---

[39] Roland GLOWINSKI, French-born mathematician, born in 1937. He worked at Université Paris VI (Pierre et Marie Curie), Paris, France, and he works now in Houston, TX.

[40] Nikolai Nikolaevich YANENKO, Russian mathematician, 1921–1984. He worked in Novosibirsk, Russia.

[41] Jean CÉA, French mathematician, born in 1932. He worked in Rennes, and in Nice, France.

[42] Mestrius PLUTARCHUS, Greek biographer, 46–120.

[43] ARCHIMEDES, Greek mathematician, 287 BCE–212 BCE. He worked in Siracusa (Syracuse), then a Greek colony, now in Italy.

[44] BCE = before common era; those who insist in linking questions of date with questions of religion may consider that it means "before Christian era".

[45] Marcus TULLIUS Cicero, Roman orator and politician, 106 BCE–43 BCE.

PENROSE[46-48] expresses the same fact that in listening to someone he actually tries to guess what the other person says, because he cannot in general follow the details of the other person's reasoning, which is the way I feel; however, he wrote that in a book whose whole argument looked quite silly to me (and Jacques-Louis LIONS also made the same comment about the book), but I read it entirely because my good friend Constantine DAFERMOS[49-51] had offered it to me, which I interpreted as meaning that I should have a critical opinion about "artificial intelligence". What had prompted that information from Laurent SCHWARTZ about Jacques-Louis LIONS was that I always insist that my proofs are not difficult (and that the importance is more in the analysis about what kind of result to look for), because I always try to simplify what I have done, so that it can be easily understood; Jacques-Louis LIONS had this quality of looking for simplifying proofs, and he sometimes asked me to find a general proof after he had obtained a particular result whose proof looked much too complicated to him to be the "right one".

I recall a remark of Jacques-Louis LIONS that a framework which is too general cannot be very deep, and he had made this comment about semi-group theory; he did not deny that the theory is useful, and the proof of the Hille[52,53]–Yosida[54] theorem is certainly more easy to perform in the abstract setting of a Banach space than in each particular situation, but the result applies to equations with very different properties that the theory cannot

---

[46] Sir Roger PENROSE, English mathematician, born in 1931. He received the Wolf Prize (in Physics!) in 1988. He has worked in London and in Oxford, England, where he held the Rouse Ball professorship.

[47] Ricardo WOLF, German-born diplomat and philanthropist, 1887–1981. He emigrated to Cuba before World War I; from 1961 to 1973 he was Cuban Ambassador to Israel, where he stayed afterwards. The Wolf foundation was established in 1976 with his wife, Francisca SUBIRANA-WOLF, 1900–1981, "to promote science and art for the benefit of mankind".

[48] Walter William Rouse BALL, English mathematician, 1850–1925. He worked in Cambridge, England.

[49] Constantine M. DAFERMOS, Greek-born mathematician, born in 1941. He worked at Cornell University, Ithaca, NY, and he works now at Brown University, Providence, RI.

[50] Ezra CORNELL, American philanthropist, 1807–1974.

[51] Nicholas BROWN Jr., American merchant, 1769–1841.

[52] Einar Carl HILLE (HEUMAN), Swedish-born mathematician, 1894–1980. He worked in Princeton, NJ, and at Yale University, New Haven, CT.

[53] Elihu YALE, American-born English philanthropist, Governor of Fort St George, Madras, India, 1649–1721.

[54] Kôsaku YOSIDA, Japanese mathematician, 1909–1990. He worked in Tokyo, Japan, where I met him during my first trip to Japan, in the fall of 1976.

distinguish. After having written an article with Jean LERAY,[55,56] who had
wanted to use the generalization of the Brouwer[57] topological degree that he
had obtained with SCHAUDER,[58] Jacques-Louis LIONS had observed that the
regularity hypotheses for applying the Leray–Schauder theory are too strong
and that a better approach is to use the Brouwer topological degree for finite-
dimensional approximations, and then pass to the limit, taking advantage of
the particular properties of the problem for performing that limiting process.
Maybe Jacques-Louis LIONS also thought that the theory of interpolation
spaces lacks depth because it is a general framework based on two arbitrary
Banach spaces, but certainly the theory is worth studying, and I decided to
teach it after the first part on Sobolev spaces for that reason.

In an issue of the Notices of the American Mathematical Society, Enrico
MAGENES had recalled the extreme efficiency of Jacques-Louis LIONS when
they worked together, in particular for the study of their famous interpolation
space $H_{00}^{1/2}(\Omega)$, but Peter LAX[59,60] had recalled an interesting encounter,
showing that early in his career Jacques-Louis LIONS was already playing
the functional analysis card against continuum mechanics. I had opposed my
former advisor on this question, because I wanted to understand more about
continuum mechanics and physics, and to develop new tools for going further
in the study of the partial differential equations of continuum mechanics or
physics, and in particular to show the limitations of the classical tools from
functional analysis, but despite our opposition on these important questions,
I have chosen to dedicate these lecture notes to Jacques-Louis LIONS, because
he played an important role in the development of the theory of interpolation
spaces, although the theory would have been quite difficult to use without the
simplifying work of Jaak PEETRE.

Jaak PEETRE wrote to me a few years ago that he had obtained the same
results on interpolation as Jacques-Louis LIONS, who had kindly proposed
that they publish an article together. It seems to me that Jaak PEETRE was

[55] Jean LERAY, French mathematician, 1906–1998. He shared the Wolf Prize in
1979 with WEIL. He worked in Nancy, France, in a prisoner of war camp in
Austria (1940–1945), and in Paris, France; he held a chair (théorie des équations
différentielles et fonctionnelles, 1947–1978) at Collège de France, Paris, France.
[56] André WEIL, French-born mathematician, 1906–1998. He received the Wolf Prize
in 1979 (shared with Jean LERAY). He worked in Aligarh, India, in Haverford,
PA, in Swarthmore, PA, in São Paulo, Brazil, in Chicago, IL, and at IAS (Institute
for Advanced Study), Princeton, NJ.
[57] Luitzen Egbertus Jan BROUWER, Dutch mathematician, 1881–1966. He worked
in Amsterdam, The Netherlands.
[58] Juliusz Pawel SCHAUDER, Polish mathematician, 1899–1943. He worked in Lwów
(then in Poland, now Lvov, Ukraine).
[59] Peter David LAX, Hungarian-born mathematician, born in 1926. He received the
Wolf Prize in 1987, and the Abel Prize in 2005. He works at NYU (New York
University), New York, NY.
[60] Niels Henrik ABEL, Norwegian mathematician, 1802–1829.

following the influence of M. RIESZ[61–65] in Lund, Sweden, while Jacques-Louis LIONS was following the path opened by the characterization of traces of functions in $W^{1,p}$ by Emilio GAGLIARDO,[66] who had also worked with Nachman ARONSZAJN.[67] The interpolation spaces studied in the joint article of Jacques-Louis LIONS and Jaak PEETRE seem to depend upon three parameters, and it is an important simplification of Jaak PEETRE to have shown that they actually only depend upon two parameters, one parameter $\theta \in (0,1)$ which they had already introduced, and a single parameter[68] $p \in [1,\infty]$. The theory of interpolation spaces as it can be taught now thus owes much more to Jaak PEETRE than to Jacques-Louis LIONS, a point which I had not emphasized enough before. In some instances, the name of Jacques-Louis LIONS appears for questions of interpolation related to his joint works with Enrico MAGENES, and that corresponds to applying the already developed theory of interpolation spaces to questions of partial differential equations.

A few years ago, I had needed the support of a few friends to find the strength to decide to revise the lecture notes which I had already written, and to complete the writing of some unfinished ones, in view of publishing them to attain a wider audience. I carried out the first revision of this course in August 2002, and I want to thank Thérèse BRIFFOD for her hospitality at that time.

I would not have been able to complete the publication of my first lecture notes and to feel able to start revising again this second set of lecture notes without the support of Lucia OSTONI, and I want to thank her for that and for much more, giving me the stability that I had lacked so much in the last

---

[61] Marcel RIESZ, Hungarian-born mathematician, 1886–1969 (the younger brother of Frederic RIESZ). He worked in Stockholm and in Lund, Sweden.

[62] Frigyes (Frederic) RIESZ, Hungarian mathematician, 1880–1956. He worked in Kolozsvár (then in Hungary, now Cluj-Napoca, Romania), in Szeged and in Budapest, Hungary. He introduced the spaces L^p in honor of LEBESGUE and the spaces $\mathcal{H}^p$ in honor of HARDY, but no spaces are named after him, and the Riesz operators have been introduced by his younger brother Marcel RIESZ.

[63] Henri Léon LEBESGUE, French mathematician, 1875–1941. He worked in Rennes, in Poitiers, and in Paris, France; he held a chair (mathématiques, 1921–1941) at Collège de France, Paris, France.

[64] Godfrey Harold HARDY, English mathematician, 1877–1947. He worked in Cambridge, in Oxford, England, holding the Savilian chair of geometry in 1920–1931, and in Cambridge again, holding the Sadleirian chair of pure mathematics (established in 1701 by Lady SADLEIR) in 1931–1942.

[65] Sir Henry SAVILE, English mathematician, 1549–1622. He founded professorships of geometry and astronomy at Oxford.

[66] Emilio GAGLIARDO, Italian mathematician, born in 1930. He worked in Pavia, Italy.

[67] Nachman ARONSZAJN, Polish-born mathematician, 1907–1980. He worked in Lawrence, KS, where I visited him during my first visit to United States, in 1971.

[68] Jaak PEETRE has even observed that one can use $p \in (0,\infty]$, and he developed the theory of interpolation for quasi-normed spaces.

twenty-five years, so that I could feel safer in resuming my research of giving a sounder mathematical foundation to 20th century continuum mechanics and physics.

I want to thank my good friends Carlo SBORDONE and Franco BREZZI for having proposed to publish my lecture notes in a series of Unione Matematica Italiana.

Milano,[69] July 2006

Luc TARTAR
Correspondant de l'Académie des Sciences, Paris
University Professor of Mathematics
Department of Mathematical Sciences
Carnegie Mellon University
Pittsburgh, PA 15213-3890
United States of America

---

[69] Two months after writing this preface, I was elected a foreign member of Istituto Lombardo Accademia di Scienze e Lettere, Milano.

Detailed Description of Lectures

a.b refers to definition, lemma or theorem # b in lecture # a, while (a.b) refers to equation # b in lecture # a.

Lecture 1, Historical background: Dirichlet principle: (1.1) and (1.2); Radon measures: (1.3).

Lecture 2, The Lebesgue measure, convolution: Convolution: 2.1 and (2.1)–(2.3); condition on supports: (2.4); operators τ_a: 2.2 and (2.5); commutation of τ_a with convolution: 2.3 and (2.6) and (2.7); an example of a C^∞ function: 2.4 and (2.8).

Lecture 3, Smoothing by convolution: Smoothing sequences: 3.1 and (3.1); convergence of smoothing sequences: 3.2; density of $C_c^\infty(R^N)$: 3.3.

Lecture 4, Truncation; Radon measures; distributions: Truncating sequences: 4.1; smoothing characteristic functions: 4.2 and (4.1); Radon measures and distributions: 4.3 and (4.2)–(4.5); derivation of distributions: 4.4 and (4.6)–(4.8).

Lecture 5, Sobolev spaces; multiplication by smooth functions: $W^{m,p}(\Omega)$: 5.1 and (5.1); $W^{m,p}(\Omega)$ is a Banach space and $H^m(\Omega)$ is a Hilbert space: 5.2 and (5.2) and (5.3); convergences in C_c^∞ and in $\mathcal{D}'$: (5.4) and (5.5); principal value of $\frac{1}{x}$: (5.6)–(5.9); multiplication by smooth functions: 5.3 and (5.10); Leibniz's formula for functions: 5.4 and (5.11) and (5.12), and for distributions: 5.5 and (5.13).

Lecture 6, Density of tensor products; consequences: Tensor product of functions: (6.1); density of tensor products: 6.2 and (6.1)–(6.4); $x_j T = 0$ for all j means $T = C\,\delta_0$: 6.3 and (6.5) and (6.6); Ω connected and $\frac{\partial T}{\partial x_j} = 0$ for all j means T constant: 6.4 and (6.7)–(6.12); $C_c^\infty(R^N)$ is dense in $W^{m,p}(R^N)$ for $1 \le p < \infty$: 6.5 and (6.13) and (6.14); $W_0^{m,p}(\Omega)$: 6.6; Sobolev's embedding theorem: 6.7.

Lecture 7, Extending the notion of support: Being 0 on an open subset: 7.1 and (7.1) and (7.2); partitions of unity: 7.2; being 0 on a union of open subsets: 7.3; support of Radon measures or of distributions: 7.4; $W_{loc}^{m,p}(\Omega)$: 7.5; product of functions in $W^{1,p}(\Omega)$ and $W^{1,q}(\Omega)$: 7.6; Lipschitz functions and $W^{1,\infty}(\Omega)$: 7.7 and 7.8.

Lecture 8, Sobolev's embedding theorem, $1 \le p < N$: A domain with a cusp, where Sobolev's embedding does not hold: 8.1; interpolation inequality for L^p: 8.2 and (8.1); $W^{1,p}(R^N) \subset L^q(R^N)$ implies $p \le q \le p_*$: 8.3 and (8.2)–(8.6); elementary solutions: 8.4; the Laplacian and its elementary solution: (8.7) and (8.8); the formula used by SOBOLEV: (8.9); $W^{1,1}(R) \subset C_0(R)$: 8.5 and (8.10)–(8.12); a lemma used by GAGLIARDO and by NIRENBERG: 8.6 and (8.13)–(8.16); estimates in various $L^q(R^N)$ spaces by the Gagliardo–Nirenberg method: 8.7 and (8.17)–(8.20).

Lecture 9, Sobolev's embedding theorem, $N \le p \le \infty$: Sobolev's embedding theorem estimates for the case $p = N$: 9.1 and (9.1)–(9.5), and for the case $p > N$: (9.6); estimates using a parametrix of the Laplacian: (9.7)–(9.9); $W^{1,p}(R^N) \subset C^{0,\gamma}(R^N)$ for $p > N$: 9.2 and (9.10)–(9.13).

Lecture 28, Bilinear and nonlinear interpolation: A nonlinear interpolation result: 28.1 and (28.1) and (28.2); a bilinear interpolation theorem of J.-L. LIONS: 28.2 and (28.3) and (28.4), and one of J.-L. LIONS & PEETRE: 28.3 and (28.5) and (28.6).

Lecture 29, Obtaining L^p by interpolation, with the exact norm: A variant of $K(t; f)$: 29.1 and (29.1), from which the L^p norm can be computed: 29.2 and (29.2)–(29.4); a generalization for Orlicz spaces: (29.5) and (29.6); a precise interpolation property: 29.3 and (29.7) and (29.8); using the nonincreasing rearrangement for evaluating norms in Lorentz spaces: 29.4 and (29.9)–(29.13).

Lecture 30, My approach to Sobolev's embedding theorem: A simple decomposition for obtaining a weak embedding theorem: (30.1)–(30.6), which one first improves by rescaling: (30.7)–(30.9), and then by applying to $\varphi_n(u)$: (30.10)–(30.19).

Lecture 31, My generalization of Sobolev's embedding theorem: The original method of SOBOLEV: (31.1), is not adapted to the case of derivatives in different $L^p(R^N)$ spaces: (31.2), but may be used if all derivatives belong to the same Lorentz space: (31.3); the case $p = 1$ is related to the isoperimetric inequality: (31.4)–(31.6), for which there is an additive and a multiplicative version: 31.1 and (31.7)–(31.12); the case of derivatives in different Lorentz spaces: 31.2 and (31.13)–(31.18), can be interpreted under a natural condition: (31.19); a case with information on various orders of derivatives in different directions: (31.20)–(31.23).

Lecture 32, Sobolev's embedding theorem for Besov spaces: Sobolev's embedding theorem for $H^s(R^N)$: 32.1; Sobolev space $W^{s,p}(R^N)$ and Besov space $B_q^{s,p}(R^N)$: 32.2 and (32.2); $W^{k,p}(R^N)$ is of class $\mathcal{H}\left(\frac{m-k}{m}\right)$ for $E_0 = W^{m,p}(R^N)$ and $E_1 = L^p(R^N)$: (32.3)–(32.5); Sobolev's embedding theorem for $W^{s,p}(R^N)$ and $B_q^{s,p}(R^N)$: (32.6) and (32.7); $\gamma_0\left((H^1(R^N), L^2(R^N))_{1/2,1}\right) = L^2(R^{N-1})$: (32.8); $\gamma_0\left(W^{1,1}(R^N)\right) = L^1(R^{N-1})$: (32.9).

Lecture 33, The Lions–Magenes space $H_{00}^{1/2}(\Omega)$: $H^{1/2}(R)$ has a kind of continuity: 33.1; $H_{00}^{1/2}(R_+)$: 33.2 and (33.1).

Lecture 34, Defining Sobolev spaces and Besov spaces for Ω: Zygmund space: (34.1); two definitions of $W^{s,p}(\Omega)$: (34.2)–(34.4).

Lecture 35, Characterization of $W^{s,p}(R^N)$: $B_\infty^{s,p}(R^N)$: 35.1 and (35.1)–(35.5); $W^{s,p}(R^N)$: 35.2 and (35.6)–(35.12).

Lecture 36, Characterization of $W^{s,p}(\Omega)$: $W^{s,p}(\Omega)$: 36.1 and (36.1)–(36.3); a bounded open set for which $W^{1,\infty}(\Omega)$ is not dense in $W^{1,p}(\Omega)$ for $p < \infty$: 36.2.

Lecture 37, Variants with BV spaces: Functions in $W^{s,p}(\Omega)$ with extension by 0 belonging to $W^{s,p}(R^N)$: 37.1; nonnegative distributions are Radon measures: 37.2; $BV(\Omega)$: 37.3; $BV(R^N) \subset L^{1^*,1}(R^N)$: 37.4 and (37.1) and (37.2); some interpolations spaces with $\mathcal{M}_b(R^N)$ and with $BV(R^N)$: 37.5 and (37.3)–(37.5).

Contents

1

Historical Background

In the mid 1930s, Sergei SOBOLEV introduced some functional spaces which have been very important in the development of partial differential equations, mostly those related to continuum mechanics or physics. They are known as *Sobolev spaces*, but others have mentioned having defined similar spaces, like FICHERA[1] and FRIEDRICHS.[2] A similar idea was used a little after by Jean LERAY in his study of *weak solutions* of the Navier–Stokes equation,[3] and he thought that the lack of regularity is related to turbulent flows, but although nobody really understands at a mathematical level what turbulence is, it is quite clear from a continuum mechanics point of view that Jean LERAY's ideas do not correspond to it; the ideas introduced later by KOLMOGOROV[4] have received more attention, but have some defects which are not emphasized enough.

The basic idea for defining a Sobolev space consists in using *weak derivatives*, as Sergei SOBOLEV or Jean LERAY did in the mid 1930s; it consists in giving a precise meaning to the statement that a function u from the *Lebesgue space* $L^p(\Omega)$[5] (for a nonempty open set Ω in R^N) has all its partial derivatives $\frac{\partial u}{\partial x_j}$ also in $L^p(\Omega)$. However, they did not define partial derivatives for every function in $L^p(\Omega)$, but only said that some of these functions have partial derivatives belonging also to $L^p(\Omega)$, and the important step of defining more general mathematical objects, which permit one to define as many derivatives

---

[1] Gaetano FICHERA, Italian mathematician, 1922–1996. He worked in Trieste and in Roma (Rome), Italy.

[2] Kurt Otto FRIEDRICHS, German-born mathematician, 1901–1982. He worked in Aachen, in Braunschweig, Germany, and at NYU (New York University), New York, NY.

[3] NAVIER introduced the equation in 1821, while STOKES introduced later the linearized version now known as the Stokes equation, which neglects inertial effects.

[4] Andrey Nikolaevich KOLMOGOROV, Russian mathematician, 1903–1987. He received the Wolf Prize in 1980. He worked in Moscow, Russia.

[5] It was F. RIESZ who introduced the $L^p(\Omega)$ spaces for $1 \leq p \leq \infty$.

as one may want for any locally integrable function, was performed by Laurent SCHWARTZ, who called his objects *distributions*. Laurent SCHWARTZ went further than the theory developed by Sergei SOBOLEV, which he did not know about, and he points out that BOCHNER[6,7] had also obtained some partial results, which he also only learnt about later. Laurent SCHWARTZ told me that some people quote GEL'FAND[8,9] for developing the theory of distributions, but that what GEL'FAND did was mostly to popularize the theory. Someone pointed out to me that WEYL[10] should be quoted for the theory too, but I have not checked that, and Laurent SCHWARTZ was not aware of his work when I last saw him.

Once distributions are defined and their basic properties obtained, one defines the Sobolev space $W^{m,p}(\Omega)$ for a nonnegative integer m, for $1 \leq p \leq \infty$ and for an open set Ω of R^N, as the space of functions $u \in L^p(\Omega)$ such that $D^\alpha u \in L^p(\Omega)$ for each multi-index $\alpha = (\alpha_1, \ldots, \alpha_N)$ with $|\alpha| = \alpha_1 + \ldots + \alpha_N \leq m$, $\alpha_j \geq 0$ for $j = 1, \ldots, N$, and $D^\alpha u = \frac{\partial^{\alpha_1}}{\partial x_1^{\alpha_1}} \ldots \frac{\partial^{\alpha_N}}{\partial x_N^{\alpha_N}} u$; one also denotes $\alpha! = \alpha_1! \ldots \alpha_N!$, $x^\alpha = x_1^{\alpha_1} \ldots x_N^{\alpha_N}$, and Laurent SCHWARTZ told me that these simplifying notations have been introduced by WHITNEY[11-13]. One should be aware of the fact that some authors, like Lars HÖRMANDER,[14,15] use D for denoting $\frac{1}{i}\frac{\partial}{\partial x}$ instead; one should not be surprised then if two books

---

[6] Salomon BOCHNER, Polish-born mathematician, 1899–1982. He worked in München (Munich), Germany, and in Princeton, NJ. He used "Zorn's lemma" seven years before ZORN.

[7] Max August ZORN, German-born mathematician, 1906–1993. He worked at UCLA (University of California at Los Angeles), Los Angeles, CA, and at University of Indiana, Bloomington, IN, where I met him in 1980.

[8] Izrail Moiseevic GEL'FAND, Russian-born mathematician, born in 1913. He received the Wolf Prize in 1978. He worked in Moscow, Russia, and at Rutgers University, Piscataway, NJ.

[9] Henry RUTGERS, American colonel, 1745–1830.

[10] Hermann Klaus Hugo WEYL, German-born mathematician, 1885–1955. He worked in Göttingen, Germany, at ETH (Eidgenössische Technische Hochschule), Zürich, Switzerland, and at IAS (Institute for Advanced Study), Princeton, NJ.

[11] Hassler WHITNEY, American mathematician, 1907–1989. He shared the Wolf Prize in 1982 with KREIN. He worked at Harvard University, Cambridge, MA, and at IAS (Institute for Advanced Study), Princeton, NJ.

[12] Mark Grigorievich KREIN, Ukrainian mathematician, 1907–1989. He shared the Wolf Prize in 1982 with WHITNEY. He worked in Moscow, in Kazan, Russia, in Odessa and in Kiev, Ukraine.

[13] John HARVARD, English clergyman, 1607–1638.

[14] Lars HÖRMANDER, Swedish mathematician, born in 1931. He received the Fields Medal in 1962, and the Wolf Prize in 1988. He worked in Stockholm, Sweden, at Stanford University, Stanford, CA, at IAS (Institute for Advanced Study), Princeton, NJ, and in Lund, Sweden.

[15] Leland STANFORD, American businessman, 1824–1893.

contain similar formulas with different constants, and one should check what is the definition of symbols like D, or how the *Fourier transform*[16],[17] is defined. The reason that Sergei SOBOLEV introduced the space $W^{1,2}(\Omega)$, which is also denoted $H^1(\Omega)$ (but should not be confused with the Hardy space,[18] which I shall denote $\mathcal{H}^1$), is that it is a natural space for solving the *Laplace equation*,[19],[20] also called the *Poisson equation*,[21] of the form $-\Delta u = f$ with boundary conditions. It was probably in relation to the *Dirichlet principle*,[22] which consists in noticing that if a function u of class C^2 *minimizes the functional* J defined by

$$J(v) = \int_\Omega |grad(v)|^2\, dx - 2 \int_\Omega f\, v\, dx \tag{1.1}$$

among all functions having a given boundary value, then u satisfies

$$-\Delta u = f \text{ in } \Omega. \tag{1.2}$$

The name for the principle was chosen by RIEMANN,[23] who had heard it from DIRICHLET, but it had been used before by GAUSS[24] and by GREEN.[25]

---

[16] Jean-Baptiste Joseph FOURIER, French mathematician, 1768–1830. He worked in Auxerre, and in Paris, France; he accompanied BONAPARTE in Egypt, he was prefect in Grenoble, France, until the fall of NAPOLÉON, and he worked in Paris again afterward. The first of three universities in Grenoble, France, Université de Grenoble I, is named after him, and the Institut Fourier is its department of mathematics.

[17] Napoléon BONAPARTE, French general, 1769–1821. He became Premier Consul after his coup d'état in 1799, was elected Consul for life in 1802, and he proclaimed himself Emperor in 1804, under the name NAPOLÉON I (1804–1814, and 100 days in 1815).

[18] The term has been introduced by F. RIESZ.

[19] Pierre-Simon LAPLACE, French mathematician, 1749–1827. He worked in Paris, France. He was made count in 1806 by NAPOLÉON and marquis in 1817 by LOUIS XVIII. In his memoirs written during his exile in St Helena, BONAPARTE wrote that he had removed LAPLACE from the office of Minister of the Interior, which he held in 1799, after only six weeks, "because he brought the spirit of the infinitely small into the government".

[20] Louis Stanislas Xavier de France, 1755–1824, comte de Provence, duc d'Anjou, was King of France from 1814 to 1824, under the name of LOUIS XVIII.

[21] Siméon Denis POISSON, French mathematician, 1781–1840. He worked in Paris, France.

[22] Johann Peter Gustav LEJEUNE DIRICHLET, German mathematician, 1805–1859. He worked in Breslau (then in Germany, now Wrocław, Poland), in Berlin and in Göttingen, Germany.

[23] Georg Friedrich Bernhard RIEMANN, German mathematician, 1826–1866. He worked in Göttingen, Germany.

[24] Johann Carl Friedrich GAUSS, German mathematician, 1777–1855. He worked in Göttingen, Germany.

[25] George GREEN, English mathematician, 1793–1841. He was a miller and never held any academic position.

WEIERSTRASS[26] had pointed out later that *the functional might not attain its minimum*, and I think that the complete solution of the Dirichlet principle was one in the famous list of problems which HILBERT proposed in 1900 at the International Congress of Mathematicians in Paris, France; the introduction of Sobolev spaces, which are Hilbert spaces, together with some developments in functional analysis, by FRÉCHET, F. RIESZ, and BANACH is one way to attack the problem, which I think was first solved by POINCARÉ. As the principle is also named after THOMSON,[27] it is possible that Sergei SOBOLEV had considered the question of *electrostatics*, a simplification of the *Maxwell equation*.[28],[29]

HADAMARD[30] introduced the notion of *well-posed problems* and he proved that there are continuous functions f for which the solution u is not of class C^2. One way to avoid this difficulty is to work with the family of Hölder spaces $C^{k,\alpha}$, where k is a nonnegative integer and $0 < \alpha \leq 1$, named after Ernst HÖLDER[31],[32] or LIPSCHITZ[33]; questions like $f \in C^{k,\alpha}$ implies $u \in C^{k+2,\alpha}$ for $0 < \alpha < 1$ were investigated by GIRAUD[34] and for Hölder continuous coefficients by SCHAUDER, but the similar statement is false for $\alpha = 1$, and an adapted space was introduced by Antoni ZYGMUND.[35] An underlying question is related to *singular integrals* acting on Hölder spaces $C^{0,\alpha}$ for $0 < \alpha < 1$, and the work of GIRAUD was extended to L^p spaces for $1 < p < \infty$

---

[26] Karl Theodor Wilhelm WEIERSTRASS, German mathematician, 1815–1897. He first taught in high schools in Münster, in Braunsberg, Germany, and then he worked in Berlin, Germany.

[27] William THOMSON, Irish-born physicist, 1824–1907. He was made baron Kelvin of Largs in 1892, and thereafter known as Lord Kelvin. He worked in Glasgow, Scotland.

[28] James CLERK MAXWELL, Scottish physicist, 1831–1879. He worked in Aberdeen, Scotland, in London, in Cambridge, England, holding the first Cavendish professorship of physics (1871–1879).

[29] Henry CAVENDISH, English physicist and chemist (born in Nice, not yet in France then), 1731–1810. He was wealthy and lived in London, England.

[30] Jacques Salomon HADAMARD, French mathematician, 1865–1963. He worked in Bordeaux, in Paris, France; he held a chair (mécanique analytique et mécanique céleste, 1909–1937) at Collège de France, Paris, France.

[31] Ernst HÖLDER, German mathematician, 1901–1990. He worked in Leipzig, and at Johannes Gutenberg-Universität, Mainz, Germany. I once saw him at a meeting at the Mathematisches Forschungsinstitut in Oberwolfach, Germany.

[32] Johannes GUTENBERG, German inventor and printer, 1397–1468. He worked in Mainz, Germany, where the university is named after him.

[33] Rudolf Otto Sigismund LIPSCHITZ, German mathematician, 1832–1903. He worked in Breslau (then in Germany, now Wrocław, Poland) and in Bonn, Germany.

[34] Georges GIRAUD, French mathematician, 1889–1943?.

[35] Antoni Szczepan ZYGMUND, Polish-born mathematician, 1900–1992. He worked in Warsaw, Poland, in Wilno (then in Poland, now Vilnius, Lithuania), and at the University of Chicago, Chicago IL.

by Alberto CALDERÓN[36] and Antoni ZYGMUND, so that for $1 < p < \infty$, $f \in L^p(\Omega)$ implies $u \in W^{2,p}_{loc}(\Omega)$; the question for adapted boundary conditions was investigated by Shmuel AGMON, Avron DOUGLIS[37] and Louis NIRENBERG,[38],[39] but the case $p = 2$ had been understood earlier, because one can use the Fourier transform, and there were simpler methods for proving regularity results in this case, by Louis NIRENBERG, or by Jaak PEETRE. In the late 1950s and early 1960s, Sobolev spaces were used in a more systematic way for solving *linear partial differential equations* from continuum mechanics or physics, with suitable boundary conditions, the *Lax–Milgram lemma*[40] being the cornerstone for the elliptic cases, but others had obtained the same result, like Mark VISHIK[41]; a generalization to *evolution problems* was worked out by Jacques-Louis LIONS, and after his development of the real methods for interpolation of Banach spaces with Jaak PEETRE, he studied the application to Sobolev spaces of noninteger order with Enrico MAGENES; the late 1960s saw the generalization to *nonlinear partial differential equations*, but using the same tools used for the linear cases.

The use of Sobolev spaces and the study of their properties was facilitated by the theory of distributions, which used some results of functional analysis developed for that purpose, and which is a natural generalization of the previously developed theory of *Radon measures*.[42] Laurent SCHWARTZ insisted that it had been good for the invention of the theory of distributions

---

[36] Alberto Pedro CALDERÓN, Argentine-born mathematician, 1920–1998. He received the Wolf Prize in 1989. He worked at Buenos Aires, Argentina, at OSU (Ohio State University), Columbus, OH, at MIT (Massachusetts Institute of Technology), Cambridge, MA, and at the University of Chicago, Chicago, IL. He kept strong ties with Argentina, as can be witnessed from the large number of mathematicians from Argentina having studied harmonic analysis, and often working now in United States.

[37] Avron DOUGLIS, American mathematician, 1918–1995. He worked at University of Maryland, College Park, MD.

[38] Louis NIRENBERG, Canadian-born mathematician, born in 1925. He received the Crafoord Prize in 1982. He works at NYU (New York University), New York, NY.

[39] Holger CRAFOORD, Swedish industrialist and philanthropist, 1908–1982. He invented the artificial kidney, and he and his wife Anna-Greta CRAFOORD, 1914–1994, established the Crafoord Prize in 1980 by a donation to the Royal Swedish Academy of Sciences, to reward and promote basic research in scientific disciplines that fall outside the categories of the Nobel Prize (which have included mathematics, geoscience, bioscience, astronomy, and polyarthritis).

[40] Arthur Norton MILGRAM, American mathematician, 1912–1960. He worked in Syracuse, NY, and in Minneapolis, MN.

[41] Mark Iosipovich VISHIK, Russian mathematician, born in 1921. He worked at the Russian Academy of Sciences, Moscow, Russia.

[42] Johann RADON, Czech-born mathematician, 1887–1956. He worked in Hamburg, in Greifswald, in Erlangen, Germany, in Breslau (then in Germany, now Wrocław, Poland) before World War II, and in Vienna, Austria, after 1947.

that Bourbaki[43,44] had chosen to define integration by the approach of Radon measures, probably following ideas from WEIL, because it would have been much more difficult for him to think of that generalization if Bourbaki had chosen the abstract measure theory[45–47] and started with *Borel measures*[48] instead, like probabilists do. Of course nothing could have been done without the development of a good theory of integration by LEBESGUE (actually first done two years earlier by W.H. YOUNG[49–52,]), and although integration is more easy to understand than differentiation if one considers that ARCHIMEDES had computed the area below a parabola (without even having at his disposal a Cartesian equation of the parabola as analytical geometry was only invented by DESCARTES[53]), while one had to wait almost two thousand years to see the invention of differential calculus by NEWTON[54] (and

---

[43] Nicolas Bourbaki is the pseudonym of a group of mathematicians, mostly French; those who chose the name certainly knew about a French general named BOURBAKI.

[44] Charles Denis Sauter BOURBAKI, French general, 1816–1897. Of Greek ancestry, he declined an offer of the throne of Greece in 1862.

[45] René DE POSSEL had left the Bourbaki group on this occasion, because he advocated the abstract measure theory.

[46] Lucien Alexandre Charles René DE POSSEL, French mathematician, 1905–1974. He worked in Marseille, in Clermont-Ferrand, in Besançon, France, in Alger (Algiers) (then in France, now the capital of Algeria), and in Paris, France. I had him as a teacher for my DEA (diplôme d'études approfondies) in numerical analysis, at Institut Blaise Pascal in Paris, in 1967–1968.

[47] Blaise PASCAL, French mathematician and philosopher, 1623–1662. The Université de Clermont-Ferrand II, Aubière, France, is named after him.

[48] Félix Edouard Justin Emile BOREL, French mathematician, 1871–1956. He worked in Lille and in Paris, France.

[49] William Henry YOUNG, English mathematician, 1863–1942. He worked in Liverpool, England, in Calcutta, India, holding the first Hardinge professorship (1913–1917), in Aberystwyth, Wales, and in Lausanne, Switzerland. There are many results attributed to him which may be joint work with his wife, Grace CHISHOLM, as they collaborated extensively; their son, Laurence, is known for his own mathematical results, and among them the introduction of *Young measures* in the Calculus of Variations, whose use in *partial differential equations* I pioneered in the late 1970s, not knowing at the time that he had introduced them (although I had first met him in 1971 in Madison WI), as I had heard about them as parametrized measures in seminars on control theory.

[50] Sir Charles HARDINGE, 1st Baron HARDINGE of Penshurst, English diplomat, 1858–1944. He was Viceroy and Governor-General of India (1910–1916).

[51] Grace Emily CHISHOLM-YOUNG, English mathematician, 1868–1944.

[52] Laurence Chisholm YOUNG, English-born mathematician, 1905–2000. He worked in Cape Town, South Africa, and at University of Wisconsin, Madison, WI.

[53] René DESCARTES, French mathematician and philosopher, 1596–1650. Université de Paris 5 is named after him.

[54] Sir Isaac NEWTON, English mathematician, 1643–1727. He held the Lucasian chair (1669–1701) at Cambridge, England.

LEIBNIZ[55] who made it more efficient), it is useful to observe that the previous understanding of integration was not good enough. Although one usually teaches first the Riemann integral, with Darboux[56] sums, there are not enough *Riemann-integrable*[57] functions in order to make some natural spaces complete, and this can be done by using the Lebesgue integral. Although the space $L^1(R)$ of *Lebesgue-integrable*[58] functions of a real variable is complete, it is not a dual, but one can consider $L^1(R)$ as a subset of the dual of $C_c(R)$, the space of continuous functions with compact support,[59] so that bounded sequences in $L^1(R)$ may approach a Radon measure (*in the weak $\star$ topology*); for example the sequence u_n defined by $u_n(x) = n$ on $(0, 1/n)$ and $u_n(x) = 0$ elsewhere converges to the Dirac[60] mass[61] at 0, and as one checks that for every $\varphi \in C_c(R)$ one has $\int_R u_n(x)\varphi(x)\,dx \to \varphi(0)$, the Dirac mass at 0 corresponds to the linear functional $\varphi \mapsto \varphi(0)$. More generally, a Radon measure μ on an open subset Ω of R^N is a linear form $\varphi \mapsto \langle \mu, \varphi \rangle$ on $C_c(\Omega)$, the space of continuous functions with compact support in Ω, such that for every compact $K \subset \Omega$, there exists a constant $C(K)$ such that

$$|\langle \mu, \varphi \rangle| \leq C(K) \max_{x \in K} |\varphi(x)| \text{ for all } \varphi \in C_c(\Omega) \text{ with support } \subset K. \quad (1.3)$$

[Taught on Monday January 17, 2000. The course met on Mondays, Wednesdays, and Fridays.]

---

[55] Gottfried Wilhelm VON LEIBNIZ, German mathematician, 1646–1716. He worked in Frankfurt, in Mainz, Germany, in Paris, France and in Hanover, Germany, but never in an academic position.

[56] Jean Gaston DARBOUX, French mathematician, 1842–1917. He worked in Paris, France.

[57] The starting point of LEBESGUE may have been his characterization of the Riemann-integrable functions, as those functions whose points of discontinuity form a set of (Lebesgue) measure zero, i.e. a set which for every $\varepsilon > 0$ can be covered by intervals whose sums of lengths is less than ε.

[58] They are actually equivalence classes, as one identifies two functions which are equal almost everywhere (abbreviated a.e.), i.e. which only differ on a set of (Lebesgue) measure 0.

[59] For a continuous function f defined on a topological space and taking values in a vector space, the support is the closure of the set of x such that $f(x) \neq 0$.

[60] Paul Adrien Maurice DIRAC, English physicist, 1902–1984. He received the Nobel Prize in Physics in 1933. He worked in Cambridge, England, holding the Lucasian chair (1932–1969).

[61] The intuition of a point mass (or charge) is obvious for anyone interested in physics, but DIRAC went much further than dealing with these objects, as he was not afraid of taking derivatives of his strangely defined "function", a quite bold move which was given a precise mathematical meaning by Laurent SCHWARTZ in his theory of distributions.

The Lebesgue Measure, Convolution

The Lebesgue measure on R^N is invariant by *translation*, by *rotation* and by *mirror symmetry*, i.e., if $a \in R^N$ and $M \in \mathcal{L}(R^N; R^N)$ belongs to the *orthogonal group* $O(N)$ (i.e., satisfies $M^T M = I$), then for a Lebesgue-measurable set A its image B by the *isometry* $x \mapsto a + M x$ is also Lebesgue-measurable and has the same measure as A. A *rigid displacement* is the particular case where M is a *rotation*, i.e., belongs to the *special orthogonal group* $SO(N)$, which is the subgroup of $O(N)$ for which the *determinant* of M is +1.

One can "construct" *nonmeasurable sets* by using the *axiom of choice*, the classical example being to start with the unit circle S^1, and to define equivalence classes, so that two points are equivalent if one can be obtained from the other by applying a rotation of an integer angle $n \in Z$; then one uses the axiom of choice in order to assert that there exists a subset A which contains exactly one element in each equivalence class, and denoting $A_n = n + A$ the subset obtained from A by a rotation of n, one finds that S^1 is partitioned into the $A_n, n \in Z$, so that if A was Lebesgue-measurable, all the A_n would have the same measure and this measure could not be > 0 because the measure of S^1 is finite, but if the measure was 0, S^1 would be a countable union of subsets of measure 0 and would have measure 0, and as A can have neither a positive measure nor a zero measure there only remains the possibility that it is not measurable (so that there is no *paradox*).

A more subtle construction was carried out in R^3 by HAUSDORFF,[1] and then used and simplified by BANACH and TARSKI[2], giving the Hausdorff–Banach–Tarski "paradox": if $N \geq 3$, if A and B are two closed bounded sets of R^N with nonempty interior, then there exists a positive integer m, a partition

---

[1] Felix HAUSDORFF, German mathematician, 1869–1942. He worked in Leipzig, in Greifswalf and in Bonn, Germany. He wrote literary and philosophical work under the pseudonym of Paul MONGRÉ.

[2] Alfred TARSKI (TEITELBAUM), Polish-born mathematician, 1902–1983. He worked in Warsaw, Poland, and at UCB (University of California at Berkeley), Berkeley, CA.

of A into (disjoint) subsets $A_1, \ldots, A_m$, a partition of B into (disjoint) subsets $B_1, \ldots, B_m$, such that for $i = 1, \ldots, m$ the subset B_i is the image of A_i by a rigid displacement; of course some of the subsets are not measurable if A and B have different measures [I have read the statement (but not seen the proof) that for $N = 2$ there does exist a finitely additive measure defined for all subsets and invariant by translation and rotation, so such a *paradoxical decomposition* cannot be performed in R^2].

Up to a multiplication by a constant, the Lebesgue measure is the only nonzero Radon measure which is invariant by translation, so that it is uniquely defined if we add the requirement that the volume of the unit cube be 1. For any locally compact[3] commutative[4] group there exists a nonzero Radon measure which is invariant by translation, unique up to multiplication by a constant, a *Haar measure*[5] of the group. For the additive group Z, a Haar measure is the counting measure; for the additive group R^N, a Haar measure is the Lebesgue measure dx, and $\frac{dt}{t}$ is a Haar measure for the multiplicative group $(0, \infty)$ (which is isomorphic to the additive group R by the *logarithm*, whose inverse is the *exponential*).

The *convolution product* can be defined for any locally compact group for which one has chosen a Haar measure, but we shall only use it for the additive group R^N. For $f, g \in C_c(R^N)$, the convolution product $h = f \star g$ is defined by

$$h(x) = \int_{R^N} f(y)g(x - y)\,dy = \int_{R^N} f(x - y)g(y)\,dy, \qquad (2.1)$$

showing that the convolution product is *commutative*. One has $f \star g \in C_c(R^N)$ and a localization of support,

$$support(f \star g) \subset support(f) + support(g). \qquad (2.2)$$

The convolution product is *associative*, i.e., for $a, b, c \in C_c(R^N)$ one has $(a \star b) \star c = a \star (b \star c)$; one has to be careful that *associativity may be lost* for some generalizations with functions (or distributions) with noncompact support.

For convolution of continuous functions with compact support, *Young's inequality*[6] holds, which asserts that for $f, g \in C_c(R^N)$ one has

---

[3] I have heard Laurent SCHWARTZ say that the result is not true for some groups which are not locally compact.

[4] In the noncommutative case one distinguishes between invariance by action of the group on the left and invariance by action of the group on the right.

[5] Alfréd HAAR, Hungarian mathematician, 1885–1933. He worked in Göttingen, Germany, in Kolozsvár (then in Hungary, now Cluj-Napoca, Romania), in Budapest and in Szeged, Hungary.

[6] William Henry YOUNG worked on so many problems with his wife, Grace CHISHOLM-YOUNG, that it is not clear if a result attributed to him was obtained jointly with his wife or not.

$$\|f \star g\|_r \leq \|f\|_p \|g\|_q \text{ if } 1 \leq p,q,r \leq \infty \text{ and } \frac{1}{r} = \frac{1}{p} + \frac{1}{q} - 1, \qquad (2.3)$$

where for $1 \leq s \leq \infty$, $\|h\|_s$ means $\|h\|_{L^s(R^N)}$. If p or q or r is 1, it is just an application of *Hölder's inequality*[7] and it is optimal, while for other cases one may prove it by applying Hölder's inequality, or *Jensen's inequality*,[8] a few times, but the constant is not optimal and the *best constant* $C(p,q)$ for which one has $\|f \star g\|_r \leq C(p,q)\|f\|_p\|g\|_q$ has been found independently by William BECKNER,[9] who used probabilistic methods, and by Elliot LIEB[10] and BRASCAMP,[11] who used nonprobabilistic methods (equality holds for some particular Gaussians). Of course, under the preceding relation between p,q,r, the convolution product extends from $L^p(R^N) \times L^q(R^N)$ into $L^r(R^N)$ with the same inequality (2.3), and this can be proven either directly or by using the density[12] of $C_c(R^N)$ in $L^p(R^N)$ for $1 \leq p < \infty$, and the (sequential) weak $\star$ density of $C_c(R^N)$ in $L^\infty(R^N)$.

Lemma 2.1. *(i) If $1 < p < \infty$, $f \in L^p(R^N)$ and $g \in L^{p'}(R^N)$ (where p' denotes the* conjugate exponent *of p, defined by $\frac{1}{p} + \frac{1}{p'} = 1$), then $f \star g \in C_0(R^N)$, the space of continuous (bounded) functions converging to 0 at infinity.*
(ii) If $f \in L^1(R^N)$ and $g \in L^\infty(R^N)$, then $f \star g$ belongs to $BUC(R^N)$, the space of bounded uniformly continuous functions.

Proof: In that case Young's inequality (2.3) follows from Hölder's inequality, which gives $\left|\int_{R^N} f(y)g(x-y)\,dy\right| \leq \|f\|_p\|g\|_{p'}$ for all x. There exists a sequence $f_n \in C_c(R^N)$ converging to f in $L^p(R^N)$ strong, and a sequence $g_n \in C_c(R^N)$ converging to g in $L^{p'}(R^N)$ strong, and as $f \star g - f_n \star g_n = f \star (g - g_n) + (f - f_n) \star g_n$, one deduces that $\|f \star g - f_n \star g_n\|_\infty \leq \|f\|_p\|g - g_n\|_{p'} + \|f - f_n\|_p\|g_n\|_{p'} \to 0$, so that $f \star g$ is the uniform limit

[7] Otto Ludwig HÖLDER, German mathematician, 1859–1937. He worked in Leipzig, Germany. He was the father of Ernst HÖLDER.
[8] Johan Ludwig William Valdemar JENSEN, Danish mathematician, 1859–1925. He never held any academic position, and worked for a telephone company.
[9] William Eugene BECKNER, American mathematician. He works at University of Texas, Austin, TX.
[10] Elliott H. LIEB, American mathematician. He worked at IBM (International Business Machines), at Yeshiva University, New York, NY, at Northeastern University, Boston, MA, at MIT (Massachusetts Institute of Technology), Cambridge, MA, and in Princeton, NJ.
[11] Herm Jan BRASCAMP, Dutch mathematical physicist. He works in Groningen, The Netherlands.
[12] Notice that I admit that this density has been proven when constructing the Lebesgue measure. Although we shall study later an explicit way of approaching functions in $L^p(R^N)$ by functions in $C_c^\infty(R^N)$, the proof will use the fact that $C_c(R^N)$ is dense, and will not be an independent proof of that result.

of the sequence $f_n \star g_n \in C_c(R^N)$, and it belongs to the closure of $C_c(R^N)$ in $L^\infty(R^N)$, which is[13,14] $C_0(R^N)$.

Using a sequence $f_n \in C_c(R^N)$ converging to f in $L^1(R^N)$ strong, one has $||f \star g - f_n \star g||_\infty \leq ||f - f_n||_1 ||g||_\infty$, so that $f \star g$ is the uniform limit of the sequence $f_n \star g$, and it is enough to show that each $f_n \star g$ is bounded uniformly continuous, as a uniform limit of such functions also belongs to the same space $BUC(R^N)$.[15] As the function f_n belongs to $C_c(R^N)$ it is uniformly continuous, so that $|f_n(a) - f_n(b)| \leq \omega_n(|a - b|)$ with $\lim_{t \to 0} \omega_n(t) = 0$. One has $(f_n \star g)(x) - (f_n \star g)(x') = \int_{R^N} (f_n(x - y) - f_n(x' - y)) g(y) \, dy$, but if the support of f_n is included in the closed ball centered at 0 with radius R_n the integral may be restricted to the set of y such that $|y - x| \leq R_n$ and $|y - x'| \leq R_n$, so that $|(f_n \star g)(x) - (f_n \star g)(x')| \leq \int_{|y-x| \leq R_n, |y-x'| \leq R_n} |f_n(x - y) - f_n(x' - y)| |g(y)| \, dy \leq \omega_n(|x - x'|) ||g||_\infty meas(B(0, R_n))$, showing that $f_n \star g$ is uniformly continuous, and it is bounded by (2.3). $\square$

Of course, the property of commutativity of the convolution product extends to the case where it is defined on $L^p(R^N) \times L^q(R^N)$ (i.e., if $p, q \geq 1$ and $\frac{1}{p} + \frac{1}{q} \geq 1$), and similarly the property of associativity of the convolution product extends to the case of functions belonging to $L^a(R^N), L^b(R^N), L^c(R^N)$, with $a, b, c \geq 1$, $\frac{1}{a} + \frac{1}{b} \geq 1$, $\frac{1}{b} + \frac{1}{c} \geq 1$, and $\frac{1}{a} + \frac{1}{b} + \frac{1}{c} \geq 2$, and can be proven directly using *Fubini's theorem*.[16]

However, one must be careful that there are other cases where the convolution products $f_1 \star f_2$, $f_2 \star f_3$, $(f_1 \star f_2) \star f_3$ and $f_1 \star (f_2 \star f_3)$ are all defined, for example if in each convolution product considered at least one of the functions has *compact support*,[17] but with $(f_1 \star f_2) \star f_3 \neq f_1 \star (f_2 \star f_3)$: let $f_1 = 1$,

---

[13] Using the *Aleksandrov one point compactification* of R^N with a point ∞, $C_c(R^N)$ is the subset of functions which are 0 in a neighborhood of ∞ and $C_0(R^N)$ is the subset of functions which are 0 at ∞.

[14] Pavel Sergeevich ALEKSANDROV, Russian mathematician, 1896–1982. He worked in Smolensk, and in Moscow, Russia.

[15] If $h_n \in BUC(R^N)$ then $||h_n||_\infty = M_n < \infty$ and $|h_n(x) - h_n(y)| \leq \omega_n(|x-y|)$ with $\lim_{\sigma \to 0} \omega_n(\sigma) = 0$; if $||h_n - h||_\infty = \varepsilon_n \to 0$, then $||h||_\infty \leq \inf_n (M_n + \varepsilon_n) < \infty$, and because $|h(x) - h(y)| \leq |h(x) - h_n(x)| + |h_n(x) - h_n(y)| + |h_n(y) - h(y)| \leq 2\varepsilon_n + \omega_n(|x-y|)$, one has $|h(x) - h(y)| \leq \omega(|x-y|)$ with $\omega(\sigma) = \inf_n (2\varepsilon_n + \omega_n(\sigma))$, so that $\lim_{\sigma \to 0} \omega(\sigma) = 0$ and ω is a modulus of uniform continuity for h.

[16] Guido FUBINI, Italian-born mathematician, 1879–1943. He worked in Catania, in Genova (Genoa), in Torino (Turin), Italy, and after 1939 in New York, NY.

[17] For a measurable function f, the definition of *support* must be different than the one for continuous functions, because a function is an equivalence class of functions equal almost everywhere, and the definition of the support of f must not change if one changes f on a set of measure 0. For an open set ω one says that $f = 0$ on ω if f is equal to 0 almost everywhere in ω; if for a family $\omega_i, i \in I$, of open sets such that one has $f = 0$ on ω_i for each $i \in I$, then $f = 0$ on $\omega = \bigcup_{i \in I} \omega_i$, and in the case where I is not countable one may use a *partition of unity* for showing that the support of f is the closed set whose complement is the largest open set on which $f = 0$.

$f_2 \in C_c(R)$ with $\int_R f_2(x)\, dx = 0$, and let f_3 be the Heaviside[18] function, defined by $f_3(x) = 0$ for $x < 0$ and $f_3(x) = 1$ for $x > 0$; one sees immediately that $f_1 \star f_2 = 0$ and $f_4 = f_2 \star f_3 \in C_c(R)$, and one has to check that f_2 can be chosen in such a way that $\int_R f_4(x)\, dx \neq 0$; if one chooses f_2 with support in $[-1, +1]$, with $\int_{-1}^{+1} f_2(y)\, dy = 0$ and $\int_{-1}^{+1}(1 - y)f_2(y)\, dy \neq 0$, then one has $f_4(x) = \int_{-1}^{x} f_2(y)\, dy$, and $\int_R f_4(x)\, dx = \int_{-1}^{+1}(1 - y)f_2(y)\, dy \neq 0$.

If one uses *locally integrable functions*, i.e., measurable functions which are integrable on every compact, denoted by $L^1_{loc}(R^N)$, then for $f, g \in L^1_{loc}(R^N)$ one can define the convolution product $f \star g$ if the following condition is satisfied:

for every compact $C \subset R^N$ there exist compact sets $A, B \subset R^N$ such that $x \in support(f), y \in support(g)$ and $x + y \in C$ imply $x \in A, y \in B$.

(2.4)

For example, if $f, g \in L^1_{loc}(R)$ vanish on $(-\infty, 0)$, then $f \star g$ is well defined and $(f \star g)(x) = \int_0^x f(y)g(x - y)\, dy$ shows that $f \star g \in L^1_{loc}(R)$ and vanishes on $(-\infty, 0)$; TITCHMARSH[19] has proven that if the support of f starts at a and the support of g starts at b, then the support of $f \star g$ starts at $a + b$, and Jacques-Louis LIONS has generalized this result to similar situations in R^N, proving that the *closed convex hull of the support* of $f \star g$ is equal to the (vector) sum of the closed convex hull of the support of f and the closed convex hull of the support of g.

Definition 2.2. *For a vector $a \in R^N$, and $f \in L^1_{loc}(R^N)$, $\tau_a f$ denotes the function defined by $(\tau_a f)(x) = f(x - a)$ for almost every $x \in R^N$.* □

This means that the graph of $\tau_a f$ is obtained from that of f by a translation of vector $(a, 0)$. Of course, one has

$$\tau_b(\tau_a f) = \tau_{a+b} f \text{ for all } a, b \in R^N, f \in L^1_{loc}(R^N).$$

(2.5)

An important property of convolution is that it *commutes with translation*; this is of course related to the fact that the Lebesgue measure is invariant by translation.

---

[18] Oliver HEAVISIDE, English engineer, 1850–1925. He worked as a telegrapher, in Denmark, and in Newcastle upon Tyne, England, and then he did research on his own, living in the south of England. He developed an *operational calculus*, which was given a precise mathematical explanation by Laurent SCHWARTZ using his theory of distributions, but we also owe him the simplified version of *Maxwell's equation* using vector calculus, which should be called the *Maxwell–Heaviside equation* (he replaced a set of 20 equations in 20 variables written by MAXWELL by a set of 4 equations in 2 variables).

[19] Edward Charles TITCHMARSH, English mathematician, 1899–1963. He worked in London, in Liverpool, and in Oxford, England, where he held the Savilian chair of geometry.

Lemma 2.3. *(i) If the convolution product of f and g is defined (i.e., if $f, g \in L^1_{loc}(R^N)$ and (2.4) holds), then*

$$\tau_a(f \star g) = (\tau_a f) \star g = f \star (\tau_a g) \text{ for every } a \in R^N. \tag{2.6}$$

(ii) If $k \geq 0$, $f \in C^k_c(R^N)$, the space of functions of class C^k with compact support and $g \in L^1_{loc}(R^N)$, then $f \star g \in C^k(R^N)$ and if $|\alpha| \leq k$ one has

$$D^\alpha(f \star g) = (D^\alpha f) \star g \text{ almost everywhere.} \tag{2.7}$$

Proof: One has $((\tau_a f) \star g)(x) = \int_{R^N} (\tau_a f)(y) g(x-y) \, dy = \int_{R^N} f(y-a) g(x-y) \, dy$, which by a change of variable in the integral is $\int_{R^N} f(y) g(x-a-y) \, dy = (f \star g)(x-a) = (\tau_a(f \star g))(x)$, showing that $(\tau_a f) \star g = \tau_a(f \star g)$, and it is also $f \star (\tau_a g)$ by commutativity of the convolution product.

If $e_1, \ldots, e_N$ is the canonical basis of R^N, then a function h has a partial derivative $\frac{\partial h}{\partial x_j}$ at x if and only if $\frac{1}{\varepsilon}(h - \tau_{\varepsilon e_j} h)$ has a limit at x when ε tends to 0 (with $\varepsilon \neq 0$, of course). If $f \in C^1_c(R^N)$, then $\frac{1}{\varepsilon}(f - \tau_{\varepsilon e_j} f)$ converges uniformly to $\frac{\partial f}{\partial x_j}$ so that if one takes the convolution product with a function $g \in L^1_{loc}(R^N)$, one finds that $\frac{1}{\varepsilon}(f - \tau_{\varepsilon e_j} f) \star g$ converges uniformly on compact sets to $\frac{\partial f}{\partial x_j} \star g$; if one defines $h = f \star g$, one has $\frac{1}{\varepsilon}(f - \tau_{\varepsilon e_j} f) \star g = \frac{1}{\varepsilon}(h - \tau_{\varepsilon e_j} h)$ by (2.6), so that the limit must be $\frac{\partial h}{\partial x_j}$ and it is equal to $\frac{\partial f}{\partial x_j} \star g$. A reiteration of this argument then gives $D^\alpha(f \star g) = (D^\alpha f) \star g$ if $|\alpha| \leq k$. $\square$

If $f \in C^\infty_c(R^N)$ (which in Laurent SCHWARTZ's notation is $\mathcal{D}(R^N)$), then $f \star g$ belongs to $C^k(R^N)$ for all k, i.e., $f \star g \in C^\infty(R^N)$ (which in Laurent SCHWARTZ's notation is $\mathcal{E}(R^N)$). We shall see that there are enough functions in $C^\infty_c(R^N)$ for approaching any function in $L^p(R^N)$ for $1 \leq p < \infty$, but just one particular function in $C^\infty_c(R^N)$ has to be constructed explicitly, and the properties of convolution will help for the rest of the argument.

Lemma 2.4. *The function ϱ defined on R^N by*

$$\varrho(x) = \begin{cases} exp(\frac{-1}{1-|x|^2}) & \text{if } |x| < 1 \\ 0 & \text{if } |x| \geq 1 \end{cases} \tag{2.8}$$

belongs to $C^\infty_c(R^N)$.

Proof: It is nonnegative, and continuous because if $|x| \to 1$ with $|x| < 1$, then $\frac{-1}{1-|x|^2} \to -\infty$ and $\varrho(x) \to 0$; obviously ϱ has for support the closed unit ball. One has $\frac{\partial \varrho}{\partial x_j} = \frac{-2x_j}{(1-|x|^2)^2}\varrho$, and by induction $D^\alpha \varrho = \frac{P_\alpha(x)}{(1-|x|^2)^{2|\alpha|}}\varrho$ for a polynomial P_α; when $|x| \to 1$ with $|x| < 1$, the exponential wins over the term $\frac{1}{(1-|x|^2)^{2|\alpha|}}$ so that $D^\alpha \varrho \to 0$, showing that all derivatives of ϱ are continuous, and therefore $\varrho \in C^\infty_c(R^N)$. $\square$

[Taught on Wednesday January 19, 2000.]

3

Smoothing by Convolution

Once one knows a nonzero function from $C_c^\infty(R^N)$, like (2.8), convolution and *scaling* help create automatically a lot of such functions, enough for approaching all functions in $L^p(R^N)$ for $1 \le p < \infty$. For doing this, the concept of a *smoothing sequence* is used.

Definition 3.1. *(i)* A smoothing sequence *is a sequence* $\varrho_n \in C_c^\infty(R^N)$ *such that* $support(\varrho_n) \to \{0\}$, *with* $\int_{R^N} |\varrho_n(x)| \, dx$ *bounded and* $\int_{R^N} \varrho_n(x) \, dx \to 1$ *as* $n \to \infty$.
(ii) A special smoothing sequence *is defined by*

$$\varrho_n(x) = n^N \varrho_1(n\,x), \tag{3.1}$$

where $\varrho_1 \in C_c^\infty(R^N)$ *is nonnegative, has integral 1 and has support in the closed unit ball centered at 0.* □

Starting from an arbitrary nonzero function $\varphi \in C_c^\infty(R^N)$, one may assume that it is nonnegative by replacing it by φ^2, that it has its support in the closed unit ball centered at 0 by replacing it by $\varphi(k\,x)$ for k large enough, and that it has integral 1 by multiplying it by a suitable constant. This gives a function ϱ_1, which is then rescaled by (3.1), so that the integral of ϱ_n is 1, and its support is in the closed ball centered at 0 with radius $\frac{1}{n}$.

Lemma 3.2. *(i) If* $1 \le p < \infty$, $f \in L^p(R^N)$ *and* ϱ_n *is a smoothing sequence, then* $f \star \varrho_n \to f$ *in* $L^p(R^N)$ *strong as* $n \to \infty$.
(ii) If $f \in L^\infty(R^N)$ *and* ϱ_n *is a smoothing sequence, then* $f \star \varrho_n \to f$ *in* $L^\infty(R^N)$ *weak* $\star$ *and in* $L_{loc}^q(R^N)$ *strong for* $1 \le q < \infty$ *as* $n \to \infty$, *i.e., for every compact* K *one has* $\int_K |f \star \varrho_n - f|^q \, dx \to 0$ *as* $n \to \infty$.

Proof: (i) There exists a sequence $f_m \in C_c(R^N)$ which converges to f in $L^p(R^N)$ strong as $m \to \infty$. One writes $f\star\varrho_n - f = (f-f_m)\star\varrho_n + (f_m\star\varrho_n - f_m) + (f_m - f)$, so that if one chooses m such that $||f-f_m||_p \le \varepsilon$, and if C is a bound for all the $L^1(R^N)$ norms of ϱ_n, one has $||(f - f_m) \star \varrho_n||_p \leq ||f - f_m||_p ||\varrho_n||_1 \leq C\,\varepsilon$, so that $||f \star \varrho_n - f||_p \leq C\,\varepsilon + ||f_m \star \varrho_n - f_m||_p + \varepsilon$, and it remains to show that for m fixed $f_m \star \varrho_n$ converges to f_m in $L^p(R^N)$ strong as $n \to \infty$. If $\int_{R^N} \varrho_n(x)\,dx = \kappa_n$, one writes $f_m \star \varrho_n - f_m = (f_m \star \varrho_n - \kappa_n f_m) + (\kappa_n - 1)f_m$, and because $\kappa_n \to 1$ the second part tends to 0 in $L^p(R^N)$ strong, and the first part is $\int_{R^N} \varrho_n(y)\big(f_m(x-y) - f_m(x)\big)\,dy$, which tends to 0 uniformly (and in $L^p(R^N)$ strong because its support stays bounded), because of the uniform continuity of f_m, and because the support of ϱ_n tends to $\{0\}$; indeed, if $support(\varrho_n)$ is included in the ball centered at 0 with radius r_n, and η_m is the modulus of uniform continuity of f_m, one has $\big|\int_{R^N} \big(f_m(x - y) - f_m(x)\big)\varrho_n(y)\,dy\big| \leq \int_{R^N} |\varrho_n(y)|\,dy\,\big(\max_{|y| \leq r_n} |f_m(x - y) - f_m(x)|\big) \leq C\,\eta_m(r_n)$.

(ii) For $f \in L^\infty(R^N)$, one wants to show that $f \star \varrho_n$ converges to f in $L^\infty(R^N)$ weak $\star$ as $n \to \infty$, and this means that for every $g \in L^1(R^N)$ one has $\int_{R^N} (f \star \varrho_n)(x)g(x)\,dx \to \int_{R^N} f(x)g(x)\,dx$. One notices that, by Fubini's theorem, one has $\int_{R^N} (f \star \varrho_n)(x)g(x)\,dx = \int\int_{R^N \times R^N} f(y)\varrho_n(x-y)g(x)\,dx\,dy = \int_{R^N} f(y)(g \star \check{\varrho}_n)(y)\,dy$, where $\check{\varrho}_n(y) = \varrho_n(-y)$ for all $y \in R^N$, so that $\check{\varrho}_n$ is a smoothing sequence and by the first part $g \star \check{\varrho}_n$ converges to g in $L^1(R^N)$ strong as $n \to \infty$, so that $\int_{R^N} f(y)(g \star \check{\varrho}_n(y)\,dy \to \int_{R^N} f(y)g(y)\,dy$ as $n \to \infty$. In order to show that $\int_K |f \star \varrho_n - f|^q\,dx \to 0$ for a compact K and $1 \leq q < \infty$, one notices that the integral only uses values of f in a ball centered at 0 with radius R_0 large enough (for the ball to contain K and $K + support(\check{\varrho}_n)$), so that if $\tilde{f}$ coincides with f inside the ball centered at 0 with radius R_0 and is 0 outside it, then the integral is $\int_K |\tilde{f} \star \varrho_n - \tilde{f}|^q\,dx$, which does converge to 0 as $n \to \infty$ because $\tilde{f} \star \varrho_n$ converges to $\tilde{f}$ in $L^q(R^N)$ as $n \to \infty$, as a consequence of the first part and of the fact that $\tilde{f}$ belongs to $L^q(R^N)$. $\qquad\square$

Of course, because $\varrho_n \in C_c^\infty(R^N)$, one has $f \star \varrho_n \in C^\infty(R^N)$ and $D^\alpha(f \star \varrho_n) = f \star (D^\alpha \varrho_n)$ for all multi-indices α, but the support of $f \star \varrho_n$ is not compact in general.

Lemma 3.3. *(i) For $1 \leq p < \infty$, the space $C_c^\infty(R^N)$ is dense in $L^p(R^N)$. (ii) $C_c^\infty(R^N)$ is (sequentially) weak $\star$ dense in $L^\infty(R^N)$.*

Proof: (i) If $1 \leq p < \infty$, $f_m \star \varrho_n \in C_c^\infty(R^N)$ because f_m has compact support. Because $f - f_m \star \varrho_n = (f - f_m) + (f_m - f_m \star \varrho_n)$, the argument used in the first part of the Lemma 3.2 shows that there are sequences m_k and n_k such that $f_{m_k} \star \varrho_{n_k}$ converges to f in $L^p(R^N)$ strong as $k \to \infty$.

(ii) For $f \in L^\infty(R^N)$, one defines g_m by $g_m(x) = f(x)$ if $|x| \leq m$ and $g_m(x) = 0$ if $|x| > m$; then, as $m \to \infty$, g_m converges to f in $L^\infty(R^N)$ weak $\star$ and $L_{loc}^q(R^N)$ strong for $1 \leq q < \infty$; one concludes by noticing that for m fixed $g_m \star \varrho_n$ converges to g_m in $L^\infty(R^N)$ weak $\star$ and $L_{loc}^q(R^N)$ strong for $1 \leq q < \infty$, and this argument uses the fact that on bounded sets of $L^\infty(R^N)$ the weak $\star$ topology is metrizable. $\qquad\square$

[Taught on Friday January 21, 2000.]

4

Truncation; Radon Measures; Distributions

For reasons which will become clear later, it is useful to define in a more general setting the truncation step used previously.

Definition 4.1. *(i) A* truncating sequence *is a sequence* $\theta_n \in C_c^\infty(R^N)$ *which is bounded in* $L^\infty(R^N)$, *and such that* $\theta_n(x) \to 1$ *for almost every* x *and* $D^\alpha \theta_n \to 0$ *in* $L^\infty(R^N)$ *strong for each multi-index* α *with* $|\alpha| \geq 1$, *as* $n \to \infty$. *(ii) A* special truncating sequence *is defined by* $\theta_n(x) = \theta_1\left(\frac{x}{n}\right)$ *with* $\theta_1 \in C_c^\infty(R^N)$, $0 \leq \theta_1(x) \leq 1$ *for all* x, *and* $\theta_1(x) = 1$ *for* $|x| \leq 1$ *(and usually with* $\theta_1(y) = 0$ *for* $|y| \geq 2$*).* $\square$

That such a θ_1 exists follows easily from the smoothing by convolution, and more precisely one has the following result.

Lemma 4.2. *Let* $0 < a < b < c$, *then there exists* $\theta \in C_c^\infty(R^N)$ *with* $0 \leq \theta(x) \leq 1$ *for all* x, *with* $\theta(x) = 1$ *if* $|x| \leq a$, $\theta(x) = 0$ *if* $|x| \geq c$, *and* $\int_{R^N} \theta(x)\,dx = \int_{|x| \leq b} dx$.

Proof: Let f be the characteristic function of the ball centered at 0 with radius b; let ε satisfy $0 < \varepsilon < \min\{b-a, c-b\}$, and let $\varrho_\varepsilon \in C_c^\infty(R^N)$ be nonnegative, with support in the ball centered at 0 with radius ε, and $\int_{R^N} \varrho_\varepsilon(x)\,dx = 1$. Then $\theta = f \star \varrho_\varepsilon$ satisfies all the desired properties, the last one coming from the fact that for two functions $f, g \in L^1(R^N)$, one has

$$\int_{R^N} (f \star g)(x)\,dx = \left(\int_{R^N} f(x)\,dx\right)\left(\int_{R^N} g(x)\,dx\right). \square \qquad (4.1)$$

Of course, if $h \in L^p(R^N)$ and θ_n is a truncating sequence, $h\theta_n$ converges almost everywhere to h and is bounded by $C|h|$, and $h\theta_n \to h$ in $L^p(R^N)$ strong if $1 \leq p < \infty$, by the Lebesgue dominated convergence theorem, and in $L^\infty(R^N)$ weak $\star$ and $L_{loc}^q(R^N)$ strong for $1 \leq q < \infty$ in the case $p = \infty$.

After these preliminaries, one defines *Radon measures* and *distributions on an open set Ω of R^N* in the following way.

Definition 4.3. *(i) A* Radon measure *μ in Ω is a linear form defined on $C_c(\Omega)$ (the space of continuous functions with compact support in Ω), $\varphi \mapsto \langle \mu, \varphi \rangle$, such that for every compact $K \subset \Omega$ there exists a constant $C(K)$ such that*

$$|\langle \mu, \varphi \rangle| \leq C(K) \|\varphi\|_\infty \text{ for all } \varphi \in C_c(\Omega) \text{ with } support(\varphi) \subset K. \quad (4.2)$$

One writes $\mu \in \mathcal{M}(\Omega)$, and the elements of $C_c(\Omega)$ are called test functions.
(ii) A distribution *S in Ω is a linear form defined on $C_c^\infty(\Omega)$ (the space of C^∞ functions with compact support in Ω), $\varphi \mapsto \langle S, \varphi \rangle$, such that for every compact $K \subset \Omega$ there exists a constant $C(K)$ and a nonnegative integer $m(K)$ such that*

$$|\langle S, \varphi \rangle| \leq C(K) \max_{|\alpha| \leq m(k)} \|D^\alpha \varphi\|_\infty \text{ for all } \varphi \in C_c^\infty(\Omega) \text{ with } support(\varphi) \subset K.$$
$$(4.3)$$

One writes $S \in \mathcal{D}'(\Omega)$, and the elements of $C_c^\infty(\Omega)$ are called test functions.
(iii) If one can take $m(K) = m_0$ for all compact $K \subset \Omega$, then the distribution is said to have order[1] $\leq m_0$. $\qquad\qquad\qquad\qquad\qquad\qquad\qquad\qquad\qquad$ □

Radon measures are distributions of order ≤ 0, and actually every distribution of order ≤ 0 is a Radon measure.[2] $L^1_{loc}(\Omega)$ denotes the space of locally integrable functions in Ω, i.e., the (equivalence classes of) Lebesgue-measurable functions which are integrable on every compact $K \subset \Omega$; it is not a Banach space but it is a *Fréchet space* (i.e., it is *locally convex, metrizable and complete*), and a sequence f_n converges to 0 in $L^1_{loc}(\Omega)$ if and only if for

---

[1] If $T \in \mathcal{D}'(\Omega)$ has a finite order, Laurent SCHWARTZ defined its order as the smallest nonnegative integer m such that T is of order $\leq m$, but I do not like his definition. The distribution $pv\frac{1}{x}$, the *principal value* of $\frac{1}{x}$, which will be defined in (5.7), is a distribution of order ≤ 1 but it is not a Radon measure; however, one sees easily that (5.7) makes sense (with natural bounds) if one uses test functions $\varphi \in C_c^{0,\alpha}(R)$, the space of Hölder continuous functions of order α which have compact support, so that I find natural to say that $pv\frac{1}{x}$ is a distribution of order $\leq \alpha$, for every $\alpha > 0$, while with the definition of Laurent SCHWARTZ it is called a distribution of order 1. The discrepancy comes from the fact that Laurent SCHWARTZ limited himself to taking test functions in spaces C_c^k of functions of class C^k with compact support, while he could as well have used spaces $C_c^{k,\alpha}$ of functions with compact support whose derivatives of order k are Hölder continuous of order α, and define distributions of order $\leq k + \alpha$.
[2] For showing this, one must observe that if T is a distribution of order 0, then one can extend the definition of $\langle T, \varphi \rangle$ to test functions $\varphi \in C_c(\Omega)$ instead of $\varphi \in C_c^\infty(\Omega)$, by choosing a special smoothing sequence ϱ_ε and then defining $\langle T, \varphi \rangle = \lim_{\varepsilon \to 0} \langle T, \varphi \star \varrho_\varepsilon \rangle$, with $\varepsilon > 0$ small enough in order to have $support(\varphi) + support(\varrho_\varepsilon) \subset \Omega$.

every compact $K \subset \Omega$ one has $\int_K |f_n(x)| \, dx \to 0$. One identifies any function $f \in L^1_{loc}(\Omega)$ with a Radon measure (and therefore with a distribution), which one usually also writes f, defined by the formula

$$\langle f, \varphi \rangle = \int_{\Omega} f(x)\varphi(x) \, dx \text{ for all } \varphi \in C_c(\Omega). \qquad (4.4)$$

It is not really such a good notation, because it relies upon having selected the Lebesgue measure dx and it would be better to call this measure (or distribution) $f \, dx$; this abuse of notation is of no consequence for open sets of R^N, and corresponds to the usual identification of $L^2(\Omega)$ with its dual, but when one deals with a *differentiable manifold* one should remember that there is no preferred *volume form* like dx.

If $a \in \Omega$, the *Dirac mass* at a, which is denoted by δ_a, is defined by

$$\langle \delta_a, \varphi \rangle = \varphi(a) \text{ for all } \varphi \in C_c(\Omega), \qquad (4.5)$$

and it is a Radon measure (and therefore a distribution). If a sequence $a_n \in \Omega$ converges to the boundary $\partial\Omega$ of Ω and c_n is an arbitrary sequence, then $\mu = \sum_n c_n \delta_{a_n}$ is a Radon measure in Ω because in the formula $\langle \mu, \varphi \rangle = \sum_n c_n \varphi(a_n)$, only a finite number of a_n belong to the compact support K of φ.

Physicists use the notation $\delta(x - a)$ instead of δ_a, and they define $\delta(x)$ as the "function" which is 0 for $x \neq 0$ and has integral 1; of course there is no such function and it is actually a measure, but after studying Radon measures and distributions one learns which formulas are right, and one can then decide quickly if a formula used by physicists can be proven easily, or if it is a questionable one, either by showing that it is false or by noticing that mathematicians do not know yet how to make sense out of the formal computations used by physicists in that particular case.

One can create a lot of distributions by taking derivatives, and it is one of the important properties of distributions that they have as many derivatives as one wants, and as locally integrable functions are measures and therefore distributions, one then has a way to define their derivatives.

Definition 4.4. *If $\alpha = (\alpha_1, \dots, \alpha_N)$ is a multi-index with $\alpha_j \geq 0$ for $j = 1, \dots, N$, then for any distribution $T \in \mathcal{D}'(\Omega)$ one defines the distribution $D^\alpha T$ by the formula*

$$\langle D^\alpha T, \varphi \rangle = (-1)^{|\alpha|} \langle T, D^\alpha \varphi \rangle \text{ for all } \varphi \in C_c^\infty(\Omega). \ \square \qquad (4.6)$$

One must first check that $D^\alpha T$ *is a distribution*, i.e., for any compact $K \subset \Omega$ one must bound $|\langle D^\alpha T, \varphi \rangle|$ for $\varphi \in C_c^\infty(\Omega)$ with $support(\varphi) \subset K$, and the bound should only involve the sup norm of a fixed finite number of derivatives of φ. As $|\langle D^\alpha T, \varphi \rangle| = |(-1)^{|\alpha|} \langle T, D^\alpha \varphi \rangle| \leq C(K) \max_{|\beta| \leq m(K)} ||D^\beta(D^\alpha \varphi)||_\infty$, and as $D^{\alpha+\beta}$ is a derivation of order $\leq m(K) + |\alpha|$, this is bounded by $C(K) \max_{|\gamma| \leq m(K)+|\alpha|} ||D^\gamma \varphi||_\infty$, so $D^\alpha T$ is a distribution. One deduces that if T is a distribution of order $\leq m_0$ then $D^\alpha T$ is a distribution of order $\leq m_0 + |\alpha|$.

One must then check that the formula is *compatible with the notion of derivative for smooth functions*, i.e., if $f \in C^1(\Omega)$, then $\frac{\partial f}{\partial x_j} \in C^0(\Omega)$, and as both f and $\frac{\partial f}{\partial x_j}$ are locally integrable they are distributions and one must check that the derivative of the distribution associated to f (which should have been denoted by $f\,dx$) is the distribution associated to $\frac{\partial f}{\partial x_j}$, and this means that one should check that for every $\varphi \in C_c^\infty(\Omega)$ one has $\int_\Omega \left(\frac{\partial f}{\partial x_j} \varphi + f \frac{\partial \varphi}{\partial x_j} \right) dx = 0$, but this is $\int_\Omega \frac{\partial (f\varphi)}{\partial x_j}\,dx$, and because $f\varphi$ has compact support, one can invoke Fubini's theorem and one may start by integrating in x_j, and then in the other variables; one has to integrate on an open set O of R a function with compact support, and O could be made of a countably infinite number of open intervals, but only a finite number of intervals have to be taken into account, and for each of these intervals one integrates the derivative of a C^1 function vanishing near the ends of the interval and the integral is then 0.

The *Heaviside function* H is defined by

$$H(x) = \begin{cases} 0 \text{ for } x < 0 \\ 1 \text{ for } x > 0, \end{cases} \tag{4.7}$$

and one has $\left\langle \frac{dH}{dx}, \varphi \right\rangle = -\left\langle H, \frac{d\varphi}{dx} \right\rangle = -\int_0^\infty \frac{d\varphi}{dx}\,dx = \varphi(0)$ for all $\varphi \in C_c^\infty(R)$, so that

$$\frac{dH}{dx} = \delta_0. \tag{4.8}$$

Let $u = -1 + 2H$, which is the *sign function*, and let D be $\frac{d}{dx}$, so that $Du = 2\delta_0$; noticing that $u^3 = u$ and $u^2 = 1$, one discovers the following "paradox", that $D(u^3) = 2\delta_0$ but $3u^2 Du = 6\delta_0$. Of course one would have been in trouble with checking if $D(u^2)$, which is 0, coincides with $2u\,Du$, because the multiplication of u by δ_0 is not defined; one can actually multiply any Radon measure by a continuous function, but u is not continuous. At this point one should remember that products are not always defined, and this question will be considered in the next lecture.

[Taught on Monday January 24, 2000.]

5

Sobolev Spaces; Multiplication by Smooth Functions

With the notion of distributions, it is now easy to give the definition of what Sobolev spaces are.

Definition 5.1. *For a nonnegative integer m, for $1 \leq p \leq \infty$ and for an open set $\Omega \subset R^N$, the Sobolev space $W^{m,p}(\Omega)$ is the space of (equivalence classes of) functions $u \in L^p(\Omega)$ such that $D^\alpha u \in L^p(\Omega)$ for all derivations D^α of length $|\alpha| \leq m$. It is a normed space equipped with the norm $||u|| = \sum_{|\alpha| \leq m} ||D^\alpha u||_p$, or the equivalent norm*

$$||u||_{m,p} = \left(\int_\Omega \left(\sum_{|\alpha| \leq m} |D^\alpha u|^p \right) dx \right)^{1/p} \text{ if } 1 \leq p < \infty \qquad (5.1)$$
$$||u||_{m,\infty} = \max_{|\alpha| \leq m} ||D^\alpha u||_\infty \text{ if } p = \infty. \quad \square$$

Lemma 5.2. *(i) For $1 \leq p \leq \infty$ and $m \geq 0$ the Sobolev space $W^{m,p}(\Omega)$ is a Banach space.*
(ii) For $p = 2$, $W^{m,2}(\Omega)$ is denoted[1] by $H^m(\Omega)$ and is a Hilbert space, for the scalar product

$$((u,v)) = \int_\Omega \left(\sum_{|\alpha| \leq m} D^\alpha u \, \overline{D^\alpha v} \right) dx. \qquad (5.2)$$

Proof: (i) Let u_n be a Cauchy[2-4] sequence in $W^{m,p}(\Omega)$, i.e., for every $\varepsilon > 0$ there exists $n(\varepsilon)$ such that for $n, n' \geq n(\varepsilon)$ one has $||u_n - u_{n'}||_{m,p} \leq \varepsilon$.

---

[1] The specialists of harmonic analysis use the same notation for Hardy spaces, which I shall denote by $\mathcal{H}^q$ when using them (for $0 < q \leq \infty$).

[2] Augustin Louis CAUCHY, French mathematician, 1789–1857. He was made baron by CHARLES X. He worked in Paris, France, went into exile after the 1830 revolution and worked in Torino, Italy, returned from exile after the 1848 revolution, and worked in Paris again. The concept of a Cauchy sequence was first introduced a few years before him by BOLZANO.

[3] Charles-Philippe de France, 1757–1836, comte d'Artois, duc d'Angoulême, pair de France, was King of France from 1824 to 1830 under the name CHARLES X.

[4] Bernhard Placidus Johann Nepomuk BOLZANO, Czech mathematician and philosopher, 1781–1848. He worked in Prague (then in Austria, now capital of the Czech Republic).

This implies that for each multi-index α with $|\alpha| \leq m$ one has $||D^\alpha u_n - D^\alpha u_{n'}||_p \leq \varepsilon$, i.e., $D^\alpha u_n$ is a Cauchy sequence in $L^p(\Omega)$, and as $L^p(\Omega)$ is complete (because one uses the Lebesgue measure, for which the *Riesz–Fischer theorem*[5] applies), one deduces that $D^\alpha u_n \to f_\alpha$ in $L^p(\Omega)$ strong as $n \to \infty$. One must then prove that $f_\alpha = D^\alpha f_0$ and that u_n tends to f_0 in $W^{m,p}(\Omega)$. For this one uses the derivative in the sense of distributions, and for all $\varphi \in C_c^\infty(\Omega)$ and all multi-indices α with $|\alpha| \leq m$ one has

$$\langle D^\alpha f_0, \varphi \rangle = (-1)^{|\alpha|} \langle f_0, D^\alpha \varphi \rangle = (-1)^{|\alpha|} \lim_{n \to \infty} \langle f_n, D^\alpha \varphi \rangle = \lim_{n \to \infty} \langle D^\alpha f_n, \varphi \rangle = \langle f_\alpha, \varphi \rangle, \tag{5.3}$$

showing that $D^\alpha f_0 = f_\alpha$, so that $f_0 \in W^{m,p}(\Omega)$ and $D^\alpha u_n \to D^\alpha f_0$ in $L^p(\Omega)$ for $|\alpha| \leq m$; by taking the limit $n' \to \infty$, one finds that $||u_n - f_0||_{m,p} \leq \varepsilon$ for $n \geq n(\varepsilon)$, i.e., $u_n \to f_0$ in $W^{m,p}(\Omega)$.

(ii) The proposed formula for the scalar product is indeed linear in u and anti-linear in v, and for $v = u$ it gives the square of the norm.   □

In the proof, one has shown some kind of *continuity* for the derivations on $\mathcal{D}'(\Omega)$. Indeed, there exists a topology on $C_c^\infty(\Omega)$ for which the *dual* is $\mathcal{D}'(\Omega)$, on which one uses the *weak ⋆ topology* (which coincides with the *weak topology*); these topologies are *not metrizable*, but it is rarely necessary to know what they are; nevertheless, it is useful to know that a *sequence* T_n *converges to* T_∞ in $\mathcal{D}'(\Omega)$ if and only if

$$\langle T_n, \varphi \rangle \to \langle T_\infty, \varphi \rangle \text{ for all } \varphi \in C_c^\infty(\Omega), \tag{5.4}$$

but as the topology is not metrizable, one should remember that the knowledge of converging sequences does not characterize what the topology is. Any derivation D^α is indeed linear continuous from $\mathcal{D}'(\Omega)$ into itself, but the argument in (5.3) has only shown that it is sequentially continuous. Although it is rarely necessary to use the precise topology on $C_c^\infty(\Omega)$ or on $\mathcal{D}'(\Omega)$, it is useful to check that all the operations which one defines are sequentially continuous, and for that one should know what the convergence of a sequence in $C_c^\infty(\Omega)$ means, i.e.,

φ_n converges to φ_∞ in $C_c^\infty(\Omega)$ if and only if there exists
a compact $K \subset \Omega$ such that $support(\varphi_n) \subset K$ for all n and
for all multi-indices α one has $||D^\alpha \varphi_n - D^\alpha \varphi_\infty||_\infty \to 0$. (5.5)

The next step is to define multiplication of distributions by smooth functions.

Laurent SCHWARTZ has shown that it is *not possible to define a product for distributions in an associative way*, and more precisely he noticed that

$$\left(pv\frac{1}{x} \cdot x \right) \cdot \delta_0 = 1 \cdot \delta_0 = \delta_0, \text{ while } pv\frac{1}{x} \cdot (x \cdot \delta_0) = pv\frac{1}{x} \cdot 0 = 0, \tag{5.6}$$

---

[5] Ernst Sigismund FISCHER, Austrian-born mathematician, 1875–1954. He worked in Brünn (then in Austria-Hungary, now Brno, Czech Republic), in Erlangen, and in Köln (Cologne), Germany.

where the *principal value* of $\frac{1}{x}$ is the distribution denoted by $pv\frac{1}{x}$, defined by

$$\left\langle pv\frac{1}{x}, \varphi \right\rangle = \lim_{n \to \infty} \int_{n\,|x|>1} \frac{\varphi(x)}{x}\, dx \text{ for all } \varphi \in C_c^\infty(R). \tag{5.7}$$

CAUCHY had already defined the notion of principal value, so formulas like (5.7) were usual and Laurent SCHWARTZ proved that it defines a distribution of order ≤ 1, because if $support(\varphi) \subset [-a, +a]$, then

$$\int_{|x|\geq \frac{1}{n}} \frac{\varphi(x)}{x}\, dx = \int_{\frac{1}{n}\leq |x|\leq a} \frac{\varphi(x)-\varphi(0)}{x}\, dx \to \int_{|x|\leq a} \frac{\varphi(x)-\varphi(0)}{x}\, dx, \tag{5.8}$$

which exists because $|\varphi(x) - \varphi(0)| \leq |x|\,||D\,\varphi||_\infty$. This distribution satisfies $x\,pv\frac{1}{x} = 1$ (and it is odd and homogeneous of degree -1); that it is not a Radon measure can be seen by constructing a sequence of functions $\varphi_k \in C_c^\infty(R^N)$ which stay uniformly bounded, keep their support in a fixed compact set and for which $\langle pv\frac{1}{x}, \varphi_k \rangle \to +\infty$; taking φ_k nonnegative with support in $[0,1]$ and $\varphi_k(x) = 1$ for $\frac{1}{k} \leq x \leq 1 - \frac{1}{k}$, one has $\langle pv\frac{1}{x}, \varphi_k \rangle \geq \int_{\frac{1}{k}}^{1-\frac{1}{k}} \frac{dx}{x}$. Some physicists write formulas where

$$\pi\,\delta(x) \cdot \pi\,\delta(x) - \frac{1}{x}\cdot\frac{1}{x} \text{ could be equal to } \frac{C}{x^2}, \tag{5.9}$$

and we know that $\delta(x)$ should be δ_0 and is a Radon measure but not a function, and as $\frac{1}{x}$ and $\frac{1}{x^2}$ are not locally integrable functions because of the singularities at 0, making distributions out of them requires some care, and the case of $\frac{1}{x}$ leads to (5.7), and similarly HADAMARD had defined the *finite part* of $\frac{1}{x^k}$, and Laurent SCHWARTZ defined by analogy a distribution $fp\frac{1}{x^k}$. Neither $\delta_0 \cdot \delta_0$ nor $pv\frac{1}{x} \cdot pv\frac{1}{x}$ can be defined, but physicists do not pretend to define these products and they notice that the same "infinities" appear in trying to define them, so that the difference seems to make sense; it is not entirely clear what such a formula could mean, and one should check why physicists play with such quantities in order to discover what mathematical statements explain their observations, but one guess is that they need results of what I call *compensated integrability* or *compensated regularity*.[6],[7] Another possibility

---

[6] What I call *compensated integrability* or *compensated regularity* is a different notion than *compensated compactness*, a term which I have introduced with François MURAT, but which some authors have used out of the correct context, for designing a result of compensated regularity. For example, in R^2, $\frac{\partial f}{\partial x}\frac{\partial g}{\partial y}$ and $\frac{\partial f}{\partial y}\frac{\partial g}{\partial x}$ are not defined if $f, g \in W^{1,p}(R^2)$ with $1 \leq p < 2$, but $a = \frac{\partial f}{\partial x}\frac{\partial g}{\partial y} - \frac{\partial f}{\partial y}\frac{\partial g}{\partial x}$ can be defined by preferring to write $b = \frac{\partial}{\partial x}\left(f\frac{\partial g}{\partial y}\right) - \frac{\partial}{\partial y}\left(f\frac{\partial g}{\partial x}\right)$ or $c = \frac{\partial}{\partial x}\left(\frac{\partial f}{\partial y}g\right) - \frac{\partial}{\partial y}\left(\frac{\partial f}{\partial x}g\right)$, because b and c are defined for $f, g \in W^{1,p}(R^2)$ with $\frac{4}{3} \leq p < 2$, by using Sobolev's embedding theorem.

[7] François MURAT, French mathematician, born in 1947. He works at CNRS (Centre National de la Recherche Scientifique) at Université de Paris VI (Pierre et Marie Curie).

is to take $\varepsilon > 0$ and to observe that as ε tends to 0, $\frac{x}{x^2+\varepsilon^2}$ converges in the sense of distributions to $pv\frac{1}{x}$, and that $\frac{\varepsilon}{x^2+\varepsilon^2}$ converges in the sense of distributions to $\pi\,\delta_0$; then $S_\varepsilon = \frac{x-\varepsilon}{x^2+\varepsilon^2}$ converges in the sense of distributions to $S_0 = pv\frac{1}{x} - \pi\,\delta_0$, and $T_\varepsilon = \frac{x+\varepsilon}{x^2+\varepsilon^2}$ converges in the sense of distributions to $T_0 = pv\frac{1}{x} + \pi\,\delta_0$, but $S_\varepsilon T_\varepsilon = \frac{x^2-\varepsilon^2}{(x^2+\varepsilon^2)^2}$ is the derivative of $\frac{-x}{x^2+\varepsilon^2}$ and converges in the sense of distributions to $U_0 = -\frac{d}{dx}\left(pv\frac{1}{x}\right)$ (which is $fp\frac{1}{x^2}$, the finite part of $\frac{1}{x^2}$); it is tempting to say that $S_0 T_0$ is U_0, which is formally what (5.9) says. Lars HÖRMANDER has actually shown that some products of distributions are defined, using his notion of *wave front set*, and indeed $S_0 T_0$ is defined, but I have not checked what his definition gives for that product.

Definition 5.3. *If $T \in \mathcal{D}'(\Omega)$ and $\psi \in C^\infty(\Omega)$, then ψT (or $T\psi$) is the distribution defined by*

$$\langle \psi T, \varphi \rangle = \langle T, \psi\varphi \rangle \text{ for all } \varphi \in C_c^\infty(\Omega). \quad \Box \qquad (5.10)$$

Notice that one only defines the product of a distribution by a function in $C^\infty(\Omega)$ (or more generally of a distribution of order $\leq m$ by a function in $C^m(\Omega)$), but first one must *check that ψT is a distribution*, and this follows from *Leibniz's formula*. Leibniz's formula in one dimension states that

$$\frac{d^n(f\,g)}{dx^n} = \sum_{m=0}^{n} \binom{n}{m} \frac{d^m f}{dx^m} \frac{d^{n-m} g}{dx^{n-m}}, \qquad (5.11)$$

where the *binomial coefficient* $\binom{n}{m}$ is $\frac{n!}{m!(n-m)!}$; it is easily proven by induction, starting from $(f\,g)' = f'g + f\,g'$. Writing a generalization of Leibniz's formula to the N-dimensional case is simplified by using a notation for multi-indices.

Definition 5.4. *If $\alpha = (\alpha_1, \ldots, \alpha_N)$, then $|\alpha|$ means $|\alpha_1| + \ldots + |\alpha_n|$, $\beta \leq \alpha$ means $\beta_j \leq \alpha_j$ for $j = 1, \ldots, N$, $\alpha!$ means $\alpha_1! \ldots \alpha_N!$, and $\binom{\alpha}{\beta}$ means $\binom{\alpha_1}{\beta_1} \ldots \binom{\alpha_N}{\beta_N} = \frac{\alpha!}{\beta!(\alpha-\beta)!}$.* $\qquad \Box$

Then, Leibniz's formula has the same form

$$D^\alpha(f\,g) = \sum_{\beta \leq \alpha} \binom{\alpha}{\beta} D^\beta f \, D^{(\alpha-\beta)} g, \qquad (5.12)$$

and it is easily proven by induction on N.

One has $||D^\alpha(\psi\varphi)||_\infty \leq \sum_{\beta \leq \alpha} \binom{\alpha}{\beta} ||D^\beta \psi||_{L^\infty(K)} ||D^{\alpha-\beta}\varphi||_\infty$ if $support(\varphi)$ is contained in a compact $K \subset \Omega$, so that one has $\max_{|\alpha| \leq m} ||D^\alpha(\psi\varphi)||_\infty \leq C(K) \max_{|\alpha| \leq m} ||D^\alpha\varphi||_\infty$; one also deduces that if T is a distribution of order $\leq m$, then ψT is also a distribution of order $\leq m$.

One must also check that the notation is *compatible with the classical multiplication*, i.e., if $f \in L^1_{loc}(\Omega)$ and T is the corresponding distribution (which should be written $f\,dx$), then the distribution S associated to ψf is indeed ψT as was just defined. This follows from the definition, as $\langle S, \varphi \rangle = \int_\Omega (\psi f) \varphi \, dx = \int_\Omega f(\psi \varphi) \, dx = \langle T, \psi \varphi \rangle = \langle \psi T, \varphi \rangle$ for all $\varphi \in C_c^\infty(\Omega)$.

The mapping $(\psi, T) \mapsto \psi T$ from $C^\infty(\Omega) \times \mathcal{D}'(\Omega)$ into $\mathcal{D}'(\Omega)$ is sequentially continuous. The space $C^\infty(\Omega)$ is a Fréchet space and $\psi_n \to \psi_\infty$ in $C^\infty(\Omega)$ as $n \to \infty$ means that for every compact $K \subset \Omega$ and for every multi-index α, $D^\alpha \psi_n \to D^\alpha \psi_\infty$ uniformly on K as $n \to \infty$. The topology of $\mathcal{D}'(\Omega)$ is more technical to describe but a sequence T_n converges to T_∞ in $\mathcal{D}'(\Omega)$ if and only if $\langle T_n, \varphi \rangle \to \langle T_\infty, \varphi \rangle$ as $n \to \infty$, for all $\varphi \in C_c^\infty(\Omega)$. Because the subspace $C_K^\infty(\Omega)$ of the functions in $C_c^\infty(\Omega)$ which have their support in K is a Fréchet space, it has *Baire's property*,[8] from which *Banach–Steinhaus theorem*[9] follows, so that if $T_n \to T_\infty$ as $n \to \infty$, then there exists a constant $C(K)$ and an integer $m(K)$ independent of n such that $|\langle T_n, \chi \rangle| \leq C(K) \max_{|\alpha| \leq m(K)} \|D^\alpha \chi\|_\infty$ for all $\chi \in C_K^\infty(\Omega)$. Therefore $\langle (\psi_n T_n - \psi_\infty T_\infty), \varphi \rangle = \langle (\psi_n - \psi_\infty) T_n, \varphi \rangle + \langle \psi_\infty (T_n - T_\infty), \varphi \rangle$, so that one has $|\langle (\psi_n T_n - \psi_\infty T_\infty), \varphi \rangle| \leq |\langle T_n, (\psi_n - \psi_\infty) \varphi \rangle| + |\langle T_n - T_\infty, \psi_\infty \varphi \rangle|$, and the first term tends to 0 because for each multi-index α, $D^\alpha((\psi_n - \psi_\infty) \varphi)$ converges uniformly to 0 by using Leibniz's formula, and the second term tends to 0 by definition.

Proposition 5.5. *For $\psi \in C^\infty(\Omega)$, $T \in \mathcal{D}'(\Omega)$ and any multi-index α, one has*

$$D^\alpha(\psi T) = \sum_{\beta \leq \alpha} \binom{\alpha}{\beta} (D^\beta \psi)(D^{\alpha - \beta} T). \tag{5.13}$$

Proof: One proves that $\frac{\partial(\psi T)}{\partial x_j} = \frac{\partial \psi}{\partial x_j} T + \psi \frac{\partial T}{\partial x_j}$ for $j = 1, \ldots, N$, because for every $\varphi \in C_c^\infty(\Omega)$ one has $\langle \frac{\partial(\psi T)}{\partial x_j}, \varphi \rangle = -\langle \psi T, \frac{\partial \varphi}{\partial x_j} \rangle = -\langle T, \psi \frac{\partial \varphi}{\partial x_j} \rangle = \langle T, -\frac{\partial(\psi \varphi)}{\partial x_j} + \varphi \frac{\partial \psi}{\partial x_j} \rangle = \langle \frac{\partial T}{\partial x_j}, \psi \varphi \rangle + \langle \frac{\partial \psi}{\partial x_j} T, \varphi \rangle = \langle \frac{\partial \psi}{\partial x_j} T + \psi \frac{\partial T}{\partial x_j}, \varphi \rangle$. Then (5.13) follows by induction. □

If $\psi \in C^\infty(\Omega)$ and $a \in \Omega$, then $\psi \delta_a = \psi(a) \delta_a$, because $\langle \psi \delta_a, \varphi \rangle = \langle \delta_a, \psi \varphi \rangle = (\psi \varphi)(a) = \psi(a) \langle \delta_a, \varphi \rangle$ for all $\varphi \in C_c^\infty(\Omega)$. In particular $x_j \delta_0 = 0$ for $j = 1, \ldots, N$.

As $\langle x \, pv\frac{1}{x}, \varphi \rangle = \langle pv\frac{1}{x}, x \varphi \rangle = \lim_{n \to \infty} \int_{|x| \geq \frac{1}{n}} \frac{x \varphi(x)}{x} \, dx = \int_R \varphi(x) \, dx = \langle 1, \varphi \rangle$ for all $\varphi \in C_c^\infty(R)$, one has $x \, pv\frac{1}{x} = 1$. Notice that $x(pv\frac{1}{x} + C \delta_0) = 1$ for all C, but $pv\frac{1}{x}$ can be shown to be the only solution T of $x T = 1$ which is odd.

[Taught on Wednesday January 26, 2000.]

---

[8] René-Louis BAIRE, French mathematician, 1874–1932. He worked in Montpellier and in Dijon, France.

[9] Hugo Dyonizy STEINHAUS, Polish mathematician, 1887–1972. He worked in Lwów (then in Poland, now Lvov, Ukraine) until 1941, and after 1945 in Wrocław, Poland.

Density of Tensor Products; Consequences

Having enlarged the class of functions by introducing distributions, some partial differential equations that had been studied before may have gained new solutions which could not be considered before, and some other partial differential equations do not get new solutions. For example, the equation $x f = 0$ a.e. in R for $f \in L^1_{loc}(R)$ only has 0 as a solution, but for distributions $f \in \mathcal{D}'(R)$, $f = c\,\delta_0$ is a solution which is a Radon measure and not a (locally integrable) function (although physicists do call the Dirac measure a "function"); it is useful to know that one has found all the distributions $T \in \mathcal{D}'(R)$ solutions of $x\,T = 0$.

Definition 6.1. *The* tensor product $f_1 \otimes f_2$ *of a real (or complex) function f_1 defined on a set X_1 and a real (or complex) function f_2 defined on a set X_2 is the real (or complex) function defined on $X_1 \times X_2$ by the formula $(f_1 \otimes f_2)(x_1, x_2) = f_1(x_1)f_2(x_2)$ for all $(x_1, x_2) \in X_1 \times X_2$.* □

For defining convolution of distributions, Laurent SCHWARTZ used tensor products of distributions, and the uniqueness of tensor products requires that one shows the *density of finite combinations of tensor products* in a suitable topology, which is what Lemma 6.2 is about.

Lemma 6.2. *If $T \in \mathcal{D}'(R^{N_1} \times R^{N_2})$ and*

$$\langle T, \varphi_1 \otimes \varphi_2 \rangle = 0 \text{ for all } \varphi_1 \in C_c^{\infty}(R^{N_1}), \varphi_2 \in C_c^{\infty}(R^{N_2}), \qquad (6.1)$$

then $T = 0$.

Proof: First one assumes that T is a Radon measure μ and let $\psi \in C_c(R^{N_1} \times R^{N_2})$ with support in K. One chooses $\eta_1 \in C_c(R^{N_1})$ equal to 1 on $B(0, R_1)$ and $\eta_2 \in C_c(R^{N_2})$ equal to 1 on $B(0, R_2)$, so that $K \subset B(0, R_1) \times B(0, R_1)$, in order to have $\psi = (\eta_1 \otimes \eta_2)\psi$. One uses Weierstrass's theorem for approaching ψ uniformly by a sequence of polynomials P_n on a compact K' containing $support(\eta_1) \times support(\eta_2)$, and because every polynomial is a finite combination of tensor products, one obtains

$$\langle \mu, \psi \rangle = \langle \mu, (\eta_1 \otimes \eta_2)\psi \rangle = \lim_{n \to \infty} \langle \mu, (\eta_1 \otimes \eta_2)P_n \rangle = 0. \tag{6.2}$$

Then, if T is a distribution satisfying (6.1), one chooses two smoothing sequences $\varrho_{1\varepsilon} \in C_c^\infty(R^{N_1}), \varrho_{2\varepsilon} \in C_c^\infty(R^{N_2})$ and one defines $\mu_\varepsilon \in \mathcal{D}'(R^{N_1} \times R^{N_2})$ by

$$\langle \mu_\varepsilon, \varphi \rangle = \langle T, \varphi \star (\varrho_{1\varepsilon} \otimes \varrho_{2\varepsilon}) \rangle \text{ for all } \varphi \in C_c^\infty(R^{N_1} \times R^{N_2}). \tag{6.3}$$

Of course, μ_ε is a Radon measure, because the sup norm of any derivative of $\varphi \star (\varrho_{1\varepsilon} \otimes \varrho_{2\varepsilon})$ can be obtained from the sup norm of φ; μ_ε satisfies condition (6.1) because if $\varphi = \psi_1 \otimes \psi_2$ then $\varphi \star (\varrho_{1\varepsilon} \otimes \varrho_{2\varepsilon})$ is a tensor product, namely $(\psi_1 \star \varrho_{1\varepsilon}) \otimes (\psi_2 \star \varrho_{2\varepsilon})$, and by the first part $\mu_\varepsilon = 0$, and then

$$\langle T, \varphi \rangle = \lim_{\varepsilon \to 0} \langle T, \varphi \star (\varrho_{1\varepsilon} \otimes \varrho_{2\varepsilon}) \rangle = 0. \quad \square \tag{6.4}$$

Lemma 6.3. *If $T \in \mathcal{D}'(R^N)$ and $x_j T = 0$ for $j = 1, \ldots, N$, then there exists $C \in R$ such that $T = C\,\delta_0$.*

Proof: For $N = 1$, let $\varphi \in C_c^\infty(R)$ satisfies $\varphi(0) = 0$, then $\varphi(x) = x\,\psi(x)$ and $\psi \in C_c^\infty(R)$; indeed, Taylor's[1] expansion formula for φ near 0 is $\varphi(x) = \varphi'(0)x + \ldots + \frac{\varphi^{(n)}(0)}{n!}x^n + o(x^n)$, so that $\psi(x) = \varphi'(0) + \ldots + \frac{\varphi^{(n)}(0)}{n!}x^{n-1} + o(x^{n-1})$ shows that one must take $\psi(0) = \varphi'(0)$, and more generally $\psi^{(n-1)}(0) = \frac{\varphi^{(n)}(0)}{n}$ for $n \geq 1$, and as Leibniz's formula gives $\varphi^{(n)}(x) = x\,\psi^{(n)}(x) + n\,\psi^{(n-1)}(x)$, the derivatives of ψ are continuous at 0. One deduces that $\langle T, \varphi \rangle = \langle T, x\,\psi \rangle = \langle x\,T, \psi \rangle = 0$. Let $\theta \in C_c^\infty(R)$ with $\theta(0) = 1$, then every function $\varphi \in C_c^\infty(R)$ may be written in the form

$$\varphi(x) = \varphi(0)\theta(x) + x\,\psi(x) \text{ for a function } \psi \in C_c^\infty(R), \tag{6.5}$$

because $\varphi - \varphi(0)\theta$ vanishes at 0, so that

$$\langle T, \varphi \rangle = \langle T, \varphi(0)\theta + x\,\psi \rangle = \varphi(0)\langle T, \theta \rangle, \tag{6.6}$$

i.e., $T = C\,\delta_0$ with $C = \langle T, \theta \rangle$.

One uses an induction on the dimension N. For $\psi \in C^\infty(R^{N-1})$, the mapping $\varphi \mapsto \langle T, \varphi \otimes \psi \rangle$ for $\varphi \in C_c^\infty(R)$ defines an element of $\mathcal{D}'(R)$ that one denotes by U_ψ and because $\langle x_1 U_\psi, \varphi \rangle = \langle U_\psi, x_1\varphi \rangle = \langle T, x_1\varphi \otimes \psi \rangle = \langle x_1 T, \varphi \otimes \psi \rangle = 0$, one deduces that $x_1 U_\psi = 0$, so that $U_\psi = C(\psi)\delta_0$. One checks that $\psi \mapsto C(\psi)$ defines a distribution $V \in \mathcal{D}'(R^{N-1})$. For $j = 2, \ldots, N$, one has $\langle T, \varphi \otimes x_j\psi \rangle = \langle x_j T, \varphi \otimes \psi \rangle = 0$, i.e., $C(x_j\psi) = 0$, which means $\langle V, x_j\psi \rangle = 0$ or $x_j V = 0$, so that by induction $V = C_*\delta_0$, and therefore $\langle T, \varphi \otimes \psi \rangle = \varphi(0)C(\psi) = C_*\varphi(0)\psi(0)$. One deduces that $T = C_*\delta_0$ by Lemma 6.2, because $T - C_*\delta_0$ vanishes on tensor products. $\square$

---

[1] Brook TAYLOR, English mathematician, 1685–1731. He worked in London, England.

Lemma 6.4. *Let Ω be a connected open set of R^N. If $T \in \mathcal{D}'(\Omega)$ satisfies $\frac{\partial T}{\partial x_j} = 0$ for $j = 1, \ldots, N$, then T is a constant, i.e., there exists C such that $\langle T, \varphi \rangle = C \int_\Omega \varphi(x)\, dx$ for all $\varphi \in C_c^\infty(\Omega)$.*

Proof: By a connectedness argument it is enough to show the result with Ω replaced by any open cube $\Omega_0 \subset \Omega$. One uses an induction on the dimension N, and one starts with the case $N = 1$, Ω_0 being an interval (a, b). One notices that if $\varphi \in C_c^\infty(a, b)$ satisfies $\int_a^b \varphi(x)\, dx = 0$, then $\varphi = \frac{d\psi}{dx}$ for a function $\psi \in C_c^\infty(a, b)$, and ψ is given explicitly by $\psi(x) = \int_a^x \varphi(t)\, dt$. One chooses $\eta \in C_c^\infty(a, b)$ such that $\int_a^b \eta(x)\, dx = 1$, and then every $\varphi \in C_c^\infty(a, b)$ can be written as $\varphi = (\int_a^b \varphi(t)\, dt)\eta + \frac{d\psi}{dx}$ for a function $\psi \in C_c^\infty(a, b)$, because the integral of $\varphi - (\int_a^b \varphi(t)\, dt)\eta$ is 0, so for all $\varphi \in C_c^\infty(a, b)$ one has

$$\langle T, \varphi \rangle = \left\langle T, \left(\int_a^b \varphi(t)\, dt\right)\eta + \frac{d\psi}{dx} \right\rangle = \left(\int_a^b \varphi(t)\, dt\right) \langle T, \eta \rangle - \left\langle \frac{dT}{dx}, \psi \right\rangle = \\ C \left(\int_a^b \varphi(t)\, dt\right), \tag{6.7}$$

with $C = \langle T, \eta \rangle$, and that means $T = C$.

Writing $\Omega_0 = \omega \times (a, b)$ where ω is a cube in R^{N-1}, then for $\varphi \in C_c^\infty(\omega)$ one defines $T_\varphi \in \mathcal{D}'(a, b)$ by

$$\langle T_\varphi, \psi \rangle = \langle T, \varphi \otimes \psi \rangle \text{ for } \psi \in C_c^\infty(a, b), \tag{6.8}$$

and one checks immediately that this indeed defines a distribution T_φ on (a, b) because the bounds on derivatives of $\varphi \otimes \psi$ only involve a finite number of derivatives of ψ and the support of $\varphi \otimes \psi$ is the product of the supports of φ and of ψ, so that it stays in a fixed compact set when the support of ψ stays in a fixed compact set, φ being kept fixed. Then

$$\left\langle \frac{d T_\varphi}{dx_N}, \psi \right\rangle = -\left\langle T_\varphi, \frac{d\psi}{dx_N} \right\rangle = -\left\langle T, \varphi \otimes \frac{d\psi}{dx_N} \right\rangle = -\left\langle T, \frac{\partial(\varphi \otimes \psi)}{\partial x_N} \right\rangle = \\ \left\langle \frac{\partial T}{\partial x_N}, \varphi \otimes \psi \right\rangle = 0, \tag{6.9}$$

so that T_φ is a constant C_φ, i.e., $\langle T, \varphi \otimes \psi \rangle = C_\varphi \int_a^b \psi(t)\, dt$ for every $\varphi \in C_c^\infty(\omega)$ and $\psi \in C_c^\infty(a, b)$. One then uses this formula to show that $\varphi \mapsto C_\varphi$ defines a distribution S on ω, as it is obviously linear and in order to obtain the desired bounds one chooses a function $\psi \in C_c^\infty(a, b)$ with $\int_a^b \psi(t)\, dt \neq 0$ and the bounds for S follow easily from the bounds for T, so that one can write

$$\langle T, \varphi \otimes \psi \rangle = \langle S, \varphi \rangle \int_a^b \psi(t)\, dt \text{ for all } \varphi \in C_c^\infty(\omega), \psi \in C_c^\infty(a, b). \tag{6.10}$$

Then for $j = 1, \ldots, N-1$ one has

$$0 = \left\langle \frac{\partial T}{\partial x_j}, \varphi \otimes \psi \right\rangle = -\left\langle T, \frac{\partial(\varphi \otimes \psi)}{\partial x_j} \right\rangle = -\left\langle T, \frac{\partial \varphi}{\partial x_j} \otimes \psi \right\rangle = \\ -\left\langle S, \frac{\partial \varphi}{\partial x_j} \right\rangle \int_a^b \psi(t)\, dt = \left\langle \frac{\partial S}{\partial x_j}, \varphi \right\rangle \int_a^b \psi(t)\, dt \tag{6.11}$$

so that $\frac{\partial S}{\partial x_j} = 0$, and by the induction hypothesis S is a constant C_*, so that one has shown

$$\langle T, \varphi \otimes \psi \rangle = C_* \left(\int_\omega \varphi(y)\,dy \right) \left(\int_a^b \psi(t)\,dt \right) = C_* \int_{\Omega_0} (\varphi \otimes \psi)(x)\,dx, \quad (6.12)$$

which implies $T = C_*$ by Lemma 6.2, because $T - C_*$ vanishes on tensor products. □

Once multiplication has been defined, and Leibniz's formula has been extended, one can prove density results.

Lemma 6.5. *For $1 \le p < \infty$, and any integer $m \ge 0$, the space $C_c^\infty(R^N)$ is dense in $W^{m,p}(R^N)$.*

Proof: Let θ_n be a special truncating sequence, i.e., $\theta_n(x) = \theta_1\left(\frac{x}{n}\right)$, with $\theta_1 \in C_c^\infty(R^N)$, $0 \le \theta(x) \le 1$ on R^N and $\theta(x) = 1$ for $|x| \le 1$. For $u \in W^{m,p}(R^N)$, one defines $u_n = \theta_n u$, and one notices that $u_n \to u$ in $W^{m,p}(R^N)$ strong as $n \to \infty$. Indeed, one has $|u_n(x)| \le |u(x)|$ almost everywhere, and $u_n(x) \to u(x)$ as $n \to \infty$, and by the Lebesgue dominated convergence theorem one deduces that $u_n \to u$ in $L^p(R^N)$ strong as $n \to \infty$. Then for $|\alpha| \le m$ one has $D^\alpha u_n = \sum_{\beta \le \alpha} \binom{\alpha}{\beta} D^\beta \theta_n D^{\alpha - \beta} u$, and the term for $\beta = 0$ converges to $D^\alpha u$ again by the Lebesgue dominated convergence theorem, while the terms for $|\beta| > 0$ contains derivatives of θ_n which converge uniformly to 0, so that one has $D^\alpha u_n \to D^\alpha u$ in $L^p(R^N)$ strong as $n \to \infty$.

Then one approaches u_n by functions in $C_c^\infty(R^N)$ by convolution with a smoothing sequence ϱ_ε for a sequence of ε converging to 0, and using a diagonal argument there is a sequence $u_n \star \varrho_{\varepsilon(n)} \in C_c^\infty(R^N)$ which converges to u in $W^{m,p}(R^N)$ strong as $n \to \infty$. The crucial point is to notice that for $|\alpha| \le m$ one has

$$D^\alpha(\varrho_\varepsilon \star u_n) = \varrho_\varepsilon \star D^\alpha u_n,$$
$$\text{which converges to } D^\alpha u_n \text{ in } L^p(R^N) \text{ strong as } \varepsilon \to 0. \quad (6.13)$$

Indeed, for any test function $\varphi \in C_c^\infty(R^N)$, one has

$$\langle D^\alpha(\varrho_\varepsilon \star u_n), \varphi \rangle = (-1)^{|\alpha|} \langle \varrho_\varepsilon \star u_n, D^\alpha \varphi \rangle = (-1)^{|\alpha|} \langle u_n, \check{\varrho}_\varepsilon \star D^\alpha \varphi \rangle =$$
$$(-1)^{|\alpha|} \langle u_n, D^\alpha(\check{\varrho}_\varepsilon \star \varphi) \rangle = \langle D^\alpha u_n, \check{\varrho}_\varepsilon \star \varphi \rangle = \langle \varrho_\varepsilon \star D^\alpha u_n, \varphi \rangle,$$

$$(6.14)$$

where $\check{f}$ is defined by $\check{f}(x) = f(-x)$. □

For $p = \infty$, the same method shows that one can approach any $u \in W^{m,\infty}(R^N)$ by a sequence $\psi_n \in C_c^\infty(R^N)$ such that for every $|\alpha| \le m$, $D^\alpha \psi_n$ converges to $D^\alpha u$ in $L^\infty(R^N)$ weak $\star$ and $L^q_{loc}(R^N)$ strong for every finite q as $n \to \infty$.

If Ω is an open set of R^N, it is not true in general that $C_c^\infty(\Omega)$ is dense in $W^{m,p}(\Omega)$, and one is led to the following definition.

Definition 6.6. *$W_0^{m,p}(\Omega)$ is the closure of $C_c^\infty(\Omega)$ in $W^{m,p}(\Omega)$.* □

If the boundary $\partial\Omega$ is Lipschitz, then the functions of $W_0^{m,p}(\Omega)$ are 0 on the boundary, as will be seen later. If $\partial\Omega$ is too small, then it may happen that $W_0^{m,p}(\Omega) = W^{m,p}(\Omega)$; this is related to the fact that the functions in $W^{m,p}(\Omega)$ are not necessarily continuous, and Sobolev's embedding theorem, given now and proven later, tells for what values of m and p functions in $W^{m,p}(\Omega)$ are automatically continuous.

Theorem 6.7. *(Sobolev's embedding theorem[2,3]) (i) If $1 \leq p < \frac{N}{m}$ then $W^{m,p}(R^N) \subset L^r(R^N)$ with $\frac{1}{r} = \frac{1}{p} - \frac{m}{N}$, but $W^{m,p}(R^N)$ is not a subspace of $L^s(R^N)$ for $s > r$.*
(ii) If $p = \frac{N}{m}$ then $W^{m,p}(R^N) \subset L^q(R^N)$ for every $q \in [p,\infty)$, but $W^{m,N}(R^N)$ is not a subspace of $L^\infty(R^N)$ if $p > 1$; however, $W^{N,1}(R^N) \subset C_0(R^N)$.
(iii) If $\frac{N}{m} < p < \infty$ then $W^{m,p}(R^N) \subset C_0(R^N)$, the space of continuous functions tending to 0 at ∞. If $\frac{N}{k} < p < \frac{N}{k-1}$ for an integer k, then $W^{k,p}(R^N) \subset C^{0,\gamma}(R^N)$, the space of Hölder continuous functions of order γ, with $\gamma = k - \frac{N}{p}$. $\square$

For example if $\Omega = R^N \setminus F$, where F is a finite number of points and $p \leq \frac{N}{m}$, then $W_0^{m,p}(\Omega) = W^{m,p}(\Omega)$ and coincides with $W^{m,p}(R^N)$, as will be shown later.

It is useful to recall that any closed set K of R^N can be the zero set of a C^∞ function, because $R^N \setminus K$ can be written as the countable union of open balls $B(z_n, r_n)$, and if $\varphi \in C_c^\infty(R^N)$ has its support equal to the closed unit ball and is positive in the open unit ball, then one considers the series $\sum_n c_n \varphi\left(\frac{x - z_n}{r_n}\right)$ and one can choose the sequence c_n such that the series converges uniformly, as well as any of its derivatives. Therefore the zero set of a smooth function can be as irregular as one may wish (among closed sets, of course).

It is useful to recall that there are open sets with thick boundary, for example if one has an enumeration of the points with rational coordinates of R^N, $z_1, \ldots, z_n, \ldots$, and for $\varepsilon > 0$ one considers $A_\varepsilon = \bigcup_n B(z_n, \varepsilon\, 2^{-n})$, then A_ε is open, has Lebesgue measure $\leq \varepsilon$ and its boundary is its complement $R^N \setminus A_\varepsilon$ which has infinite Lebesgue measure.

[Taught on Friday January 28, 2000.]

---

[2] The part of (iii) concerning Hölder continuity seems to be an improvement due to MORREY.

[3] Charles Bradfield Jr. MORREY, American mathematician, 1907–1980. He worked at UCB (University of California at Berkeley), Berkeley, CA.

Extending the Notion of Support

As I recalled in footnote 59 of Lecture 1, for a continuous function u from a topological space X into a vector space, the support of u, denoted by *support(u)*, is the closure of the set of points $x \in X$ such that $u(x) \neq 0$. As mentioned in footnote 17 of Lecture 2, one needs a different definition for locally integrable functions, and one can actually define the support of any Radon measure or any distribution in a similar way, by characterizing the complement of the support as the largest open set on which u is 0; this leads to the following definition.

Definition 7.1. *A Radon measure $\mu \in \mathcal{M}(\Omega)$ is said to be 0 on an open subset $\omega \subset \Omega$ if*

$$\langle \mu, \varphi \rangle = 0 \text{ for all } \varphi \in C_c(\omega). \tag{7.1}$$

A distribution $T \in \mathcal{D}'(\Omega)$ is said to be 0 on an open subset $\omega \subset \Omega$ if

$$\langle T, \varphi \rangle = 0 \text{ for all } \varphi \in C_c^\infty(\omega). \; \square \tag{7.2}$$

For a Radon measure, being 0 on ω as in (7.1) or considering it as a distribution and being 0 on ω as in (7.2) coincide (by smoothing by convolution, as in the second part of the proof of Lemma 7.3). In order to define the support of a Radon measure or a distribution, one must deduce that being 0 on a family of open sets implies being 0 on its union, and this is done by using a *partition of unity.*

Lemma 7.2. *Let F be a closed set of R^N and let U_i, $i \in I$, be an open covering of F. Then for each $i \in I$ there exists $\theta_i \in C_c^\infty(U_i)$ with $0 \leq \theta_i \leq 1$ on R^N and $\sum_{i \in I} \theta_i = 1$ on an open set V containing F, the sum being locally finite, i.e., for each $x \in \bigcup_{i \in I} U_i$ there exists an open set W_x containing x such that only a finite number of θ_j are not identically 0 on W_x.*

Proof: Let $\varrho_1 \in C_c^\infty(R^N)$, with *support*$(\varrho_1) \subset \overline{B(0,1)}$, such that $\varrho_1 \geq 0$ and $\int_{R^N} \varrho_1(x)\,dx = 1$, and for $\varepsilon > 0$ let $\varrho_\varepsilon(x) = \varepsilon^{-N}\varrho_1\left(\frac{x}{\varepsilon}\right)$. For each $x \in F$ there exists $i(x) \in I$ such that $x \in U_{i(x)}$ and $0 < r(x) \leq 1$ such that

$B(x, 4r(x)) \subset U_{i(x)}$. For $n \geq 1$, let $F_n = \{x \in F \mid n - 1 \leq |x| \leq n\}$, and as F_n is compact and covered by the open balls $B(x, r(x))$ for $x \in F_n$, it is covered by a finite number of them, with centers $y \in G_n$, G_n being a finite subset of F_n. One chooses $\varepsilon_n = \min_{y \in G_n} r(y) > 0$, and for $y \in G_n$ one denotes by χ^y the *characteristic function* of $B(y, 2r(y))$ and $\alpha_n^y = \varrho_{\varepsilon_n} \star \chi^y$, so that $\alpha_n^y \in C_c^\infty(B(y, 3r(y))) \subset C_c^\infty(U_{i(y)})$ and $\alpha_n^y = 1$ on $B(y, r(y))$, and therefore $\beta_n = \sum_{y \in G_n} \alpha_n^y \geq 1$ on the open set $V_n = \bigcup_{y \in G_n} B(y, r(y))$ which contains F_n (and $\beta_n \geq 0$ elsewhere).

For $j \in I$, let η_j be the sum of all α_n^y for which $i(y) = j$; there might be an infinite number of such y, but the sum is locally finite and $\eta_j \in C^\infty(U_j)$ because when $n - 1 \leq |y| \leq n$ the function α_n^y is 0 outside $B(y, 3)$, because of the choice $r(y) \leq 1$; if F is compact, there are only a finite number of terms, so that $\eta_j \in C_c^\infty(U_j)$ in this case. Similarly, let $\zeta = \sum_{j \in I} \eta_j$, the sum being also locally finite and equal to $\sum_n \beta_n$, so that $\zeta \geq 1$ on $V = \bigcup_n V_n$. Choose $\psi \in C^\infty(R^N)$ such that $\psi = 0$ on $\overline{V}$ and $\psi > 0$ on $R^N \setminus \overline{V}$. For $j \in I$, let $\theta_j = \frac{\eta_j}{\zeta + \psi}$, which is C^∞ because $\zeta + \psi$ does not vanish (as $\zeta \geq 1$ and $\psi = 0$ on $\overline{V}$ and $\psi > 0$ and $\zeta \geq 0$ outside $\overline{V}$), so that $support(\theta_j) \subset support(\eta_j) \subset U_j$. One has $\sum_{j \in I} \theta_j = \frac{\zeta}{\zeta + \psi}$, which is 1 on V as $\psi = 0$ on V. $\square$

Lemma 7.3. *If a Radon measure $\mu \in \mathcal{M}(\Omega)$ or a distribution $T \in \mathcal{D}'(\Omega)$ is 0 on $\omega_i \subset \Omega$ for $i \in I$, then it is 0 on $\bigcup_{i \in I} \omega_i$.*

Proof: Let $\omega = \bigcup_{i \in I} \omega_i$ and $\varphi \in C_c^\infty(\omega)$ with support K. There is a finite number of functions $\theta_i \in C_c^\infty(\omega_i)$ with $\sum_i \theta_i = 1$ on K, so that $\varphi = \sum_{i \in I} \theta_i \varphi$, and as $\theta_i \varphi \in C_c^\infty(\omega_i)$ one has $\langle T, \theta_i \varphi \rangle = 0$ and by summing in i one deduces that $\langle T, \varphi \rangle = 0$.

If $\psi \in C_c(\omega)$, then for a smoothing sequence ϱ_n one defines $\varphi_n = \psi \star \varphi_n$, and for n large enough the support of all the φ_n stays in a fixed compact set K' of Ω; considering μ as a distribution, the preceding result shows that $\langle \mu, \varphi_n \rangle = 0$ for n large enough, but as $\varphi_n \to \psi$ uniformly on K' one has $\langle \mu, \psi \rangle = \lim_{n \to \infty} \langle \mu, \varphi_n \rangle = 0$. $\square$

Definition 7.4. *For a Radon measure $\mu \in \mathcal{M}(\Omega)$, the* support, *denoted by* $support(\mu)$, *is the closed set which is the complement of the largest open set where μ is 0. For a distribution $T \in \mathcal{D}'(\Omega)$, the* support, *denoted by* $support(T)$, *is the closed set which is the complement of the largest open set where T is 0.* $\square$

The definitions make sense because the largest open set is the union of all open sets where μ is 0, or where T is 0 (it is empty if no such open set exists). The second part of the proof of Lemma 7.3 shows that for $\mu \in \mathcal{M}(\Omega)$, the two definitions of being 0 coincide, so the two definitions of support coincide.

Partitions of unity will be useful for studying how functions in Sobolev spaces behave near the boundary of an open set. There are properties of Sobolev spaces which do depend upon the smoothness of the boundary $\partial\Omega$, but for some other properties the boundary plays no role, and these properties are said to be local, and they may be expressed for larger spaces.

Definition 7.5. *For an open set $\Omega \subset R^N$, an integer $m \geq 0$ and $1 \leq p \leq \infty$, the space $W_{loc}^{m,p}(\Omega)$ is the space of distributions $T \in \mathcal{D}'(\Omega)$ such that for every $\varphi \in C_c^\infty(\Omega)$ one has $\varphi T \in W^{m,p}(\Omega)$.*   □

One checks immediately that the space $L_{loc}^1(\Omega)$, described previously as the set of (equivalence classes of) Lebesgue-measurable functions T such that for any compact $K \subset \Omega$ one has $\chi_K T \in L^1(\omega)$, where χ_K is the characteristic function of K, is identical with the space described in Definition 7.5 for $m = 0$ and $p = 1$, which is the space of distributions T such that $\varphi T \in L^1(\Omega)$ for every $\varphi \in C_c^\infty(\Omega)$.

Of course, $W_{loc}^{m,p}(\Omega)$ is a Fréchet space, but (assuming $\Omega \neq \emptyset$) it is not a Banach space.

Sobolev's embedding theorem implies that for $1 \leq p < N$ one has $W^{1,p}(R^N) \subset L^{p^*}(R^N)$ with $\frac{1}{p^*} = \frac{1}{p} - \frac{1}{N}$, and one deduces that for any open set $\Omega \subset R^N$ one has $W^{1,p}(\Omega) \subset L_{loc}^{p^*}(\Omega)$. Indeed, for $\varphi \in C_c^\infty(\Omega)$ and $u \in W^{1,p}(\Omega)$, the function φu also belongs to $W^{1,p}(\Omega)$ and is 0 outside the support of φ, and by extending it by being 0 outside Ω, one finds a function $\widetilde{\varphi u} \in W^{1,p}(R^N) \subset L^{p^*}(R^N)$, showing that one has $\varphi u \in L^{p^*}(\Omega)$. Without a Lipschitz boundary, $W^{1,p}(\Omega)$ may not be a subspace of $L^{p^*}(\Omega)$.

Lemma 7.6. *(i) If $1 \leq p, q, r \leq \infty$ and $\frac{1}{r} = \frac{1}{p} + \frac{1}{q}$, then for $u \in W^{1,p}(\Omega)$ and $v \in W^{1,q}(\Omega)$ one has $u\,v \in W^{1,r}(\Omega)$ (and $||u\,v||_{1,r} \leq C||u||_{1,p}||v||_{1,q}$).*
(ii) If $1 \leq p, q, s < N$ and $\frac{1}{s} = \frac{1}{p} + \frac{1}{q} - \frac{1}{N}$, then for $u \in W^{1,p}(\Omega)$ and $v \in W^{1,q}(\Omega)$ one has $u\,v \in W_{loc}^{1,s}(\Omega)$.

Proof: The first part is a consequence of applying Hölder's inequality to the formula $\frac{\partial(u\,v)}{\partial x_j} = \frac{\partial u}{\partial x_j} v + u \frac{\partial v}{\partial x_j}$. To prove the formula, one must show that $-\int_\Omega u\,v \frac{\partial \varphi}{\partial x_j}\,dx = \int_\Omega \left(\frac{\partial u}{\partial x_j} v + u \frac{\partial v}{\partial x_j}\right)\varphi\,dx$ for every $\varphi \in C_c^\infty(\Omega)$. One chooses $\theta \in C_c^\infty(\Omega)$ with $\theta = 1$ on the support K of φ, and the formula to be proven does not change if one replaces u by $\theta\,u$ (as the derivative of θ vanishes on the support of φ), but as $\theta\,u$ extended by 0 outside Ω is a function of $W^{1,p}(R^N)$, one may approach it by a sequence $w_n \in C_c^\infty(R^N)$, and as the formula is true for u replaced by w_n one just lets n tend to ∞ and each term converges to the right quantity.

The second part is similar and consists in using the fact that on the support K of φ one has $u \in L^{p^*}(K)$ and $v \in L^{q^*}(K)$ (by the remark preceding Lemma 7.6, which relies on Sobolev's embedding theorem).   □

Definition 7.7. *For a nonempty set $A \subset R^N$, the space of Lipschitz continuous functions on A, denoted by $Lip(A)$, is the space of bounded (continuous) functions u on A such that there exists M with $|u(a) - u(b)| \leq M\,|b - a|$ for all $a, b \in A$; the space of locally Lipschitz continuous functions on A, denoted by $Lip_{loc}(A)$, is the space of (continuous) functions u on A such that for every* compact $K \subset A$ there exists $M(K)$ with $|u(a) - u(b)| \leq M(K)|b - a|$ for all $a, b \in K$, i.e., the restriction of u to K belongs to $Lip(K)$.   □

Of course, $Lip_{loc}(A)$ is a Fréchet space, and it is only a Banach space when A is compact, in which case it coincides with $Lip(A)$.

Lemma 7.8. *(i)* $W^{1,\infty}(R^N) = Lip(R^N)$.
(ii) If Ω is an open subset of R^N, then $Lip(\Omega) \subset W^{1,\infty}(\Omega)$, and $W^{1,\infty}(\Omega) \subset L^{\infty}(\Omega) \cap Lip_{loc}(\Omega)$.
(iii) If $u \in W^{1,\infty}(\Omega)$ and $||grad(u)||_\infty \leq K$, then one has $|u(x) - u(y)| \leq K\, d_\Omega(x,y)$, where d_Ω is the geodesic distance from x to y in Ω, the shortest length of a smooth path connecting x to y in Ω.

Proof: (i) If $u \in W^{1,\infty}(R^N)$, and ϱ_n is a special smoothing sequence, then $u_n = \varrho_n \star u \in C^\infty(R^N)$, $||u_n||_\infty \leq ||u||_\infty$ and $||\frac{\partial u_n}{\partial x_j}||_\infty \leq ||\frac{\partial u}{\partial x_j}||_\infty \leq ||grad(u)||_\infty$ for $j = 1, \ldots, N$, and as this inequality applies to any direction (not only the N directions of the canonical basis) it implies that $|grad(u_n)| \leq ||grad(u)||_\infty$ in R^N, so that $|u_n(x) - u_n(y)| \leq |x - y|\,||grad(u)||_\infty$ for all $x, y \in R^N$; as a subsequence u_m of u_n converges almost everywhere to u as $m \to \infty$, one deduces that $|u(x) - u(y)| \leq |x - y|\,||grad(u)||_\infty$ for almost every $x, y \in R^N$, i.e., u is Lipschitz continuous with Lipschitz constant $||grad(u)||_\infty$. Conversely, if $u \in Lip(R^N)$ and $u_n = u \star \varrho_n \in C^\infty(R^N)$, then for any $h \in R^N$ one has $u_n - \tau_{sh}u_n = (u - \tau_{sh}u) \star \varrho_n$, implying $||u_n - \tau_{sh}u_n||_\infty \leq ||u - \tau_{sh}u||_\infty \leq K\,s\,|h|$, where K is the Lipschitz constant of u, so that after dividing by s and letting s tend to 0 one deduces that $||grad(u_n)||_\infty \leq K$ and then letting n tend to ∞ gives $||grad(u)||_\infty \leq K$.

(ii) The preceding argument is valid if $u \in Lip(\Omega)$, as $u \star \varrho_n$ is well defined at a short distance from the boundary.

(iii) The passage from a bound on $||grad(u)||_\infty$ to a bound on $|u(x) - u(y)|$ for $u \in C^\infty(\Omega)$ relies on the fact that the segment $[xy]$ is included in Ω, and it can be replaced by the sum of the lengths of segments for a polygonal path joining x to y and staying inside Ω, and the infimum of these quantities is the geodesic distance $d_\Omega(x,y)$.   □

If Ω is the open subset of R^2 defined in polar coordinates by $-\pi < \theta < \pi$ and $r > 1$, then the function $u = \theta$ satisfies $\frac{\partial u}{\partial x} = -\frac{y}{r^2}$ and $\frac{\partial u}{\partial y} = \frac{x}{r^2}$, so that $u \in W^{1,\infty}(\Omega)$, but for $\varepsilon > 0$ small the points with Cartesian coordinates $(-2, -\varepsilon)$ and $(-2, +\varepsilon)$ are at Euclidean[1,2] distance 2ε while the difference in values of θ is converging to 2π as $\varepsilon \to 0$ (the geodesic distance tends to $2\sqrt{3} + \frac{4\pi}{3}$).

[Taught on Monday January 31, 2000.]

---

[1] EUCLID of Alexandria, "Egyptian" mathematician, about 325 BCE–265 BCE. It is not known where he was born, but he worked in Alexandria, Egypt, shortly after it was founded by ALEXANDER the Great, in 331 BCE.

[2] Alexandros Philippou Makedonon, 356 BCE–323 BCE, was King of Macedon as ALEXANDER III, and is referred to as ALEXANDER the Great, in relation with the large empire that he conquered.

8

Sobolev's Embedding Theorem, $1 \leq p < N$

One has $H^1(R^2) = W^{1,2}(R^2) \subset L^p(R^2)$ for every $p \in [2,\infty)$ by Sobolev's embedding theorem, but the similar property does not hold for all open sets Ω, and a Lipschitz boundary will be assumed.

Lemma 8.1. *Let $\Omega = \{(x,y) \mid 0 < x < 1, 0 < y < x^2\} \subset R^2$. Then $H^1(\Omega) \not\subset L^p(\Omega)$ for $p > 6$.*

Proof: One checks for which value $\alpha \in R$ the function $u(x) = x^\alpha$ belongs to $L^q(\Omega)$ and as $\int_0^1 x^{\alpha q+2}\,dx = \int_\Omega x^{\alpha q}\,dx\,dy$, one finds that the condition is $\alpha q + 2 > -1$. Applying this remark to u and $\frac{\partial u}{\partial x}$ and $q = 2$, one finds that $u \in H^1(\Omega)$ if and only if $2(\alpha - 1) + 2 > -1$, i.e., $\alpha > -\frac{1}{2}$, and $u \in L^p(\Omega)$ if and only if $\alpha \frac{p}{6} > -\frac{1}{2}$, i.e., $H^1(\Omega)$ is not a subset of $L^p(\Omega)$ for $p > 6$. $\quad\square$

Of course, the same limitations occur for other *cusps* on the boundary, like for $\Omega = \{(x,y) \mid 0 < x < 1, 0 < y < x^\gamma\}$, with $\gamma > 1$.

A part of Sobolev's embedding theorem asserts that for $1 \leq p < N$ one has $W^{1,p}(R^N) \subset L^{p^*}(R^N)$ with $\frac{1}{p^*} = \frac{1}{p} - \frac{1}{N}$, or $p^* = \frac{Np}{N-p}$, and one deduces then that $W^{1,p}(R^N) \subset L^q(R^N)$ for every $q \in [p, p^*]$ by the following application of Hölder's inequality.

Lemma 8.2. *If $1 \leq p_0 < p_\theta < p_1 \leq \infty$, then*

$$\|u\|_{p_\theta} \leq \|u\|_{p_0}^{1-\theta}\|u\|_{p_1}^\theta \text{ for all } u \in L^{p_0}(\Omega) \cap L^{p_1}(\Omega)$$
$$\text{with } \theta \in (0,1) \text{ defined by } \frac{1}{p_\theta} = \frac{1-\theta}{p_0} + \frac{\theta}{p_1}. \tag{8.1}$$

Proof: One applies Hölder's inequality $\|f g\|_1 \leq \|f\|_q \|g\|_{q'}$ with $f = |u|^{(1-\theta)p_\theta}$ and $g = |u|^{\theta p_\theta}$, with $q = \frac{p_0}{(1-\theta)p_\theta}$ and $q' = \frac{p_1}{\theta p_\theta}$, which are conjugate exponents. $\quad\square$

The preceding result is not restricted to the Lebesgue measure, and the restriction that $p_0 \geq 1$ is not necessary (although the notation $\|v\|_r$ is not a norm for $0 < r < 1$).

Sobolev's embedding theorem is natural if one considers the question of *scaling*.

Lemma 8.3. *If $1 \le q < p$ or $1 \le p < N$ and $q > p^*$, defined by $\frac{1}{p^*} = \frac{1}{p} - \frac{1}{N}$, there is no finite constant C such that $||u||_q \le C\,||u||_{1,p}$ for all $u \in C_c^\infty(R^N)$.*

Proof: For $\lambda > 0$, one applies the inequality

$$||u||_q \le C\,||u||_p + C\,||grad(u)||_p \text{ for } u \in C_c^\infty(R^N) \qquad (8.2)$$

to the function v defined by

$$v(x) = u\left(\frac{x}{\lambda}\right) \text{ for } x \in R^N, \qquad (8.3)$$

and one notices that for $1 \le r < \infty$ one has

$$||v||_r = \left(\int_{R^N} \left|u\left(\frac{x}{\lambda}\right)\right|^r dx\right)^{1/r} = \left(\int_{R^N} |u(y)|^r \lambda^N dy\right)^{1/r} = \lambda^{N/r}||u||_r$$
$$||grad(v)||_r = \lambda^{-1+N/r}||grad(u)||_r, \qquad (8.4)$$

so that, if (8.2) is true one deduces

$$\lambda^{N/q}||u||_q \le C\,\lambda^{N/p}||u||_p + C\,\lambda^{-1+N/p}||grad(u)||_p \text{ for } u \in C_c^\infty(R^N), \quad (8.5)$$

i.e., an inequality of the form

$$||u||_q \le C\,\lambda^\alpha ||u||_p + C\,\lambda^\beta ||grad(u)||_p \text{ for all } \lambda > 0$$
$$\text{with } \alpha = \frac{N}{p} - \frac{N}{q}, \beta = \frac{N}{p^*} - \frac{N}{q}. \qquad (8.6)$$

If one had $\alpha < 0$ and $\beta < 0$, then by letting λ tend to ∞ one would deduce the contradiction $||u||_q = 0$ for all $u \in C_c^\infty(R^N)$; this corresponds to the case $q < p$. Similarly, if one had $\alpha > 0$ and $\beta > 0$ one would deduce the same contradiction by letting λ tend to 0; this corresponds to the case $p < N$ and $q > p^*$.   □

If the inequality is true for $q = p^*$, then the same argument shows that one has $||u||_{p^*} \le C\,||grad(u)||_p$ for all $u \in C_c^\infty(R^N)$. However, this is not a proof that the inequality is true, as for example the inequality $||u||_\infty \le C\,||grad(u)||_N$ implies no contradiction by the preceding scaling argument, but it is not true for $N > 1$.

One reason why one cannot deduce by a scaling argument that the limiting case of the Sobolev's embedding theorem does not hold for $p = N$, is that in the larger context of the Lorentz[1,2] spaces all the spaces $L^{p,q}(R^N)$ scale in the same way for different values of $q \in [1, \infty]$. If all the partial derivatives of u are estimated in $L^{N,1}(R^N)$ it does provide a bound for the sup norm

---

[1] George Gunther LORENTZ, Russian-born mathematician, born in 1910. He worked in Toronto, Ontario (Canada), at Wayne State University, Detroit, MI, in Syracuse, NY, and at University of Texas, Austin, TX.
[2] Anthony WAYNE, American general, 1745–1796.

of u, while for any $q > 1$ there exist unbounded functions v with all partial derivatives in $L^{N,q}(R^N)$.

The method of Sergei SOBOLEV for proving his famous embedding theorem for $W^{1,p}(R^N)$ in the case $1 < p < N$ was based on properties of an *elementary solution* of the *Laplacian*

$$\Delta = \sum_{i=1}^{N} \frac{\partial^2}{\partial x_i^2}. \tag{8.7}$$

Definition 8.4. *If $P(\xi) = \sum_{\alpha} a_{\alpha} \xi^{\alpha}$ is a polynomial in $\xi \in R^N$ (with constant coefficients), and $P(D)$ denotes the differential operator $P(D) = \sum_{\alpha} a_{\alpha} D^{\alpha}$, an elementary solution of $P(D)$ is any distribution E such that $P(D)E = \delta_0$.*   □

Elementary solutions are not unique, but a particular elementary solution may often be selected by using symmetry arguments, and in the case of Δ, one finds a radial function of the form

$$E = \begin{cases} C_N r^{2-N} \text{ for } N \geq 3 \\ C_2 \log r \text{ for } N = 2 \\ \frac{r}{2} \text{ for } N = 1, \end{cases} \tag{8.8}$$

with $r^2 = \sum_{i=1}^{N} x_i^2$. Anticipating the properties of convolution with distributions, one has

$$u = u \star \Delta E = \sum_{i=1}^{N} \frac{\partial u}{\partial x_i} \star \frac{\partial E}{\partial x_i}, \tag{8.9}$$

and Sobolev's embedding theorem would be a consequence of Young's inequality (2.3) if one had $\frac{\partial E}{\partial x_i} \in L^{N'}(R^N)$, but as these derivatives are of the order of r^{1-N} this fails to be the case; however, Sergei SOBOLEV proved that Young's inequality (2.3) still holds if instead of a function in $L^q(R^N)$ one uses the function $r^{-N/q}$. This line of argument does not work for $p = 1$, and that case was proven by Louis NIRENBERG, but a more important generalization was noticed later by Jaak PEETRE, using the theory of interpolation that he had developed in parallel with Jacques-Louis LIONS; actually, the particular result of interpolation in Lorentz spaces is also a consequence of a result obtained by O'NEIL[3], who was extending a result of HARDY and LITTLEWOOD[4] about *nonincreasing rearrangements*.

A second method for proving Sobolev's embedding theorem was developed independently by Emilio GAGLIARDO and by Louis NIRENBERG, but the same

---

[3] Richard Charles O'NEIL, American mathematician. He has worked at Rice University, Houston, TX, and in Albany, NY.

[4] John Edensor LITTLEWOOD, English mathematician, 1885–1977. He worked in Manchester and in Cambridge, England, where he held the newly founded Rouse Ball professorship (1928–1950).

idea has also been used by Olga LADYZHENSKAYA.[5,6] Of course, one always proves inequalities for smooth functions with compact support, and then one extends the results to Sobolev spaces by density.

Lemma 8.5. $W^{1,1}(R) \subset C_0(R)$ *and*

$$||f||_\infty \leq \frac{1}{2} \left|\left|\frac{df}{dx}\right|\right|_1 \quad for \ all \ f \in W^{1,1}(R). \tag{8.10}$$

Proof: One takes $f \in C_c^\infty(R)$ and by density (8.10) holds for $f \in W^{1,1}(R)$. From

$$f(x) = \int_{-\infty}^x \frac{df}{dx}(y)\,dy = -\int_x^\infty \frac{df}{dx}(y)\,dy, \tag{8.11}$$

one deduces $|f(x)| \leq \int_\infty^x |\frac{df}{dx}|\,dy$ and $|f(x)| \leq \int_x^\infty |\frac{df}{dx}|\,dy$, and adding gives $2|f(x)| \leq \int_R |\frac{df}{dx}|\,dx$; varying x gives (8.10). The constant $\frac{1}{2}$ cannot be improved by taking f increasing and then decreasing. Every element $f \in W^{1,1}(R)$ is a limit of a sequence $f_n \in C_c^\infty(R)$ and f_n converges uniformly, because

$$||f_n - f_m||_{C_0(R)} = ||f_n - f_m||_\infty \leq \frac{1}{2}\left|\left|\frac{df_n}{dx} - \frac{df_m}{dx}\right|\right|_1 \to 0. \ \square \tag{8.12}$$

Lemma 8.6. *For $N \geq 2$, and $i = 1, \ldots, N$, let f_i be a measurable function independent of x_i, and assume that $f_i \in L^{N-1}$ in its $N-1$ variables, i.e.,*

$$\frac{\partial f_i}{\partial x_i} = 0; \ f^i = f_i\Big|_{x_i=0} \in L^{N-1}(R^{N-1}), \tag{8.13}$$

then

$$F = \prod_{i=1}^N f_i \in L^1(R^N) \ and \ ||F||_1 \leq \prod_{i=1}^N ||f^i||_{N-1}. \tag{8.14}$$

Proof: For $N = 2$ one has $F(x_1, x_2) = f^1(x_2)f^2(x_1)$ and $||F||_1 = ||f^1||_1||f^2||_1$. For $N \geq 3$, let

$$g_i = \left(\int_R |f_i|^{N-1}dx_N\right)^{1/(N-2)}, i = 1, \ldots, N-1; \ G = \prod_{i=1}^{N-1} g_i, \tag{8.15}$$

so g_i is independent of x_i and x_N, $g^i \in L^{N-2}$ in its $N-2$ arguments, and $||g^i||_{N-2} \leq ||f^i||_{N-1}^{(N-1)/(N-2)}$; by *induction* one has $G \in L^1$ in the arguments $x_1, \ldots, x_{N-1}$. By Hölder's inequality, one has

[5] Olga Aleksandrovna LADYZHENSKAYA, Russian mathematician, 1922–2004. She worked at the Steklov Mathematical Institute, in Leningrad, USSR, then St Petersburg, Russia.
[6] Vladimir Andreevich STEKLOV, Russian mathematician, 1864–1926. He worked in Kharkov, and in St Petersburg (then Petrograd, USSR), Russia.

$\int_R |F| \, dx_N \le \prod_{i=1}^{N-1} g_i^{(N-2)/(N-1)} f_N = G^{(N-2)/(N-1)} f^N,$

$\int_{R^N} |F| \, dx \le \left(\int_{R^{N-1}} |G| \, dx_1 \dots dx_{N-1} \right)^{\frac{N-2}{N-1}} ||f^N||_{N-1} \le \prod_{i=1}^{N} ||f^i||_{N-1}. \ \Box$

$$(8.16)$$

Lemma 8.7. *For $a > 1$ and $1 < p < \infty$, one has*

$$\left(\int_{R^N} |u|^{N a/(N-1)} \, dx \right)^{(N-1)/N} \le \frac{a}{2} \left(\prod_{i=1}^{N} \left| \left| \frac{\partial u}{\partial x_i} \right| \right|_p \right)^{1/N} \left(\int_{R^N} |u|^{(a-1)p'} \, dx \right)^{1/p'}$$

for all $u \in C_c^\infty(R^N)$.

$$(8.17)$$

For $p = 1$ one has

$$||u||_{1^*} \le \frac{1}{2} \left(\prod_{i=1}^{N} \left| \left| \frac{\partial u}{\partial x_i} \right| \right|_1 \right)^{1/N} \quad \text{for all } u \in C_c^\infty(R^N). \tag{8.18}$$

Proof: Applying Lemma 8.5 with $f = |u|^a$ for $u \in C_c^\infty(R^N)$ (because $|u|^a$ is of class C^1 and Lemma 8.5 only requires the function to belong to $W^{1,1}(R)$), one obtains $|u(x)|^a \le \frac{a}{2} \int_R |u|^{a-1} \left| \frac{\partial u}{\partial x_i} \right| dx_i = f_i^{N-1}$, where one has chosen

$$f_i = \left(\frac{a}{2} \int_R |u|^{a-1} \left| \frac{\partial u}{\partial x_i} \right| dx_i \right)^{1/N-1} \quad \text{for } i = 1, \dots, N, \tag{8.19}$$

and then Lemma 8.6 implies

$$\int_{R^N} |u(x)|^{N a/(N-1)} \, dx \le \left| \left| \prod_{i=1}^{N} f_i \right| \right|_1 \le \prod_{i=1}^{N} ||f^i||_{N-1} =$$

$$\left(\frac{a}{2} \right)^{N/(N-1)} \prod_{i=1}^{N} \left(\int_{R^N} |u|^{a-1} \left| \frac{\partial u}{\partial x_i} \right| dx \right)^{1/(N-1)} \le \tag{8.20}$$

$$\left(\frac{a}{2} \right)^{N/(N-1)} \prod_{i=1}^{N} \left| \left| \frac{\partial u}{\partial x_i} \right| \right|_p^{1/(N-1)} \left(\int_{R^N} |u|^{(a-1)p'} \, dx \right)^{N/(N-1)p'},$$

and taking the power $(N-1)/N$ gives (8.17). For $p = 1$, one takes $a = 1$ and one uses the first line of (8.20). $\qquad \Box$

Sobolev's embedding theorem $W^{1,p}(R^N) \subset L^{p^*}(R^N)$ follows in the case $1 \le p < N$ by choosing a such that $\frac{N a}{N-1} = (a-1)p'$, and this common value appears to be p^*. This proof does not give the best constants.

[Taught on Wednesday February 2, 2000.]

Sobolev's Embedding Theorem, $N \leq p \leq \infty$

If one writes $\Lambda = \left(\prod_{i=1}^{N} \left|\left|\frac{\partial u}{\partial x_i}\right|\right|_p\right)^{1/N}$, and one defines Φ by

$$\Phi(a) = \int_{R^N} |u|^{N a/(N-1)} \, dx, \qquad (9.1)$$

then the case $p = N$ of (8.17) is

$$\Phi(a) \leq \left(\frac{a \Lambda}{2}\right)^{N/(N-1)} \Phi(a-1) \text{ for all } a > 1, \text{ where } \Lambda = \left(\prod_{i=1}^{N} \left|\left|\frac{\partial u}{\partial x_i}\right|\right|_N\right)^{1/N}. \qquad (9.2)$$

One deduces that a bound on $||u||_{1,N}$, which gives a bound on $\Phi(N-1) = ||u||_N^N$ and a bound on Λ, implies a bound for $||u||_q$ for all $q > N$, and more precisely

$$\Phi(a) \leq \left(\frac{a!}{(N-1)!}\right)^{N/(N-1)} \left(\frac{\Lambda}{2}\right)^{(a+1-N)N/(N-1)} ||u||_N^N \text{ for all integers } a \geq N. \qquad (9.3)$$

After taking the logarithm by using *Stirling's formula*,[1-3] $a! \approx \left(\frac{a}{e}\right)^a \sqrt{2\pi a}$, one finds

$$\limsup_{a \to \infty} \frac{1}{a} \left(\Phi(a)\right)^{(N-1)/(N a)} \leq \frac{\Lambda}{2e}, \text{ or } \limsup_{q \to \infty} \frac{||u||_q}{q} \leq \frac{(N-1)\Lambda}{2 N e}. \qquad (9.4)$$

Lemma 9.1. *For* $u \in W^{1,N}(R^N)$, *there exists* $\varepsilon > 0$ *depending on* $||grad(u)||_N$ *such that* $e^{\varepsilon |u|}$ *is locally integrable.*

---

[1] It is an improvement by STIRLING of a formula obtained by DE MOIVRE.

[2] James STIRLING, Scottish-born mathematician, 1692–1770. He worked in London, England, and then as a manager in a mining company in Scotland.

[3] Abraham DE MOIVRE, French-born mathematician, 1667–1754. He moved to London, England, but could not obtain an academic position.

Proof: One chooses $\kappa > \frac{(N-1)\Lambda}{2N\,e}$ and $\varepsilon > 0$ such that $\varepsilon\,\kappa\,e < 1$. By (9.4) one has $||u||_q \leq \kappa\,q$ for $q \geq q_c$ with q_c large enough $\geq N$, so that

$$\int_{R^N} \sum_{q=q_c}^{\infty} \frac{(\varepsilon\,|u|)^q}{q!}\,dx = \sum_{q=q_c}^{\infty} \int_{R^N} \frac{(\varepsilon\,|u|)^q}{q!}\,dx \leq \sum_{q=q_c}^{\infty} \frac{(\varepsilon\,\kappa\,q)^q}{q!} < \infty. \qquad (9.5)$$

For a compact K, estimating $\int_K e^{\varepsilon\,|u|}\,dx$ requires also a bound of $\int_K |u|^r\,dx$ for $0 \leq r < q_c$, which by Hölder's inequality and (9.3) can be estimated in terms of $||u||_N$ and $meas(K)$. □

Lemma 9.1 was obtained in a different way by Louis NIRENBERG and Fritz JOHN[4] as a property of the space $BMO(R^N)$ (bounded mean oscillation[5]), which they had introduced in part for studying the limiting case $p = N$ of Sobolev's embedding theorem.

For the case $p > N$, one notices that when $(a-1)p'$ and $\frac{N\,a}{N-1}$ are equal they take the negative value $-\frac{N\,p}{p-N}$, so that if one writes $q_k = -\frac{N\,p}{p-N} + \alpha\,\beta^k$ with $\beta = \frac{N(p-1)}{p(N-1)} > 1$, then the choice of a giving $(a-1)p' = q_k$ gives $\frac{N\,a}{N-1} = q_{k+1}$; one chooses $\alpha = \frac{p^2}{p-N}$ so that $q_0 = p$. Using $a \leq \frac{q_k}{p'} \leq \frac{\alpha}{p'}\beta^k$, one finds that

$$\int_{R^N} |u|^{q_{k+1}}\,dx \leq \left(\frac{\alpha\,\Lambda\,\beta^k}{2p'}\right)^{N/(N-1)} \left(\int_{R^N} |u|^{q_k}\,dx\right)^{\beta}. \qquad (9.6)$$

This gives an estimate of $||u||_{q_k}$ for all k in terms of $||u||_{1,p}$ (which gives a bound on $||u||_p$ and on Λ) and k; as $||u||_{\infty} = \lim_{r\to\infty}||u||_r$ one must show that $||u||_{q_k}$ is bounded independently of k. By homogeneity of the formula, one has $q_{k+1} = \frac{N}{N-1} + q_k\beta$, so that if one puts $|u| = \frac{\alpha\,\Lambda}{2p'}|v|$, the formula becomes $\int_{R^N} |v|^{q_{k+1}}\,dx \leq \beta^{k\,N/(N-1)}\left(\int_{R^N} |v|^{q_k}\,dx\right)^{\beta}$. Writing $f(k) = \log\left(\int_{R^N}|v|^{q_k}\,dx\right)$, one deduces $f(k+1) \leq A\,k + \beta\,f(k)$ with $A = \frac{N\log\beta}{N-1}$, and by induction this gives $f(k) \leq A\left((k-1) + (k-2)\beta + \ldots + 2\beta^{k-3} + \beta^{k-2}\right) + \beta^k f(0)$ for $k \geq 2$, and as $q_k = -\frac{N\,p}{p-N} + \alpha\,\beta^k$ one finds that $\frac{f(k)}{q_k} \to \frac{1}{\alpha}\left(f(0) + \frac{A}{(\beta-1)^2}\right)$, giving a bound for $||u||_{\infty}$ in terms of $||u||_{1,p}$.

A different way to obtain a bound in $L^{\infty}(R^N)$, following Sergei SOBOLEV method is to replace the elementary solution E of Δ by a *parametrix*,[6] which instead of solving $\Delta\,E = \delta_0$ satisfies

$$\Delta\,F = \delta_0 + \psi \text{ with } \psi \in C_c^{\infty}(R^N), \qquad (9.7)$$

---

[4] Fritz JOHN, German-born mathematician, 1910–1994. He worked at University of Kentucky, Lexington, KY, and at NYU (New York University), New York, NY.
[5] $BMO(R^N)$ is the space of locally integrable function for which there exists C such that $\int_Q |u - u_Q|\,dx \leq C\,meas(Q)$ for every cube Q, denoting by u_Q the average of u on Q, i.e., $meas(Q)u_Q = \int_Q u(x)\,dx$.
[6] The word has been coined by HADAMARD.

and it is natural to take

$$F = \theta \, E \text{ with } \theta \in C_c^\infty(R^N) \text{ equal to 1 in a small ball around 0.} \quad (9.8)$$

The derivatives $\frac{\partial F}{\partial x_j}$ are $O(r^{1-N})$ near 0 as the partial derivatives of E, but F has compact support, so that $grad(F) \in L^q(R^N)$ for $1 \leq q < \frac{N}{N-1}$, and in particular for $q = p'$ if $p > N$, so one deduces

$$u = u \star (\Delta \, F - \psi) = \sum_{i=1}^N \frac{\partial u}{\partial x_i} \star \frac{\partial F}{\partial x_i} - u \star \psi$$
$$\text{so that } ||u||_\infty \leq \sum_{i=1}^n ||\frac{\partial u}{\partial x_i}||_p ||\frac{\partial F}{\partial x_i}||_{p'} + ||u||_p ||\psi||_{p'}. \quad (9.9)$$

I suppose that it is the way Sergei SOBOLEV had proven that $W^{1,p}(R^N) \subset L^\infty(R^N)$ for $N < p < \infty$; he might have known that by density of $C_c^\infty(R^N)$ in $W^{1,p}(R^N)$ one has actually $W^{1,p}(R^N) \subset C_0(R^N)$, and he must have known the argument of scaling that leads to (9.10), but probably not the argument for obtaining (9.11), as the Hölder continuity property is attributed to MORREY.

Lemma 9.2. *For $p > N$, one has $W^{1,p}(R^N) \subset C^{0,\gamma}(R^N)$ with $\gamma = 1 - \frac{N}{p}$, and*

$$||u||_\infty \leq C \, ||u||^{1-\theta} ||grad(u)||_p^\theta \text{ with } \theta = \frac{N}{p} \text{ for all } u \in W^{1,p}(R^N), \quad (9.10)$$

and

$$|u(x) - u(y)| \leq C \, |x - y|^\gamma ||grad(u)||_p \text{ for all } u \in W^{1,p}(R^N), \quad (9.11)$$

Proof: Whatever the way one has obtained a bound $||u||_\infty \leq A \, ||u||_p + B \, ||grad(u)||_p$ when $p > N$, the scaling argument implies (9.10): one applies the inequality to $v(x) = u(\frac{x}{\lambda})$, and that gives $||u||_\infty \leq A \, |\lambda|^{N/p} ||u||_p + B \, |\lambda|^{-1+N/p} ||grad(u)||_p$, and then one chooses the best $\lambda > 0$, and that gives (9.10).

Integrating $\frac{d}{dt} u(x - t \, h)$ from 0 to 1 one obtains

$$|u(x - h) - u(x)| \leq |h| \int_0^1 |grad(u)|(x + t \, h) \, dt, \quad (9.12)$$

and taking the norm in $L^p(R^N)$ of both sides (and using the triangle inequality) one obtains

$$||\tau_h u - u||_p \leq |h| \, ||grad(u)||_p, \quad (9.13)$$

and as $||grad(\tau_h u - u)||_p \leq 2||grad(u)||_p$, applying (9.10) to $\tau_h u - u$ gives (9.11). $\square$

That Sobolev's embedding theorem cannot be improved is shown by constructing counter-examples. For instance, if $\varphi \in C_c^\infty(R^N)$ is equal to 1 in a small ball around 0, then $r^\alpha \varphi \in L^p(R^N)$ is equivalent to $p\alpha + N - 1 > -1$, i.e., $\alpha > -\frac{N}{p}$ and $r^\alpha \varphi \in W^{1,p}(R^N)$ is equivalent to $\alpha - 1 > -\frac{N}{p}$; if $1 \leq p < N$

and $q > p^*$ one can choose α such that $\alpha - 1 > -\frac{N}{p}$ but $\alpha < -\frac{N}{q}$, giving a function in $W^{1,p}(R^N)$ which does not belong to $L^q(R^N)$.

For the case $p = N$, one considers functions $|\log r|^\beta \varphi$, and one finds that $|\log r|^\beta \varphi \in W^{1,N}(R^N)$ if and only if $|\log r|^{\beta-1} r^{-1}\varphi \in L^N(R^N)$, i.e., if and only if $N(\beta - 1) < -1$, and there exists $\beta > 0$ satisfying this inequality if $N > 1$ (for $N = 1$, Lemma 8.5 has shown that $W^{1,1}(R) \subset C_0(R)$).

As mentioned before, Sobolev's embedding theorem can be made more precise by using Lorentz spaces, $L^{p,q}$, which increase with q, with $L^{p,p} = L^p$ and $L^{p,\infty}$ a space introduced by MARCINKIEWICZ,[7] which is the space of (equivalence classes of) measurable functions satisfying $\int_\omega |f|\, dx \le C\, meas(\omega)^{1/p'}$ for every measurable set ω. Jaak PEETRE has proven that for $1 < p < N$ one has $W^{1,p}(R^N) \subset L^{p,p^*}(R^N)$, and I proved the case $p = 1$, extending the result of Louis NIRENBEG that $W^{1,1}(R^N) \subset L^{1^*}(R^N)$. For $p = N$, the result of Fritz JOHN and Louis NIRENBEG using $BMO(R^N)$ was improved by Neil TRUDINGER,[8] who proved that if $u \in W^{1,N}(R^N)$, then for every $C > 0$ one has $e^{C|u|^{N/(N-1)}} \in L^1_{loc}(R^N)$, but that is not true for all functions in $BMO(R^N)$ (because $C\log r$ belongs to $BMO(R^N)$). The result was extended by Haïm BREZIS[9] and Stephen WAINGER[10] who proved that if u has all its partial derivatives in the space $L^{N,q}(R^N)$, with $1 < q < \infty$, then $e^{C|u|^{q'}} \in L^1_{loc}(R^N)$ for every $C > 0$.

Questions about the best constants in Sobolev's embedding theorems have been investigated by Thierry AUBIN[11] and by Giorgio TALENTI[12]; a good class of functions for finding the optimal constants are those of the form $\frac{1}{(1+a\,r^2)^k}$.

The preceding results can be extended to functions having derivatives $\frac{\partial u}{\partial x_j} \in L^{p_j}(R^N)$, not all p_j being equal (it occurs naturally if one coordinate denotes time and the others denote space, for example). In 1978, I visited Trento, Italy, and heard a talk on this subject by Alois KUFNER,[13] who followed the natural approach of Emilio GAGLIARDO[14] or Louis NIRENBERG,

---

[7] Józef MARCINKIEWICZ, Polish mathematician, 1910–1940. He worked in Wilno, then in Poland, now Vilnius, Lithuania. He died during World War II, presumably executed by the Soviets with thousands of other Polish officers.

[8] Neil Sidney TRUDINGER, Australian mathematician, born in 1942. He works at Australian National University, Canberra, Australia.

[9] Haïm R. BREZIS, French mathematician, born in 1944. He works at Université Paris VI (Pierre et Marie Curie), Paris (and it seems at Rutgers University, Piscataway, NJ.).

[10] Stephen WAINGER, American mathematician, born in 1936. He works at University of Wisconsin, Madison, WI.

[11] Thierry AUBIN, French mathematician, born in 1942. He worked in Lille, and at Université Paris VI (Pierre et Marie Curie), Paris, France.

[12] Giorgio G. TALENTI, Italian mathematician, born in 1940. He works in Firenze (Florence), Italy.

[13] Alois KUFNER, Czech mathematician. He works in Prague, Czech Republic.

[14] Just after the talk, I met Emilio GAGLIARDO, whom I had first met the week before in Pavia, and learnt that he was also teaching in Trento; he was no longer

as the method of Sergei SOBOLEV cannot be used in this case (at least, I do not see how one could use it), but I learnt afterwards that it had been obtained earlier by TROISI.[15] I have obtained a generalization of all these methods for the case where the partial derivatives belong to different Lorentz spaces, by a different method (the methods that have been described do not seem to be sufficient for proving such a general result).

[Taught on Friday February 4, 2000.]

---

interested in the ideas that he had introduced in the past, and he continued the explanations that he had given me a few days before on his favorite subject, applying mathematics to music.

[15] Mario TROISI, Italian mathematician, born in 1934. He worked in Salerno, Italy.

10

Poincaré's Inequality

For $1 \leq p < \infty$, one usually takes the norm of the space $W^{1,p}(R^N)$ to be

$$\left(\int_{R^N} |u|^p \, dx + \int_{R^N} |grad(u)|^p \, dx \right)^{1/p}, \qquad (10.1)$$

but one should notice that adding $\int_{R^N} |u|^p \, dx$ and $\int_{R^N} |grad(u)|^p \, dx$ is a strange practice, which mathematicians follow almost all the time, and which makes physicists wonder if mathematicians know what they are talking about, because they ignore the question of units. In real problems x usually denotes the space variables, which are measured in units of length (L), while t denotes the time variable, measured in units of time (T), and if one considers the *wave equation*

$$\frac{\partial^2 u}{\partial t^2} - c^2 \sum_{j=1}^{N} \frac{\partial^2 u}{\partial x_j^2} = 0, \qquad (10.2)$$

c is a characteristic velocity, measured in units $L\,T^{-1}$, and the equation is consistent as each of the terms of the equation is measured in units $U\,T^{-2}$, whatever the unit U for u is (u could be a vertical displacement if one looks at small waves on the surface of a lake or a swimming pool and $N = 2$ in that case, or a pressure if one looks at propagation of sound in the atmosphere, or in the ocean, or in the ground, and $N = 3$ in that case). For nonlinear equations, like the *Burgers equation*[1]

$$\frac{\partial u}{\partial t} + u \frac{\partial u}{\partial x} = 0, \qquad (10.3)$$

the dimension of u must be that of a velocity $L\,T^{-1}$, but some physicists prefer to introduce a characteristic velocity c and write it $\frac{\partial u}{\partial t} + c\,u\,\frac{\partial u}{\partial x} = 0$ and in that case u has no dimension.

---

[1] Johannes Martinus BURGERS, Dutch-born mathematician, 1895–1981. He worked at University of Maryland, College Park, MD.

Mathematicians studying equations from continuum mechanics or physics should be careful about the question of units, and as Sobolev spaces were originally introduced for studying solutions of partial differential equations from continuum mechanics or physics, this question does occur naturally in studying them. The quantities $\int_{R^N} |u|^p \, dx$ and $\int_{R^N} |grad(u)|^p \, dx$ are not measured in the same units, the first term having dimension $U^p L^N$ and the second one having dimension $U^p L^{N-p}$, and it would be more natural when dealing with physical problems to use a norm like

$$\left(\int_{R^N} |u|^p \, dx + L_0^p \int_{R^N} |grad(u)|^p \, dx \right)^p, \qquad (10.4)$$

where L_0 is a characteristic length, but as was already noticed when the argument of scaling was used in relation with Sobolev's embedding theorem, one can start from an inequality written without paying attention to units and then deduce from it a better one which does take into account this question.

An important remark is that for some open sets Ω and some particular subspaces of $W^{1,p}(\Omega)$ one can avoid adding the terms $\int_\Omega |u|^p \, dx$, because Poincaré's inequality[2] holds.

Definition 10.1. *If* $1 \leq p \leq \infty$ *and* Ω *is a nonempty open subset of* R^N, *one says that Poincaré's inequality holds for a subspace* V *of* $W^{1,p}(\Omega)$ *if there exists a constant* C *such that one has* $||u||_p \leq C \, ||grad(u)||_p$ *for all* $u \in V$. $\qquad \Box$

Of course, C then has the dimension of a length, and if there is no characteristic length that one can attach to Ω, then one expects that Poincaré's inequality does not hold.

Lemma 10.2. *(i) If* $meas(\Omega) < \infty$ *and if the constant function 1 belongs to a subspace* V *of* $W^{1,p}(\Omega)$, *then Poincaré's inequality does not hold on* V.
(ii) Poincaré's inequality does not hold on $W_0^{1,p}(\Omega)$ *if* Ω *contains arbitrarily large balls, i.e., if there exists a sequence* $r_n \to \infty$ *and points* $x_n \in \Omega$ *such that* $B(x_n, r_n) \subset \Omega$.
(iii) If Ω *is included in a strip of width* d, *i.e., there exists* $\xi \in R^N$ *with* $|\xi| = 1$ *and* $\Omega \subset \{x \in R^N \mid \alpha < (\xi.x) < \beta\}$ *and* $d = \beta - \alpha$, *then* $||u||_p \leq C_0 \, d \, ||grad(u)||_p$ *for all* $u \in W_0^{1,p}(\Omega)$, *where* C_0 *is a universal constant, i.e., independent of which* Ω *is used.*
(iv) If $p = \infty$, *Poincaré's inequality holds on* $W_0^{1,\infty}(\Omega)$ *if and only if there exists* $C < \infty$ *such that for all* $x \in \Omega$ *one has* $dist(x, \partial\Omega) \leq C$, *where dist is the Euclidean distance.*
(v) If $meas(\Omega) < \infty$, *then Poincaré's inequality holds for* $W_0^{1,p}(\Omega)$ *for* $1 \leq p \leq \infty$, *and one has* $||u||_p \leq C(p) meas(\Omega)^{1/N} ||grad(u)||_p$ *for all* $u \in W_0^{1,p}(\Omega)$.
(vi) If the injection of V *into* $L^p(\Omega)$ *is compact, then Poincaré's inequality holds on a subspace* V *of* $W^{1,p}(\Omega)$ *if and only if the constant function 1 does not belong to* V.

---

[2] I have been told that this kind of inequality was introduced by POINCARÉ in his work on tides.

Proof: (i) If $meas(\Omega) < \infty$, then $1 \in W^{1,p}(\Omega)$, but as $grad(1) = 0$ one must have $C = 0$, which is incompatible with $1 \in V$.

(ii) Let $\varphi \in C_c(R^N)$ with $\varphi \neq 0$ and $support(\varphi) \subset B(0,1)$, then one defines u_n by $u_n(x) = \varphi\left(\frac{x - x_n}{r_n}\right)$, which belongs to $C_c^\infty(\Omega)$, and one has $||u_n||_p = r_n^{N/p}||\varphi||_p$ and $||grad(u_n)||_p = r_n^{-1+N/p}||grad(\varphi)||_p$; if the inequality was true one would have $1 \leq \frac{C}{r_n}$, so that if Poincaré's inequality holds on $W_0^{1,p}(\Omega)$ it gives an upper bound for the size of balls included in Ω.

(iii) One starts from the case $N = 1$, where one has $\max_{x \in R} |v(x)| \leq \frac{1}{2}\int_R \left|\frac{dv}{dx}\right| dx$ for all $v \in C_c^\infty(R)$, so if for an interval $I = (\alpha, \beta)$ one has $u \in C_c^\infty(I)$, one deduces that $||u||_p \leq ||u||_\infty d^{1/p}$ and as $\int_I \left|\frac{du}{dx}\right| dx \leq \left|\left|\frac{du}{dx}\right|\right|_p d^{\frac{1}{p'}}$, one deduces that $||u||_p \leq \frac{d}{2}\left|\left|\frac{du}{dx}\right|\right|_p$ for all $u \in C_c^\infty(I)$ (this argument does not give the best constant $C(p)$ for $1 \leq p < \infty$). One deduces the case of the strip by applying the preceding inequality in an orthogonal basis whose last vector is $e_N = \xi$, so that the strip is defined by $\alpha < x_N < \beta$, and for each choice of $x' = (x_1, \ldots, x_{N-1})$ one has $\int_\alpha^\beta |u(x', x_N)|^p dx_N \leq 2^{-p}d^p \int_\alpha^\beta \left|\frac{\partial u}{\partial x_N}(x', x_N)\right|^p dx_N$, and one integrates then this inequality in x' in order to obtain Poincaré's inequality in the case $1 \leq p < \infty$.

(iv) In the case $p = \infty$, the condition is necessary because of (ii), and it is sufficient because for each $x \in \Omega$ there exists $z \in \partial\Omega$ with $|x - z| \leq C$, and if $u \in C_c^\infty(\Omega)$ there exists y on the segment $[x, z]$ and outside the support of u such that $|u(x)| = |u(x) - u(y)| \leq C ||grad(u)||_\infty$; then the same inequality extends to $W_0^{1,\infty}(\Omega)$.

(v) If $p = \infty$, it follows from (iv). If $1 \leq p < \infty$, one chooses $q < N$ such that $1 \leq q \leq p < q^*$, and one uses Sobolev's embedding theorem $||u||_{q^*} \leq C ||grad(u)||_q$ for all $u \in C_c^\infty(R^N)$, and Hölder's inequality: $||u||_p \leq ||u||_{q^*} \cdot meas(\Omega)^\alpha$ with $\alpha = \frac{1}{p} - \frac{1}{q^*}$ and $||grad(u)||_q \leq ||grad(u)||_p meas(\Omega)^\beta$ with $\beta = \frac{1}{q} - \frac{1}{p}$, so that $\alpha + \beta = \frac{1}{N}$. Without the precise estimate for the constant it can also be proven by the compactness argument used in (vi). There is a different proof for the case $p = 2$ based on Fourier transform, and also a proof of the compactness property using Fourier transform, and they will be shown later.

(vi) The necessity that 1 should not belong to V follows from (i). That this condition is sufficient is the consequence of what I call the *equivalence lemma* (Lemma 11.1), by taking $E_1 = \overline{V}$, $A = grad$ and $E_2 = \left(L^p(\Omega)\right)^N$ and B the injection into $E_3 = L^p(\Omega)$. □

Of course, if $\Omega_1 \subset \Omega_2$ and Poincaré's inequality holds for $W_0^{1,p}(\Omega_2)$, then it holds for $W_0^{1,p}(\Omega_1)$, because each function of $u \in W_0^{1,p}(\Omega_1)$ can be extended by 0 and gives a function $\widetilde{u} \in W_0^{1,p}(\Omega_2)$.

[Taught on Monday February 7, 2000.]

11

The Equivalence Lemma; Compact Embeddings

Questions of equivalence of norms play an important role in the theoretical part of numerical analysis, because *interpolation formulas* or *quadrature formulas* are used on a triangulation made with small elements and it is important to know how the errors behave in terms of the size of these elements. In 1974, I was told about a classical result named after James BRAMBLE[1] and one of his students named HILBERT[2]; I discovered later that it is based on results by Jacques DENY[3] and Jacques-Louis LIONS. I developed a more general framework, which also generalizes a different type of result by Jaak PEETRE, which I had seen mentioned in a footnote, in a book by Jacques-Louis LIONS and Enrico MAGENES in order to prove the *Fredholm alternative*[4] for elliptic boundary value problems. I call my framework (Lemma 11.1) the *equivalence lemma*, and at a theoretical level it is useful in order to obtain many variants of Poincaré inequalities in various subspaces of Sobolev spaces, but it requires enough regularity of the boundary in order to satisfy the hypothesis of compactness.

Lemma 11.1. *(equivalence lemma) Let E_1 be a Banach space, E_2, E_3 normed spaces (with $\| \cdot \|_j$ denoting the norm of E_j), and let $A \in \mathcal{L}(E_1, E_2)$, $B \in \mathcal{L}(E_1, E_3)$ such that one has:*
(a) $\|u\|_1 \approx \|A u\|_2 + \|B u\|_3$
(b) B is compact.
Then one has the following properties:
(i) The kernel of A is finite dimensional.

---

[1] James H. BRAMBLE, American mathematician. He has worked at University of Maryland, College Park, MD, at Cornell University, Ithaca, NY, and at Texas A&M University, College Station TX.

[2] Stephen R. HILBERT, American mathematician. He works in Ithaca, NY.

[3] Jacques DENY, French mathematician, born in 1918. He worked at Université Paris-Sud XI, Orsay, France, where he was my colleague from 1975 to 1982.

[4] Erik Ivar FREDHOLM, Swedish mathematician, 1866–1927. He worked in Stockholm, Sweden.

(ii) The range of A is closed.

(iii) There exists a constant C_0 such that if F is a normed space and $L \in \mathcal{L}(E_1, F)$ satisfies $L u = 0$ whenever $A u = 0$, then one has $||L u||_F \leq C_0 ||L|| \, ||A u||_2$ for all $u \in E_1$.

(iv) If G is a normed space and $M \in \mathcal{L}(E_1, G)$ satisfies $M u \neq 0$ whenever $A u = 0$ and $u \neq 0$, then $||u||_1 \approx ||A u||_2 + ||M u||_G$.

Proof: (i) On $X = ker(A)$, the closed unit ball for $|| \cdot ||_1$ is compact. Indeed if $||u_n||_1 \leq 1$, then $B u_n$ stays in a compact of E_3 by (b) so a subsequence $B u_m$ converges in E_3, so that it is a Cauchy sequence in E_3, and as $A u_m = 0$ it is a Cauchy sequence in E_2 and therefore (a) implies that u_m is a Cauchy sequence in E_1, which converges as E_1 is a Banach space. By a theorem of F. RIESZ, X must be finite dimensional.

(ii) As a consequence of the Hahn[5]–Banach theorem, X being finite dimensional has a topological supplement Y, i.e., $X \cap Y = \{0\}$ and there exist $\pi_X \in \mathcal{L}(E_1, X)$ and $\pi_Y \in \mathcal{L}(E_1, Y)$ such that $e = \pi_X(e) + \pi_Y(e)$ for all $e \in E_1$, and in particular Y is closed as it is the kernel of π_X, so that Y is a Banach space.

One shows then that there exists $\alpha > 0$ such that $||A u||_2 \geq \alpha ||u||_1$ for all $u \in Y$. Indeed, if it was not true there would exist a sequence $y_n \in Y$ with $||y_n||_1 = 1$ and $A y_n \to 0$, and again taking a subsequence such that $B y_m$ converges in E_3 one finds that y_m would be a Cauchy sequence in Y and its limit $y_\infty \in Y$ would satisfy $A y_\infty = 0$, i.e., $y_\infty \in X$, so that $y_\infty = 0$, contradicting the fact that $||y_\infty||_1 = \lim_m ||y_m||_2 = 1$.

Then if $f_n \in R(A)$ satisfies $f_n \to f_\infty$ in E_2, one has $f_n = A e_n = A(\pi_X e_n + \pi_Y e_n) = A \pi_Y e_n$. If one writes $y_n = \pi_Y e_n$, one has $\alpha ||y_n - y_m||_1 \leq ||A y_n - A y_m||_2 = ||f_n - f_m||_2$ so that y_n is a Cauchy sequence in Y and its limit y_∞ satisfies $A y_\infty = f_\infty$, showing that $R(A)$ is closed.

(iii) As A is a bijection from Y onto $R(A)$ it has an inverse D, and as one considers $R(A)$ equipped with the norm of E_2, $D \in \mathcal{L}(R(A), Y)$ with $||D|| \leq \frac{1}{\alpha}$ by the previously obtained inequality (it shows that $R(A)$ is a Banach space, although one has not assumed that E_2 is a Banach space, and the closed graph theorem has not been used). With this definition of D one has $y = D A y$ for all $y \in Y$, and in particular $D A e = \pi_Y e$ for all $e \in E_1$ because $A e = A \pi_Y e$. From the hypothesis $L u = 0$ for $u \in X$, one has $L e = L \pi_Y e = L D A e$ for all $e \in E_1$, so that $||L e||_F \leq ||L|| \, ||D|| \, ||A e||_2$, and therefore C_0 may be taken to be the norm of D in $\mathcal{L}(R(A), Y)$.

(iv) One has $||A e||_2 + ||M e||_G \leq (||A|| + ||M||)||e||_1$, and if the norms were not equivalent one could find a sequence $e_n \in E_1$ with $||e_n||_1 = 1$ and $||A e_n||_2 + ||M e_n||_G \to 0$. As before, a subsequence e_m would be such that $B e_m$ is a Cauchy sequence in E_3, and as $A e_m \to 0$ in E_2, e_m would be a Cauchy sequence in E_1, converging to a limit e_∞, which would satisfy the contradictory properties $||e_\infty||_1 = 1$, $A e_\infty = 0$ and $M e_\infty = 0$. □

---

[5] Hans HAHN, Austrian mathematician, 1879–1934. He worked in Vienna, Austria.

Other applications of the equivalence lemma 11.1 will be encountered later, but a crucial hypothesis is the compact injection assumption without which the result may be false (but it is not always false); for example, taking $1 \leq p < \infty$ and $E_1 = W^{1,p}(R)$, $A = \frac{d}{dx}$, $E_2 = E_3 = L^p(R)$ and B the injection of $W^{1,p}(R)$ into $L^p(R)$ (which is not compact), then the range of A is not closed, and its closure is $L^p(R)$ if $p > 1$, and the subspace of functions in $L^1(R)$ with integral 0 if $p = 1$.

A compactness result, attributed to RELLICH[6],[7] and to KONDRAŠOV,[8] asserts that if Ω is a bounded open set of R^N with a continuous boundary $\partial\Omega$, then the injection of $W^{1,p}(\Omega)$ into $L^p(\Omega)$ is compact, and it can be deduced from a result associated with the names of M. RIESZ, FRÉCHET and KOLMOGOROV, but it can be proven easily from the result for $W_0^{1,p}(\Omega')$ proven below, once *extension properties* have been studied (Lemma 12.4 and 12.5). For unbounded open sets, or for bounded sets with nonsmooth boundaries, the situation is not as simple, but for the case of $W_0^{1,p}(\Omega)$ the smoothness of the boundary is not important.

Most compactness theorems use in some way the basic results of ARZELÁ[9] and ASCOLI[10]: if u_n is a bounded sequence of real continuous functions on a separable compact metric space X, then for each $x \in X$ one can extract a subsequence u_m such that $u_m(x)$ converges, and by a diagonal argument this can be achieved for all x in a countable dense subspace; this is extended to other points if one assumes that the sequence is equicontinuous at every point, a way to say that at any point y the functions are continuous in the same way, i.e., for every $\varepsilon > 0$ there exists $\delta > 0$, depending upon y and ε but not upon n, such that $d(y,z) \leq \delta$ implies $|u_n(y) - u_n(z)| \leq \varepsilon$ for all n. In order to cover many applications to weak convergence or weak $\star$ convergence like the Banach–Alaoglu[11] theorem, one also uses a maximality argument like Zorn's lemma, hidden in the proof of Tikhonov's[12] theorem, that any product of compact spaces is compact.

For functions in Sobolev spaces, which are not necessarily continuous but can be approached by smooth functions, one needs to control precisely the error, and some smoothness properties of the boundary will be needed if one works with $W^{1,p}(\Omega)$, but the following result is only concerned with $W_0^{1,p}(\Omega)$.

---

[6] Franz RELLICH, German mathematician, 1906–1955. He worked at Georg-August-Universität, Göttingen, Germany.

[7] Georg Augustus, 1683–1760. Duke of Brunswick-Lüneburg (Hanover), he became King of Great Britain and Ireland in 1727, under the name of GEORGE II.

[8] Vladimir Iosifovich KONDRAŠOV, Russian mathematician, 1909–1971.

[9] Cesare ARZELÁ, Italian mathematician, 1847–1912. He worked in Palermo, and in Bologna, Italy.

[10] Giulio ASCOLI, Italian mathematician, 1843–1896. He worked in Milano (Milan), Italy.

[11] Leonidas ALAOGLU, Canadian-born mathematician.

[12] Andreĭ Nikolaevich TIKHONOV, Russian mathematician, 1906–1993. He worked in Moscow, Russia.

Lemma 11.2. *(i) If Ω is an unbounded open subset of R^N such that there exists $r_0 > 0$ and a sequence $x_n \in \Omega$ converging to infinity with $B(x_n, r_0) \subset \Omega$, then the injection of $W_0^{1,p}(\Omega)$ into $L^p(\Omega)$ is not compact.*
(ii) If $1 \leq p \leq \infty$ and Ω is an open set with $meas(\Omega) < \infty$, then the injection of $W_0^{1,p}(\Omega)$ into $L^p(\Omega)$ is compact.

Proof: (i) Let $\varphi \in C_c(R^N)$, with $support(\varphi) \subset B(0, r_0)$, and $\varphi \neq 0$, then $u_n = \tau_{x_n}\varphi \in W_0^{1,p}(\Omega)$, $||u_n||_{1,p}$ is constant, but no subsequence converges strongly in $L^p(\Omega)$ because u_n converges to 0 in $L^p(\Omega)$ weak (weak $\star$ if $p = \infty$) and it cannot converge strongly to 0 as its norm stays constant, and not 0.

(ii) One starts with the case where Ω is bounded. For a bounded sequence $u_n \in W_0^{1,p}(\Omega)$ one wants to show that it belongs to a compact set of $L^p(\Omega)$, i.e., there is a subsequence which converges strongly in $L^p(\Omega)$. To prove that property it is enough to show that for every $\varepsilon > 0$ one can find a compact set K_ε of $L^p(\Omega)$ such that each u_n is at a distance at most $C\varepsilon$ of K_ε, i.e., one can decompose $u_n = v_{n,\varepsilon} + w_{n,\varepsilon}$, with $||w_{n,\varepsilon}||_p \leq C\varepsilon$ and $v_{n,\varepsilon} \in K_\varepsilon$; indeed for a subsequence one has $\limsup_{m,m'\to\infty} ||u_m - u_{m'}||_p \leq 2C\varepsilon$, so that a diagonal subsequence is a Cauchy sequence.

To do this, one extends the functions u_n by 0 outside Ω (still calling them u_n instead of $\widetilde{u_n}$), so that one has a bounded sequence $u_n \in W^{1,p}(R^N)$ with support in a fixed bounded set of R^N. For a special smoothing sequence $\varrho_\varepsilon(x) = \frac{1}{\varepsilon^N} \varrho_1(\frac{x}{\varepsilon})$ (with $\varrho_1 \in C_c^\infty(R^N)$, $support(\varrho_1) \subset B(0,1)$, $\varrho_1 \geq 0$ and $\int_{R^N} \varrho_1 \, dx = 1$), one takes $v_{n,\varepsilon} = u_n \star \varrho_\varepsilon$. This gives $||v_{n,\varepsilon}||_\infty \leq ||u_n||_p ||\varrho_\varepsilon||_{p'}$ and $||\frac{\partial v_{n,\varepsilon}}{\partial x_j}||_\infty \leq ||\frac{\partial u_n}{\partial x_j}||_p ||\varrho_\varepsilon||_{p'}$, i.e., for $\varepsilon > 0$ fixed $v_{n,\varepsilon}$ stays in a bounded set of Lipschitz functions, and keeps its support in a fixed compact set of R^N, and a subsequence converges uniformly on R^N, so that the sequence of restrictions to Ω converges strongly in $L^\infty(\Omega)$, and therefore in $L^p(\Omega)$, by Arzelá–Ascoli theorem. In order to estimate $||u_n - v_{n,\varepsilon}||_p$, one notices that $(u_n - u_n \star \varrho_\varepsilon)(x) = \int_{R^N} \varrho_\varepsilon(y)(u_n(x) - u_n(x-y)) \, dy$, i.e., $u_n - u_n \star \varrho_\varepsilon = \int_{R^N} \varrho_\varepsilon(y)(u_n - \tau_y u_n) \, dy$, so that $||u_n - u_n \star \varrho_\varepsilon||_p \leq \int_{R^N} \varrho_\varepsilon(y)||u_n - \tau_y u_n||_p \, dy$, but as one has $||u_n - \tau_y u_n||_p \leq |y| \, ||grad(u_n)||_p$ and $\int_{R^N} |y| \varrho_\varepsilon(y) \, dy = A\varepsilon$, one deduces $||u_n - u_n \star \varrho_\varepsilon||_p \leq AB\varepsilon$, where B is an upper bound for $||grad(u_n)||_p$ for all n.

If Ω is unbounded but has finite measure, one chooses $r_0 < r_1 < \infty$ such that the measure of $\Omega \setminus B(0, r_0)$ is $< \eta$, and one chooses $\theta \in C_c^\infty(R^N)$ such that $\theta = 1$ on $B(0, r_0)$ and $support(\theta) \subset B(0, r_1)$. The sequence $u_n - \theta u_n$ is bounded in $W^{1,p}(R^N)$ and is 0 outside $\Omega \setminus B(0, r_0)$ (one should avoid using its support, which is closed and could be very big if $\partial\Omega$ is thick).

For $p = \infty$, the maximum distance from a point of $\Omega \setminus B(0, r_0)$ to its boundary is at most $C(N)\eta^{1/N}$, and as one can take a common Lipschitz constant for all the functions $u_n - \theta u_n$, one deduces that they are uniformly small in $L^\infty(\Omega)$, and as θu_n stays in a bounded set of $W_0^{1,p}(\Omega \cap B(0, r_1))$, it remains in a compact of $L^p(\Omega \cap B(0, r_1))$ by applying the result for the case of bounded open sets.

For $1 \leq p < \infty$, one bounds the norm of $\|u_n - \theta\,u_n\|_p$ by using Sobolev's embedding theorem, choosing $q < N$ such that $1 \leq q \leq p < q^* < \infty$, and as $u_n - \theta\,u_n$ is bounded in $W^{1,q}(R^N)$ and therefore in $L^{q^*}(R^N)$, one has $\|u_n - \theta\,u_n\|_p \leq \|u_n - \theta\,u_n\|_{q^*} meas\big(\Omega \setminus B(0,r_0)\big)^{\alpha}$ with $\alpha = \frac{1}{p} - \frac{1}{q^*} > 0$, proving the desired uniform small bound for $\|u_n - \theta\,u_n\|_p$.

For $p = 2$, one can give a different proof, using the Fourier transform. □

It is now time to start studying the many questions where the regularity of the boundary plays a role: approximation by smooth functions, compactness, extension to the whole space, traces on the boundary.

[Taught on Wednesday February 9, 2000.]

Regularity of the Boundary; Consequences

For approaching functions in $W^{1,p}(\Omega)$ but not in $W_0^{1,p}(\Omega)$, one needs smooth functions whose support intersects the boundary $\partial\Omega$.

Definition 12.1. $\mathcal{D}(\overline{\Omega})$ *denotes the space of functions which are the restrictions to Ω of functions in $C_c^\infty(R^N)$.* ☐

For some nice open sets Ω, $\mathcal{D}(\overline{\Omega})$ is dense in $W^{1,p}(\Omega)$ for $1 \leq p < \infty$, but this is not true for some open sets, like a disc in the plane from which one removes a closed segment $[a, b]$, the intuitive reason being that functions in $\mathcal{D}(\overline{\Omega})$ are continuous across the segment, while there are functions in $W^{1,p}(\Omega)$ which are discontinuous across it (and for giving a mathematical meaning to this idea, one will have to define a notion of *trace on the boundary*). The preceding example is one where the open set is not locally on one side of the boundary, and will be ruled out for the moment, but one should remember that there are applications where one must consider open sets of this type, in the study of crack propagation, or in the scattering of waves by a thin plate, for example.

Definition 12.2. *(i) An open set Ω of R^N is said to have a continuous boundary, if for every $z \in \partial\Omega$, there exists $r_z > 0$, an orthonormal basis $e_1, \ldots, e_N$, and a continuous function F of $x' = (x_1, \ldots, x_{N-1})$ such that $\{x \in \Omega \mid |x - z| < r_z\} = \{x \in R^N \mid |x - z| < r_z, x_N > F(x')\}$.*
(ii) An open set Ω of R^N is said to have a Lipschitz boundary, if the same property holds with F being a Lipschitz continuous function. ☐

These conditions are usually referred to as Ω having the *segment property* or the *cone property*, and more precise notions are used. Notice that $e_1, \ldots, e_N$ and F vary with the point z, that the origin of the coordinate system may also change with z, and that Ω is locally on only one side of the boundary.

With the preceding definition, assuming $a < b$, the open set $\{(x, y) \in R^2 \mid x > 0, a x^2 < y < b x^2\}$ is an open set with continuous boundary if $a b < 0$, but not if $a b \geq 0$.

Lemma 12.3. *Let Ω be a bounded open set with continuous boundary. If $1 \leq p < \infty$, then $\mathcal{D}(\overline{\Omega})$ is dense in $W^{1,p}(\Omega)$.*

If $u \in W^{1,\infty}(\Omega)$, there exists a sequence $u_n \in \mathcal{D}(\overline{\Omega})$ such that $u_n \rightharpoonup u$ in $L^\infty(\Omega)$ weak $\star$ and $L^q(\Omega)$ strong for $1 \leq q < \infty$, and for $j = 1, \ldots, N$, $\frac{\partial u_n}{\partial x_j} \rightharpoonup \frac{\partial u}{\partial x_j}$ in $L^\infty(\Omega)$ weak $\star$ and $L^q(\Omega)$ strong for $1 \leq q < \infty$.

Proof: Ω being bounded, $\partial\Omega$ is compact, and as it is covered by the open balls $B(z, r_z)$ for $z \in \partial\Omega$, it is covered by a finite number of them, with centers $z_1, \ldots, z_m$. There exists $\varepsilon > 0$ such that $\bigcup_{i=1}^m B(z_i, r_{z_i})$ contains all the points at distance $\leq \varepsilon$ from $\partial\Omega$, so that $\overline{\Omega}$ is covered by the open sets $B(z_1, r_{z_1}), \ldots, B(z_m, r_{z_m})$ and $U = \{x \in \Omega \mid dist(x, \partial\Omega) > \varepsilon\}$. Let $\theta_1, \ldots, \theta_m, \zeta$ be a partition of unity associated to this covering, so that $\theta_i \in C_c^\infty(B(z_i, r_{z_i})), i = 1, \ldots, m, \zeta \in C_c^\infty(U)$ and $\sum_{i=1}^m \theta_i + \zeta = 1$ on $\overline{\Omega}$, so that every $u \in W^{1,p}(\Omega)$ can be decomposed as $\sum_{i=1}^m \theta_i u + \zeta u$. As ζu has support in U, it can be approached by functions in $C_c^\infty(\Omega)$ by smoothing by convolution. For each i, in order to approximate $v_i = \theta_i u$ by functions in $\mathcal{D}(\overline{\Omega})$, one chooses the set of orthogonal directions which gives a continuous equation for the boundary, and one uses the fact that for a rigid displacement f (i.e., $f(x) = Ax + a$ for a special orthogonal matrix A and a vector a), and $f(\omega) = \omega'$, if $\varphi \in W^{1,p}(\omega')$ and ψ is defined by $\psi(x) = \varphi(f(x))$ then one has $\psi \in W^{1,p}(\omega)$. One studies the case of a special domain Ω_F defined by $\Omega_F = \{x \in R^N \mid x_N > F(x')\}$ with F (uniformly) continuous (as one actually only uses functions which have their support in a fixed compact set, one only needs F continuous, and because it is uniformly continuous on compact sets, one may change F far away in order to make it uniformly continuous). If $v \in W^{1,p}(\Omega_F)$ and v has compact support, one wants to approach it by functions from $\mathcal{D}(\overline{\Omega_F})$, and for this one translates it down, i.e., for $h > 0$ one defines

$$v_h(x', x_N) = v(x', x_N + h), \tag{12.1}$$

and then one truncates v_h and one regularizes the result by convolution.

In doing the truncation, one regularizes F by convolution, and because F is uniformly continuous one can obtain in this way a function $G \in C^\infty(R^{N-1})$ such that

$$F - \frac{h}{6} \leq G \leq F + \frac{h}{6} \text{ on } R^{N-1}; \tag{12.2}$$

one chooses $\eta \in C^\infty(R)$ such that

$$\eta(t) = 0 \text{ for } t \leq \frac{-2}{3}; \ \eta(t) = 1 \text{ for } t \geq \frac{-1}{3}, \tag{12.3}$$

and one truncates v_h by defining

$$w_h(x) = v_h(x)\eta\left(\frac{x_N - G(x')}{h}\right), \tag{12.4}$$

so that $w_h(x) = v_h(x)$ if $x_N \geq F(x') - \frac{h}{6}$ because $x_N \geq G(x') - \frac{h}{3}$ and $w_h(x) = 0$ if $x_N \leq F(x') - \frac{5h}{6}$ because $x_N \leq G(x') - \frac{2h}{3}$. Because of the truncation one has

$$\frac{\partial w_h}{\partial x_j}(x', x_N) = \frac{\partial v}{\partial x_j}(x', x_N + h)\eta\left(\frac{x_N - G(x')}{h}\right) -$$

$$v(x', x_N + h)\eta'\left(\frac{x_N - G(x')}{h}\right)\frac{\frac{\partial G}{\partial x_j}(x')}{h} \text{ for } j < N$$

$$\frac{\partial w_h}{\partial x_N}(x', x_N) = \frac{\partial v}{\partial x_N}(x', x_N + h)\eta\left(\frac{x_N - G(x')}{h}\right) +$$

$$v(x', x_N + h)\eta'\left(\frac{x_N - G(x')}{h}\right)\frac{1}{h}, \tag{12.5}$$

because the partial derivatives of v_h might have another part which is supported on $x_N = F(x') - h$ but this part is killed by the term $\eta\left(\frac{x_N - G(x')}{h}\right)$. Therefore if W_h is the restriction of w_h to Ω_F, one has $\frac{\partial W_h}{\partial x_j}(x', x_N) = \frac{\partial v}{\partial x_j}(x', x_N + h)$ in Ω_F for $j \leq N$, so that W_h converges to v in $W^{1,p}(\Omega_F)$ strong if $p < \infty$. In order to approach W_h by functions in $\mathcal{D}(\overline{\Omega_F})$, one approaches w_h by convolutions by smooth functions and one restricts them to Ω_F. □

There is another way to do the preceding two steps of truncation and regularization in one single step, and the idea is to do a convolution of v with a sequence of nonnegative smoothing functions $\varrho_n \in C_c^\infty(R^N)$ with integral 1 whose support shrinks to $\{0\}$. However, unless F is Lipschitz continuous, one must not use a special smoothing sequence, and one lets the support of ϱ_n shrink to $\{0\}$ in a special way. If $K = support(\varrho)$, the convolution $(v \star \varrho)(x) = \int_{R^N} v(x - y)\varrho(y)\, dy$ is only defined if $x - K \subset \Omega_F$, and one wants this set of x to contain Ω_F, or better to contain $\Omega_{F-\eta}$ for some $\eta > 0$; in doing this, one assumes that $|y'| \leq \varepsilon$ for $y \in K$, so that $|F(x') - F(y')| \leq \omega(\varepsilon)$ where ω is the modulus of uniform continuity of F, and one asks that $y \in K$ implies $y_N \leq -\eta - \omega(\varepsilon)$.

This method also applies in some different situations, like for $\Omega = \{(x, y) \mid x > 0, 0 < y < x^2\}$, which is not an open set with continuous boundary with our definition, but $\mathcal{D}(\overline{\Omega})$ is dense, and in order to show that the cusp at 0 is not a problem one notices that if one translates Ω by a vector $(-a, -b)$ with $a > 0$ and $b > a^2$, then one obtains an open set Ω' which contains $\overline{\Omega}$, giving room for translating (by small amounts) and smoothing by convolution.

Lemma 12.4. *Let Ω be a bounded open set with Lipschitz boundary. Then there exists a linear continuous extension P from $W^{1,p}(\Omega)$ into $W^{1,p}(R^N)$ for $1 \leq p \leq \infty$ (an extension is characterized by the property that $P\,u|_\Omega = u$ for every $u \in W^{1,p}(\Omega)$).*

Proof: One constructs the extension for the dense subspace $\mathcal{D}(\overline{\Omega})$, and using a partition of unity, it is enough to construct the extension for Ω_F, where F is Lipschitz continuous. One defines $P\,u$ by

$$P\,u(x', x_N) = \begin{cases} u(x', x_N) \text{ if } x_N > F(x') \\ u(x', 2F(x') - x_N) \text{ if } x_N < F(x'). \end{cases} \tag{12.6}$$

In that way Pu is continuous at the interface $\partial\Omega_F$ and one has

$$\frac{\partial Pu}{\partial x_j}(x', x_N) = \frac{\partial u}{\partial x_j}(x', x_N) \text{ if } x_N > F(x') \text{ and } j \leq N$$

$$\frac{\partial Pu}{\partial x_j}(x', x_N) = \frac{\partial u}{\partial x_j}(x', 2F(x') - x_N) + 2\frac{\partial F}{\partial x_j}(x')\frac{\partial u}{\partial x_N}(x', 2F(x') - x_N)$$

$$\text{if } x_N < F(x') \text{ and } j < N$$

$$\frac{\partial Pu}{\partial x_N}(x', x_N) = -\frac{\partial u}{\partial x_N}(x', 2F(x') - x_N) \text{ if } x_N < F(x'),$$

(12.7)

and one verifies on these formulas that P is indeed linear continuous.   □

The extension constructed in Lemma 12.4 is the same whatever p is, but this method does not apply for showing that there exists a continuous extension from $W^{m,p}(\Omega)$ into $W^{m,p}(R^N)$ for $m \geq 2$ because higher-order derivatives of F might not exist. STEIN[1] has constructed a different extension which maps $W^{m,p}(\Omega)$ into $W^{m,p}(R^N)$ for all $m \geq 0$ and all $p \in [1, \infty]$, but we shall only consider a simpler one which can be used for open sets with smooth boundary, and the idea is shown for $\Omega = R_+^N = \{x \in R^N \mid x_N > 0\}$.

Lemma 12.5. *There is a linear continuous extension from* $W^{m,p}(R_+^N)$ *into* $W^{m,p}(R^N)$, *defined by*

$$Pu(x', x_N) = \begin{cases} u(x', x_N) & \text{if } x_N > 0 \\ \sum_{j=1}^m \alpha_j u(x', -j\,x_N) & \text{if } x_N < 0, \end{cases}$$

(12.8)

with suitable coefficients $\alpha_j, j = 1, \ldots, m$.

Proof: Using the techniques already presented, one shows that $\mathcal{D}(\overline{R_+^N})$ is dense into $W^{m,p}(R_+^N)$. In order to check that the definition defines a continuous operator, one must show that derivatives up to order $m - 1$ are continuous on $x_N = 0$. As for smooth functions taking tangential derivatives (i.e., not involving $\frac{\partial}{\partial x_N}$) commutes with restricting to $x_N = 0$, it is enough to check that $\frac{\partial^k Pu}{\partial x_N^k}$ is continuous for $k = 0, \ldots, m - 1$. One finds the condition to be

$$\sum_{j=1}^m \alpha_j(-j)^k = 1 \text{ for } k = 0, \ldots, m - 1,$$

(12.9)

and as this linear system has a Vandermonde[2] matrix, it is invertible and the coefficients $\alpha_j, j = 1, \ldots, m$ are defined in a unique way.   □

The extension property of Lemma 12.4 does not necessarily hold for open sets which only have a Hölder continuous boundary of order $\theta < 1$. This can be checked in the plane for the open set $\Omega = \{(x, y) \mid 0 < x, -x^{1/\theta} < y <$

---

[1] Elias M. STEIN, Belgian-born mathematician, born in 1931. He received the Wolf Prize in 1999. He worked at The University of Chicago, Chicago, IL, and in Princeton, NJ.

[2] Alexandre-Théophile VANDERMONDE, French mathematician, 1735–1796.

$x^{1/\theta}$}, by showing that $H^1(\Omega)$ is not (continuously) embedded in all $L^p(\Omega)$ for $2 \leq p < \infty$, which would be the case if a continuous extension existed, because Sobolev's embedding theorem asserts that $H^1(R^2)$ is (continuously) embedded in all $L^p(R^2)$ for $2 \leq p < \infty$.

For $\varphi \in C_c^\infty(R^2)$ with $\varphi(0) = 1$ one defines ψ by $\psi(x) = x^\alpha \varphi(x)$, and one checks for what values of α the function ψ belongs to $L^p(\Omega)$ or to $H^1(\Omega)$. The function ψ belongs to $L^p(\Omega)$ if and only if $\int_0^1 x^{p\,\alpha+1/\theta}\,dx < \infty$, i.e., $p\alpha + \frac{1}{\theta} > -1$, and ψ belongs to $H^1(\Omega)$ if and only $2(\alpha - 1) + \frac{1}{\theta} > -1$; the (excluded) critical value $\alpha_C = \frac{1}{2} - \frac{1}{2\theta} = \frac{\theta-1}{2\theta}$ (which is < 0) corresponds to the (excluded) critical value $p_C = -\frac{1}{\alpha_C} - \frac{1}{\theta\,\alpha_C} = \frac{2(1+\theta)}{1-\theta}$, so that $H^1(\Omega)$ is not embedded in $L^p(\Omega)$ for $p > p_C$.

[Taught on Friday February 11, 2000.]

Traces on the Boundary

For an open set with a continuous boundary, there is a notion of restriction to the boundary (called a trace) for functions of $W^{1,p}(\Omega)$, which is easily derived for the case of Ω_F, with F continuous.

Lemma 13.1. *For $u \in \mathcal{D}(\overline{\Omega_F})$ and $1 \leq p < \infty$, one has*

$$\left\|u\big(x', F(x')\big)\right\|_{L^p(R^{N-1})} \leq \left(\frac{p}{2}\right)^{1/p} \left\|\frac{\partial u}{\partial x_N}\right\|_{L^p(\Omega_F)}^{1/p} \|u\|_{L^p(\Omega_F)}^{(p-1)/p}. \tag{13.1}$$

Proof: If $v \in C_c^\infty(R)$ one has

$$2|v(0)|^p = \int_{-\infty}^0 \frac{d(|v|^p)}{dt}\,dt - \int_0^\infty \frac{d(|v|^p)}{dt}\,dt \leq p \int_R |v|^{p-1}|v'|\,dt \leq p\,\|v\|_p^{p-1}\|v'\|_p. \tag{13.2}$$

One applies this inequality to $v(t) = u\big(x', t + F(x')\big)$ and then one integrates in x', using Hölder's inequality. $\qquad\square$

For $p = \infty$, the functions of $W^{1,\infty}(\Omega_F)$ are locally uniformly continuous (each function is an equivalence class and one element of the equivalence class is continuous), and the trace is just the restriction to the boundary.

Definition 13.2. *The linear continuous operator of trace on the boundary, defined by extension by (uniform) continuity of the operator of restriction defined for $\mathcal{D}(\overline{\Omega})$ will be denoted by γ_0.* $\qquad\square$

Notice that γ_0 is not defined as the restriction to the boundary, because the boundary has measure 0, and the restriction to a set of measure 0 is not defined for functions which are not smooth enough.

Lemma 13.3. *If $1 \leq p, q, r \leq \infty$ and $\frac{1}{r} = \frac{1}{p} + \frac{1}{q}$, then for $u \in W^{1,p}(\Omega_F)$ and $v \in W^{1,q}(\Omega_F)$ one has $uv \in W^{1,r}(\Omega_F)$ and $\gamma_0(uv) = \gamma_0 u\,\gamma_0 v$.*

Proof: The formula is true if $u, v \in \mathcal{D}(\overline{\Omega_F})$ and as both sides of the equality use continuous mappings on $W^{1,p}(\Omega_F) \times W^{1,q}(\Omega_F)$, the formula is true by density. $\square$

Using a partition of unity one can then define a notion of trace on the boundary for $u \in W^{1,p}(\Omega)$ if Ω has a continuous boundary, but one should be careful that the definition depends upon the choice of the partition of unity and the choice of local orthonormal bases. One should be aware that the usual area measure on the boundary, i.e., the $(N-1)$-dimensional Hausdorff measure, is not defined for F continuous; for F Lipschitz continuous it is $\sqrt{1 + |\nabla F(x')|^2}\, dx'$, and it has the important property of being invariant by rigid displacements (rotations and translations). Using the invariance by rotation of the $(N-1)$-dimensional Hausdorff measure, one can show that for a bounded open set with a Lipschitz boundary the trace does not depend upon the choice of the partition of unity or the choice of local orthonormal bases.

Some notion of trace can be defined for other open sets, for example some which are not even locally on one side of their boundary. For example, let Ω be the open set of R^2 defined in polar coordinates by $0 < r < 1$ and $0 < \theta < 2\pi$, i.e., the open unit disc slit on the nonnegative x axis. One can apply Lemma 13.1 to the open subsets Ω_+ defined by $0 < r < 1$ and $0 < \theta < \pi$ and Ω_- defined by $0 < r < 1$ and $\pi < \theta < 2\pi$, so that one can define two traces on the piece of the boundary corresponding to $y = 0$, $0 < x < 1$, one from the side of Ω_+ and one from the side of Ω_-; these two traces are not necessarily the same for $u \in W^{1,p}(\Omega)$, although they are the same for functions in $\mathcal{D}(\overline{\Omega})$ (and $\mathcal{D}(\overline{\Omega})$ is not dense in $W^{1,p}(\Omega)$ in this case); there is actually a compatibility condition at 0 between the traces on the two sides if $p > 2$.

An important result is to identify $W_0^{1,p}(\Omega)$, which is by definition the closure of $C_c^\infty(\Omega)$ in $W^{1,p}(\Omega)$, as the kernel of γ_0, and this makes use of the simple form[1] of *Hardy's inequality* (13.3).

Lemma 13.4. *(Hardy's inequality) For $p > 1$,*

$$\text{if } f \in L^p(R_+) \text{ and } g(t) = \frac{1}{t}\int_0^t f(s)\,ds \text{ for } t > 0, \text{ then}$$
$$g \in L^p(R_+) \text{ and } ||g||_p \le \frac{p}{p-1}||f||_p. \tag{13.3}$$

Proof: By density, it is enough to prove the result for $f \in C_c^\infty(R_+)$, for which g is 0 near 0 and decays in $\frac{C}{t}$ for large t (so that g does not belong to $L^1(R_+)$ when $f \in L^1(R_+)$ and $\int_0^\infty f(t)\,dt \ne 0$). One has

$$t\,g'(t) + g(t) = f(t), \tag{13.4}$$

and one multiplies by $p\,|g|^{p-1}sign(g)$ and integrates on $(0,\infty)$, and obtains

$$p\int_0^\infty f\,|g|^{p-1}sign(g)\,dt = \int_0^\infty t(|g|^p)'\,dt + p\int_0^\infty |g|^p\,dt = (p-1)\int_0^\infty |g|^p\,dt \tag{13.5}$$

---

[1] A more general form of Hardy's inequality will be seen at Lemma 22.5.

observing that $\int_0^\infty t(|g|^p)' \, dt = -\int_0^\infty |g|^p \, dt$ because $t \, |g(t)|^p \to 0$ as $t \to \infty$ (because $p > 1$); one concludes by using Hölder's inequality. $\qquad \square$

One shows that the constant is optimal, although no nonzero function gives an equality in (13.3) (because equality in Hölder's inequality requires $f = \lambda \, g$ for some $\lambda > 0$, and then (13.4) implies $g(t) = C \, t^{\lambda-1}$ for $t > 0$, which does not belong to $L^p(R_+)$), by considering the case

$$f(t) = \begin{cases} \alpha \, t^{\alpha-1} & \text{for } 0 < t < T \\ 0 & \text{for } t > T \end{cases} \quad g(t) = \begin{cases} t^{\alpha-1} & \text{for } 0 < t < T \\ \frac{T^\alpha}{t} & \text{for } t > T \end{cases}, \qquad (13.6)$$

for $\frac{p-1}{p} < \alpha < 1$ and letting T tend to $+\infty$ shows that the constant cannot be replaced by $\left(\frac{1}{\alpha}\right)^{1/p}$ (and that no such inequality is true for $p = 1$ by letting α tend to 0).

Another proof of Hardy's inequality (13.3) uses Young's inequality (2.3) for convolution, noticing that $(0, \infty)$ is a multiplicative group with Haar measure $\frac{dt}{t}$, then one has $g(t) = \int_0^t f(s) \frac{s}{t} \frac{ds}{s}$, i.e., g is the convolution product of f with the function h defined by $h(t) = 0$ for $0 < t < 1$ and $h(t) = \frac{1}{t}$ for $1 < t < \infty$ (so that $h \in L^1\left(R_+; \frac{dt}{t}\right)$).

Lemma 13.5. *For F continuous, and $1 < p < \infty$, one has*

$$\left\| \frac{u(x', x_N) - \gamma_0 u(x')}{x_N - F(x')} \right\|_{L_p(\Omega_F)} \leq \frac{p}{p-1} \left\| \frac{\partial u}{\partial x_N} \right\|_{L_p(\Omega_F)} \qquad \text{for all } u \in W^{1,p}(\Omega_F).$$

$$(13.7)$$

Proof: For $u \in C_c^\infty(\overline{\Omega_F})$, one applies Hardy's inequality (13.3) to $f(t) = \frac{\partial u}{\partial x_N}(x', F(x') + t)$, one takes the power p and then integrates in x'. Then one extends the inequality to $u \in W^{1,p}(\Omega_F)$ by a density argument, valid for $p < \infty$. $\qquad \square$

For $p = +\infty$, (13.7) is true if one replaces $\frac{p}{p-1}$ by 1.

Lemma 13.6. *If F is Lipschitz continuous and $p > 1$, then $W_0^{1,p}(\Omega_F)$ is the subspace of $u \in W^{1,p}(\Omega_F)$ satisfying $\gamma_0 u = 0$.*

Proof: If $u \in W_0^{1,p}(\Omega_F)$ then there exists a sequence $\varphi_n \in C_c^\infty(\Omega_F)$ such that $\varphi_n \to u$ in $W^{1,p}(\Omega_F)$; as γ_0 is continuous from $W^{1,p}(\Omega_F)$ to $L^p(R^{N-1})$, $\gamma_0 u$ is the limit of $\gamma_0 \varphi_n$, and is 0 because each φ_n is 0 near $\partial\Omega_F$ and γ_0 is the restriction to the boundary for functions in $\mathcal{D}(\overline{\Omega_F})$.

Conversely, for $u \in W^{1,p}(\Omega_F)$ satisfying $\gamma_0 u = 0$, one must approach u by a sequence from $C_c^\infty(\Omega_F)$. First one truncates at ∞, i.e., one chooses $\theta \in C_c^\infty(R^N)$ such that $\theta(x) = 1$ for $|x| \leq 1$ and one approaches u by u_n defined by $u_n(x) = u(x)\theta\left(\frac{x}{n}\right)$, and one has $\gamma_0 u_n = 0$ by Lemma 13.3, and u_n converges to u (using Lebesgue dominated convergence theorem). One may then assume that the support of u is bounded.

Then one wants to truncate near the boundary, and for this one uses Lemma 13.5, which implies

$$\frac{u}{x_N - F(x')} \in L^p(\Omega_F). \qquad (13.8)$$

Let $\eta \in C^\infty(R)$ with $\eta(t) = 0$ for $t \leq 1$ and $\eta(t) = 1$ for $t \geq 2$. One approaches u by u_n defined by

$$u_n(x) = u(x)\eta\left(n(x_N - F(x'))\right). \qquad (13.9)$$

The sequence u_n converges to u in $L^p(\Omega_F)$ strong by the Lebesgue dominated convergence theorem (if $p = \infty$ the convergence is in $L^\infty(\Omega_F)$ weak $\star$ and L^q_{loc} strong for all $q < \infty$, of course). Similarly $\frac{\partial u_n}{\partial x_j}$ has a term $\frac{\partial u}{\partial x_j}\eta\left(n(x_N - F(x'))\right)$ which converges to $\frac{\partial u}{\partial x_j}$, but also another term $n\,u\,\eta'\left(n(x_N - F(x'))\right)w_j$ with $w_j \in L^\infty(\Omega_F)$, as it is $-\frac{\partial F}{\partial x_j}$ if $j < N$ and 1 if $j = N$. This last term tends to 0 by the Lebesgue dominated convergence theorem because one may write it $\frac{u}{x_N - F(x')}\zeta_n(x_N - F(x'))w_j$ with $\zeta_n(t) = n\,t\,\eta'(nt)$, and ζ_n is bounded by $\sup_{t \in (1,2)} t\,|\eta'(t)|$ and $\zeta_n(t)$ tends to 0 for every $t > 0$. One may then assume that u has its support bounded and bounded away from the boundary.

The last part is to regularize by convolution.   □

Lemma 13.7. *If Ω is bounded with Lipschitz boundary and $p > 1$, then $W_0^{1,p}(\Omega)$ is the subspace of $u \in W^{1,p}(\Omega)$ satisfying $\gamma_0 u = 0$.*

Proof: One uses a partition of unity and a local change of orthonormal basis and one applies the preceding result.   □

[Taught on Monday February 14, 2000.]

14

Green's Formula

An important result is *Green's formula*, or simply *integration by parts*, for which one needs the boundary to be smooth enough so that a *normal* is defined almost everywhere (for the $(N-1)$-dimensional Hausdorff measure).

Definition 14.1. *If Ω is an open set with Lipschitz boundary, ν denotes the unit exterior normal.* □

For the case of Ω_F defined as $\Omega_F = \{(x', x_N) \mid x_N > F(x')\}$ for a Lipschitz function F, one has

$$\nu_j\big(x', F(x')\big) = \frac{\frac{\partial F}{\partial x_j}(x')}{\sqrt{1 + |\nabla F(x')|^2}} \text{ for } j < N; \quad \nu_N\big(x', F(x')\big) = \frac{-1}{\sqrt{1 + |\nabla F(x')|^2}}. \tag{14.1}$$

Lemma 14.2. *If F is Lipschitz continuous, $u \in W^{1,p}(\Omega_F)$, $v \in W^{1,p'}(\Omega_F)$, then one has*

$$\int_{\Omega_F} \left(u \frac{\partial v}{\partial x_N} + \frac{\partial u}{\partial x_N} v \right) dx = \int_{\partial \Omega_F} \gamma_0 u \, \gamma_0 v \, \nu_N \, dH^{N-1}, \text{ i.e., } -\int_{R^{N-1}} \gamma_0 u \, \gamma_0 v \, dx'. \tag{14.2}$$

Proof: For $u, v \in \mathcal{D}(\overline{\Omega_F})$ and each $x' \in R^{N-1}$, one has

$$\int_{F(x')}^{\infty} \left(u \frac{\partial v}{\partial x_N} + \frac{\partial u}{\partial x_N} v \right) dx_N = \int_{F(x')}^{\infty} \frac{\partial(u\,v)}{\partial x_N} dx_N = -u\big(x', F(x')\big) v\big(x', F(x')\big). \tag{14.3}$$

Integrating this equality in x' shows that the formula is true for $u, v \in \mathcal{D}(\overline{\Omega_F})$, and as both sides of the equality are continuous functionals if one uses the topologies of $W^{1,p}(\Omega_F)$ and $W^{1,p'}(\Omega_F)$, the lemma is proven (in the case where $p = 1$ or $p = \infty$, one first approaches the function in $W^{1,\infty}(\Omega_F)$ and its partial derivatives in $L^{\infty}(\Omega_F)$ weak $\star$ and the other function in $W^{1,1}(\Omega_F)$ strong). □

Lemma 14.3. *If F is Lipschitz continuous, $u \in W^{1,p}(\Omega_F)$, $v \in W^{1,p'}(\Omega_F)$, then for every $j = 1, \ldots, N$ one has*

$$\int_{\Omega_F} \left(u \frac{\partial v}{\partial x_j} + \frac{\partial u}{\partial x_j} v \right) dx = \int_{\partial \Omega_F} \gamma_0 u \, \gamma_0 v \, \nu_j \, dH^{N-1}. \tag{14.4}$$

Proof: The case $j = N$ has been proven. If $j < N$ and if one makes only x_j vary, the intersection with Ω_F can be an arbitrary open subset of R, i.e., a countable union of open intervals, so in order to avoid this technical difficulty one uses a new orthogonal basis, with last vector $e'_N = \frac{1}{\sqrt{1+\varepsilon^2}}(e_N + \varepsilon \, e_j)$ (where $e_1, \ldots, e_N$ is the initial basis). Of course $\varepsilon > 0$ is taken small enough, so that using y to denote coordinates in the new basis, the open set can be written as $y_N > G(y')$ with G Lipschitz continuous. Therefore the preceding lemma shows that $\int_{\Omega_G} \left(u \frac{\partial v}{\partial y_N} + \frac{\partial u}{\partial y_N} v \right) dy = \int_{\partial \Omega_G} \gamma_0 u \, \gamma_0 v \, (\nu . e'_N) \, dH^{N-1}$. One observes that $\frac{\partial u}{\partial y_N} = \frac{1}{\sqrt{1+\varepsilon^2}} \frac{\partial u}{\partial x_N} + \frac{\varepsilon}{\sqrt{1+\varepsilon^2}} \frac{\partial u}{\partial x_j}$ and similarly $(\nu . e'_N) = \frac{1}{\sqrt{1+\varepsilon^2}} \nu_N + \frac{\varepsilon}{\sqrt{1+\varepsilon^2}} \nu_j$, so that after multiplication by $\sqrt{1+\varepsilon^2}$, one obtains a relation of order 1 in ε; the equality for $\varepsilon = 0$ is true as it is the preceding lemma, so that the equality of the coefficients of ε gives the desired relation. $\square$

Lemma 14.4. *For any bounded open set Ω with Lipschitz boundary one has $\int_{\Omega} \left(u \frac{\partial v}{\partial x_j} + \frac{\partial u}{\partial x_j} v \right) dx = \int_{\partial \Omega} \gamma_0 u \, \gamma_0 v \, \nu_j \, dH^{N-1}$ for $j = 1, \ldots, N$, for all $u \in W^{1,p}(\Omega)$, $v \in W^{1,p'}(\Omega)$.*

Proof: One uses a partition of unity and the fact that one has a formulation invariant by rigid displacements, for example by writing the formula $\int_{\Omega} \left(u \, (\nabla v . e) + (\nabla u . e) \, v \right) dx = \int_{\partial \Omega} \gamma_0 u \, \gamma_0 v \, (\nu . e) \, dH^{N-1}$ for all vectors e (of course, the fact that the Lebesgue measure dx is also invariant by rigid displacements is also used). $\square$

It should be noticed that even for a C^∞ function F, the set of x' such that $F(x') = \lambda$ can be a general closed set; actually, for any closed set $K \subset R^{N-1}$, there exists $G \in C^\infty$, with $G \geq 0$ and $\{x' \mid G(x') = 0\} = K$. In constructing G, one notices that the complement of K is a countable union of open balls $R^{N-1} \setminus K = \bigcup_n B(M_n, r_n)$ (for example for each point $M \notin K$ and M with rational coordinates, one keeps the largest open ball centered at M which does not intersect K). One chooses a function $\varphi \in C_c^\infty(R^{N-1})$ such that $\{x \mid \varphi(x) \neq 0\} = B(0, 1)$, and one may assume $\varphi \geq 0$. One defines G by $G(x) = \sum_n c_n \varphi\left(\frac{x - M_n}{r_n}\right)$ with all $c_n > 0$, and G satisfies the desired property if one chooses the c_n converging to 0 fast enough so that the series converges uniformly, as well as any of its derivatives (so that $D^\alpha G \in C_0(R^{N-1})$ for any multi-index α).

However, there is a result of SARD[1] which says that for most λ the set is not that bad; for example, if $F \in C^1(R)$, then except for a set of λ with

---

[1] Arthur SARD, American mathematician, 1909–1980. He worked in New York, NY.

measure 0, at any point x with $F(x) = \lambda$ one has $F'(x) \neq 0$, so these points are isolated.

The next question is to identify the range of γ_0, for a bounded open set Ω with Lipschitz boundary for example. This was done by Emilio GAGLIARDO, but although γ_0 is surjective from $W^{1,1}(\Omega)$ onto $L^1(\partial\Omega)$, it is not so for $p > 1$, and the image of $W^{1,p}(\Omega)$ by γ_0 is not $L^p(\partial\Omega)$ for $p > 1$. Actually for $p = \infty$ it is $W^{1,\infty}(\partial\Omega)$, the space of Lipschitz continuous functions on the boundary, and as p varies from 1 to ∞ one goes from traces having no derivatives in L^1 to traces having one derivative in L^∞, and one can guess that for $1 < p < \infty$ the traces have $1 - \frac{1}{p}$ derivatives in L^p, if one finds a way to express what this means. The characterization of the traces is simpler in the case $p = 2$ because one can use the Fourier transform, which will be studied for that reason.

Before doing that, one now has a simple way to prove that the injection of $W^{1,p}(\Omega)$ into $L^p(\Omega)$ is compact in the case of bounded open sets with Lipschitz boundary, the case of continuous boundary being left for later.

Lemma 14.5. *If Ω is bounded with Lipschitz boundary, then the injection of $W^{1,p}(\Omega)$ into $L^p(\Omega)$ is compact.*

Proof: One uses a *continuous extension* P from $W^{1,p}(\Omega)$ to $W^{1,p}(R^N)$, as constructed in Lemma 12.4. Let $\eta \in C_c^\infty(R^N)$ such that $\eta(x) = 1$ for $x \in \overline{\Omega}$, and let Ω' be a bounded open set containing $support(\eta)$. Then if a sequence u_n is bounded in $W^{1,p}(\Omega)$, the sequence of extensions $P\,u_n$ is bounded in $W^{1,p}(R^N)$, and the sequence of truncated functions $\eta(P\,u_n)$ is bounded in $W_0^{1,p}(\Omega')$, so that a subsequence $\eta(P\,u_m)$ belongs to a compact of $L^p(\Omega')$, and the sequence of restrictions to Ω, which is u_m, belongs to a compact of $L^p(\Omega)$. □

[Taught on Wednesday February 16, 2000.]

15

The Fourier Transform

A decomposition in *Fourier series* of a scalar function f in one variable consists in writing

$$f(x) = \sum_{n \in Z} c_n e^{2i \pi n x/T} \text{ for some (complex) coefficients } c_n, n \in Z. \quad (15.1)$$

If the series converges the function f must be periodic with period T.

A decomposition in *Fourier integral* of a scalar function f in one variable consists in writing

$$f(x) = \int_R \hat{f}(\xi) e^{2i \pi x \xi} d\xi \text{ for some function } \hat{f}. \quad (15.2)$$

The generalization to functions of N variables leads to the following definition. Having learnt the theory from Laurent SCHWARTZ, I use his notation, but most mathematicians do not put the coefficient 2π in the integral, and some different powers of π occur then in their formulas, in the argument of the exponentials or multiplying some integrals, and one should be aware of the fact that *different books may use different constants*.

Definition 15.1. *For $f \in L^1(R^N)$, the Fourier transform of f is the function $\mathcal{F}f$ (or $\hat{f}$) defined on (the dual of) R^N by*

$$\mathcal{F}f(\xi) = \int_{R^N} f(x) e^{-2i \pi (x.\xi)} dx. \quad (15.3)$$

One also defines $\overline{\mathcal{F}}$ by

$$\overline{\mathcal{F}}f(\eta) = \int_{R^N} f(y) e^{+2i \pi (y.\eta)} dy. \quad \Box \quad (15.4)$$

Of course, $\overline{\mathcal{F}}$ is defined so that $\overline{\mathcal{F}f} = \overline{\mathcal{F}}\overline{f}$.

From a scaling point of view, one should remember that, if using L to denote a length unit for measuring x and U to denote a unit for measuring u, then ξ scales as L^{-1} (because the exponential must be computed on a number, i.e., with no dimension), and $\mathcal{F}u$ scales as $U\,L^N$.

The Fourier transform $\mathcal{F}$ maps $L^1(R^N)$ into $C_0(R^N)$, but one important property is that it extends as an *isometry* from $L^2(R^N)$ to itself, with inverse $\overline{\mathcal{F}}$.

Another important property is that it *transforms derivation into multiplication*, or more generally[1] it *transforms convolution into product*, and one can check easily the following properties:

$$\text{if } f, g \in L^1(R^N), \text{ one has } f \star g \in L^1(R^N) \text{ and } \mathcal{F}(f \star g) = \mathcal{F}f\,\mathcal{F}g, \quad (15.5)$$

$$\text{if } f \in L^1(R^N) \text{ and } x_j f \in L^1(R^N), \text{ one has } \frac{\partial(\mathcal{F}f)}{\partial \xi_j} = \mathcal{F}(-2i\pi\,x_j f), \quad (15.6)$$

$$\text{if } f \in L^1(R^N) \text{ and } \frac{\partial f}{\partial x_j} \in L^1(R^N), \text{ one has } \mathcal{F}\frac{\partial f}{\partial x_j} = 2i\,\pi\,\xi_j \mathcal{F}f. \quad (15.7)$$

The definition of Fourier transform extends immediately to Radon measures with *finite total mass*, i.e., those for which

$$\text{there exists } C \text{ such that } |\langle \mu, \varphi \rangle| \le C\,||\varphi||_\infty \text{ for all } \varphi \in C_c(R^N), \quad (15.8)$$

(and the *total mass* of μ is the smallest value of C). If (15.8) holds, the linear mapping $\varphi \mapsto \langle \mu, \varphi \rangle$ extends in a unique way to the Banach space $C_b(R^N)$ of continuous bounded functions on R^N equipped with the sup norm $||\cdot||_\infty$, with the same inequality (15.8) being true for all $\varphi \in C_b(R^N)$. Then one defines μ by

$$\mu(\xi) = \langle \mu, e^{-2i\,\pi(\cdot\,.\,\xi)} \rangle \text{ for all } \xi \in R^N, \quad (15.9)$$

and the Lebesgue dominated convergence theorem shows that μ is continuous, so that one has $\mu \in C_b(R^N)$. It may be useful to recall that $C_b(R^N)$ is not separable.[2]

In order to define the Fourier transform for some distributions, Laurent SCHWARTZ introduced a particular (Fréchet) space of *rapidly decaying C^∞ functions*.

---

[1] Laurent SCHWARTZ has generalized the definition of convolution product to pairs of distributions if the condition (2.4) holds for their supports, and $\frac{\partial T}{\partial x_j}$ appears then to be the convolution of T with $\frac{\partial \delta_0}{\partial x_j}$, so derivations are indeed particular cases of convolutions.

[2] One way to show that a normed space E is not separable is to exhibit an uncountable collection of elements $e_i \in E$, $i \in I$, such that there exists $\alpha > 0$ with $||e_i - e_j||_E \ge \alpha$ whenever $i, j \in I$ with $i \ne j$. For $E = C_b(R^N)$, one chooses $\varphi_0 \in C_c(R)$ with $0 \le \varphi_0 \le 1$, $\varphi_0(0) = 1$ and $support(\varphi_0) \subset B\left(0, \frac{1}{2}\right)$, and one considers the collection of functions f_η defined by $f_\eta(x) = \sum_{m \in Z^N} \eta(m)\varphi_0(x - m)$, where η is any mapping from Z^N into $\{-1, +1\}$, so for $\eta_1 \ne \eta_2$ one has $||f_{\eta_1} - f_{\eta_2}||_E = 2$, and the cardinal of the collection is $2^{\aleph_0}$.

Definition 15.2. *The space of* rapidly decaying C^∞ *functions* $\mathcal{S}(R^N)$ *is*

$$\mathcal{S}(R^N) = \{u \in C^\infty(R^N) \mid P\,D^\alpha u \in L^\infty(R^N) \\ \text{for all polynomials } P \text{ and all multi-indices } \alpha\}. \quad \Box \qquad (15.10)$$

By repeated use of (15.6) and (15.7), one checks easily that $\mathcal{F}$ maps $\mathcal{S}(R^N)$ into itself, and that *Plancherel's formula*[3]

$$\int_{R^N} (\mathcal{F}f)g\,dx = \int_{R^N} f(\mathcal{F}g)\,dx \text{ for all } f,g \in \mathcal{S}(R^N) \qquad (15.11)$$

holds.[4] Laurent SCHWARTZ used (15.11) for choosing the definition of Fourier transform on the dual $\mathcal{S}'(R^N)$ of $\mathcal{S}(R^N)$, which he called the space of *tempered distributions*.

Definition 15.3. *A tempered distribution is an element* $T \in \mathcal{S}'(R^N)$*; one defines* $\mathcal{F}T \in \mathcal{S}'(R^N)$ *by*

$$\langle \mathcal{F}T, \varphi \rangle = \langle T, \mathcal{F}\varphi \rangle \text{ for all } \varphi \in \mathcal{S}(R^N). \quad \Box \qquad (15.12)$$

One finds easily that the analog of (15.6) and (15.7) holds, i.e., that for $T \in \mathcal{S}'(R^N)$ one has

$$\frac{\partial(\mathcal{F}T)}{\partial \xi_j} = \mathcal{F}(-2i\pi\,x_j T) \text{ for } j = 1,\dots,N, \\ \mathcal{F}\frac{\partial T}{\partial x_j} = 2i\,\pi\,\xi_j \mathcal{F}T \text{ for } j = 1,\dots,N. \qquad (15.13)$$

Lemma 15.4. *(i) If* $\varphi_0 \in \mathcal{S}(R^N)$ *is defined by* $\varphi_0(x) = e^{-\pi|x|^2}$*, then* $\mathcal{F}\varphi_0 = \varphi_0$.
(ii) $\mathcal{F}\delta_0 = 1$*, and* $\mathcal{F}1 = \delta_0$.

Proof: (i) One notices that $\frac{\partial \varphi_0}{\partial x_j} = -2\pi x_j\,\varphi_0$, so that by (15.6) and (15.7) one has $\frac{\partial \mathcal{F}\varphi_0}{\partial \xi_j} = -2\pi \xi_j\,\mathcal{F}\varphi_0$ for $j = 1,\dots,N$, so that $\mathcal{F}\varphi_0(\xi) = C\,e^{-\pi|\xi|^2}$. One finds that $C = 1$ by using the definition (15.3) for $\xi = 0$ and using $\int_R e^{-\pi x^2}\,dx = 1$.

(ii) One may deduce that $\mathcal{F}\delta_0 = 1$ by using (15.9), if one notices that this definition coincides with that of (15.12). One has $\delta_0 \in \mathcal{S}'(R^N)$ because $\mathcal{S}(R^N) \subset C_0(R^N)$, and noticing that $x_j\delta_0 = 0$ gives $\frac{\partial(\mathcal{F}\delta_0)}{\partial x_j} = 0$, one deduces that $\mathcal{F}\delta_0$ is a constant C by Lemma 6.4, which is 1 by using φ_0 in (15.12). Similarly, $1 \in \mathcal{S}'(R^N)$ because $\mathcal{S}(R^N) \subset L^1(R^N)$, and noticing that $\frac{\partial 1}{\partial x_j} = 0$

---

[3] Michel PLANCHEREL, Swiss mathematician, 1885–1967. He worked in Genève (Geneva), in Fribourg, and at ETH (Eidgenössische Technische Hochschule), Zürich, Switzerland.

[4] The left side of (15.11) is $\int_{R^N} \left(\int_{R^N} e^{-2i\,\pi(x.\xi)} f(x)\,dx \right) g(\xi)\,d\xi$, and Fubini's theorem applies because $\int\int_{R^N \times R^N} |f(x)|\,|g(\xi)|\,dx\,d\xi < \infty$, so that one may exchange the order of integrations; after that, one exchanges the names of x and ξ.

gives $\xi_j \mathcal{F}1 = 0$, one deduces that $\mathcal{F}1 = C\,\delta_0$ by Lemma 6.3, and the constant C is 1 by using φ_0 in (15.12).   □

There are other classical methods for computing Fourier transforms, like changing paths of integration in the complex plane, but the observations of Laurent SCHWARTZ used in the proof of Lemma 15.3 are often more easy to apply.

Following the example of DIRAC, physicists write $\int e^{-2i\pi(x.\xi)}\,dx = \delta(\xi)$, but Laurent SCHWARTZ's extension of Fourier transform has shown that this is a loose way of saying $\mathcal{F}1 = \delta_0$, which means

$$\int_{R^N}\left(\int_{R^N} e^{-2i\pi(x.\xi)}\varphi(x)\,dx\right)d\xi = \varphi(0) \text{ for all } \varphi \in \mathcal{S}(R^N), \qquad (15.14)$$

but *Fubini's theorem does not apply* to the integral on the left.

Lemma 15.5. *(i) $\mathcal{F}$ is an isomorphism from $\mathcal{S}(R^N)$ onto itself, with inverse $\overline{\mathcal{F}}$.*

(ii) $\mathcal{F}$ is an isomorphism from $\mathcal{S}'(R^N)$ onto itself, with inverse $\overline{\mathcal{F}}$.
(iii) $\mathcal{F}$ is an isometry from $L^2(R^N)$ onto itself, with inverse $\overline{\mathcal{F}}$.

Proof: (i) For $\varphi \in \mathcal{S}(R^N)$, one uses (15.14) for $\tau_{-y}\varphi$, which gives

$$\varphi(y) = \int_{R^N}\!\left(\int_{R^N} e^{-2i\pi(x.\xi)}\varphi(x+y)\,dx\right)d\xi = \int_{R^N}\!\left(\int_{R^N} e^{-2i\pi(z-y.\xi)}\varphi(z)\,dz\right)d\xi$$
$$= \int_{R^N} e^{2i\pi(y.\xi)}\mathcal{F}\varphi(\xi)\,d\xi = \overline{\mathcal{F}}(\mathcal{F}\varphi)(y),$$
$$(15.15)$$

i.e., $\overline{\mathcal{F}}(\mathcal{F}\varphi) = \varphi$, showing that $\overline{\mathcal{F}}$ is a left inverse of $\mathcal{F}$, but taking the complex conjugate gives $\mathcal{F}\overline{\mathcal{F}\overline{\varphi}} = \overline{\mathcal{F}}(\mathcal{F}\varphi) = \overline{\varphi}$, showing that $\overline{\mathcal{F}}$ is a right inverse of $\mathcal{F}$.

(ii) By (15.12), for $T \in \mathcal{S}'(R^N)$ and $\varphi \in \mathcal{S}(R^N)$

$$\langle \overline{\mathcal{F}}\mathcal{F}T, \varphi \rangle = \langle \mathcal{F}T, \overline{\mathcal{F}}\varphi \rangle = \langle T, \mathcal{F}\overline{\mathcal{F}}\varphi \rangle = \langle T, \varphi \rangle, \qquad (15.16)$$

showing that $\overline{\mathcal{F}}\mathcal{F}T = T$, and similarly $\mathcal{F}\overline{\mathcal{F}}T = T$.

(iii) One uses Plancherel's formula (15.11) with $f \in \mathcal{S}(R^N)$ and $g = \overline{\mathcal{F}f} = \overline{\mathcal{F}}\bar{f}$, and because $\mathcal{F}g = \mathcal{F}\overline{\mathcal{F}}\bar{f} = \bar{f}$, one obtains

$$\int_{R^N}|\mathcal{F}f|^2\,d\xi = \int_{R^N}|f|^2\,dx \text{ for all } f \in \mathcal{S}(R^N) \qquad (15.17)$$

and because $\mathcal{S}(R^N)$ is dense in $L^2(R^N)$ (as it contains $C_c^\infty(R^N)$ which is dense) (15.17) is true for all $f \in L^2(R^N)$.   □

Of course, one cannot define the Fourier transform of an arbitrary distribution on R^N, as one cannot even define the Fourier transform for all smooth functions independently of what their growth at ∞ is, and keep all the known properties. For example, in one dimension the function $exp(x) = e^x$ satisfies $\frac{dexp}{dx} = exp$ and if $\mathcal{F}exp$ was defined with the property (15.7) one would have $2i\pi\xi\mathcal{F}exp = \mathcal{F}exp$, so that the support of $\mathcal{F}exp$ would be included in the zero set of $2i\pi\xi - 1$, which is empty[5-7], so that one would have $\mathcal{F}exp = 0$, a useless extension, incompatible with keeping $\mathcal{F}$ an isomorphism.

Lemma 15.6. *For* $u \in H^1(R^N)$ *one has*

$$\int_{R^N} |u|^2\, dx = \int_{R^N} |\mathcal{F}u|^2\, d\xi$$

$$\int_{R^N} |grad(u)|^2\, dx = \int_{R^N} 4\pi^2|\xi|^2|\mathcal{F}u|^2\, d\xi. \tag{15.18}$$

Proof: The first part is (15.17), and the second follows from $\mathcal{F}\frac{\partial u}{\partial x_j} = 2i\pi\xi_j\mathcal{F}u$, and again (15.17) and then summing in j. □

Lemma 15.6 suggests a natural extension for defining Sobolev spaces with noninteger order.

Definition 15.7. *For a real* $s \geq 0$, $H^s(R^N) = \{u \in L^2(R^N) \mid |\xi|^s\mathcal{F}u \in L^2(R^N)\}$.
For a real $s < 0$, $H^s(R^N) = \{u \in \mathcal{S}'(R^N) \mid (1+|\xi|^2)^{s/2}\mathcal{F}u \in L^2(R^N)\}$. □

The Fourier transform is an isometry on $L^2(R^N)$, and because it satisfies $\mathcal{F}D^\alpha u = (2i\pi)^{|\alpha|}\xi^\alpha\mathcal{F}u$ for all $u \in \mathcal{S}'(R^N)$ and all multi-indices α, one sees that if s is a nonnegative integer m, Definition (15.15) for $H^s(R^N)$ gives the same space as $W^{m,2}(R^N)$.

For $s < 0$, the space $H^s(R^N)$ is not a subset of $L^2(R^N)$, and in order to use $\mathcal{F}$ one starts from an element of $\mathcal{S}'(R^N)$, but one cannot use $(1+|\xi|)^s\mathcal{F}u$ because $(1+|\xi|)^s$ is not a C^∞ function.

Of course, $H^{-s}(R^N)$ is the dual of $H^s(R^N)$, but the notation for an open set Ω is different.

Definition 15.8. *For an open set* $\Omega \subset R^N$, *and a positive integer* m, *one denotes by* $H_0^m(\Omega)$ *the closure of* $C_c^\infty(\Omega)$ *in* $H^m(\Omega)$, *and* $H^{-m}(\Omega)$ *the dual of* $H_0^m(\Omega)$. □

---

[5] GELFAND has extended the Fourier transform, with values into a much larger space than $\mathcal{D}'(R^N)$, a space of analytic functionals $\mathcal{A}'(C^N)$, the dual space of a space of analytic functions $\mathcal{A}(C^N)$. For $T \in \mathcal{D}'$, the expected relation $\langle\mathcal{F}T, \varphi\rangle = \langle T, \mathcal{F}\varphi\rangle$ shows that one must take test functions φ such that $\mathcal{F}\varphi \in C^\infty(R^N)$, i.e., $\varphi = \overline{\mathcal{F}}\psi$ with $\psi \in C_c^\infty(R^N)$; it is easy to see that if $\psi \in C_c^\infty(R^N)$ then $\mathcal{F}\psi$ and $\overline{\mathcal{F}}\psi$ extend to an analytic function on C^N satisfying a special growth condition $|\mathcal{F}\psi(\xi+i\eta)| \leq ||\psi||_{L^1}e^{2\pi\varrho|\eta|}$ if the support of ψ is included in $\overline{B(0,\varrho)}$; the Paley–Wiener theorem, extended by Laurent SCHWARTZ to the case of distributions, gives a more precise characterization of what $\mathcal{F}\psi$ can be. One should pay attention that because a nonzero analytic function cannot have compact support, there is no notion of support for analytic functionals. With this extension of the Fourier transform, one then has $\mathcal{F}exp = C\,\delta_{z_0}$ with $z_0 = \frac{1}{2i\pi}$.

[6] Raymond Edward Alan Christopher PALEY, English mathematician, 1907–1933.

[7] Norbert G. WIENER, American mathematician, 1894–1964. He worked at MIT (Massachusetts Institute of Technology), Cambridge, MA.

Lemma 15.9. $H^{-1}(\Omega) = \{T \in \mathcal{D}'(\Omega) \mid T = f_0 - \sum_{j=1}^{N} \frac{\partial f_j}{\partial x_j}, f_0, \ldots, f_N \in L^2(\Omega)\}$. *If Poincaré's inequality holds for $H_0^1(\Omega)$, then every $g_0 \in L^2(\Omega)$ can be written as $\sum_{j=1}^{N} \frac{\partial g_j}{\partial x_j}$, with $g_1, \ldots, g_N \in L^2(\Omega)$.*

Proof: The mapping $u \mapsto (u, \frac{\partial u}{\partial x_1}, \ldots, \frac{\partial u}{\partial x_N})$ is an isometry of $H_0^1(\Omega)$ onto a closed subspace of $L^2(\Omega)^{N+1}$, so a linear continuous form L on $H_0^1(\Omega)$ is transported onto this subspace and extended to a linear continuous form on $L^2(\Omega)^{N+1}$, so that there exist $f_0, \ldots, f_N \in L^2(\Omega)$ such that $L(\varphi) = \int_\Omega (f_0 \varphi + \sum_{j=1}^{N} f_j \frac{\partial \varphi}{\partial x_j}) \, dx$ for all $\varphi \in H_0^1(\Omega)$, or equivalently for all $\varphi \in C_c^\infty(\Omega)$, but this means that $L(\varphi) = \langle T, \varphi \rangle$ with $T = f_0 - \sum_{j=1}^{N} \frac{\partial f_j}{\partial x_j}$, and because $C_c^\infty(\Omega)$ is dense in $H_0^1(\Omega)$, one has $L = T$.

If Poincaré's inequality holds for $H_0^1(\Omega)$, the mapping $u \mapsto (\frac{\partial u}{\partial x_1}, \ldots, \frac{\partial u}{\partial x_N})$ is an isometry of $H_0^1(\Omega)$ onto a closed subspace of $L^2(\Omega)^N$, and every linear continuous form L on $H_0^1(\Omega)$ has the form $L(\varphi) = \sum_{j=1}^{N} g_j \frac{\partial \varphi}{\partial x_j} \, dx$ for all $\varphi \in H_0^1(\Omega)$, with $g_1, \ldots, g_N \in L^2(\Omega)$, and in particular if $g_0 \in L^2(\Omega)$ one can write $\int_\Omega g_0 \varphi \, dx$ in this way. $\qquad\square$

Lemma 15.10. *For any $s \in R$, $C_c^\infty(R^N)$ is dense in $H^s(R^N)$.*

Proof: Consider the space $\mathcal{F}H^s$ of functions in L^2 with the weight $(1 + |\xi|^2)^s$, i.e., $\int_{R^N} (1 + |\xi|^2)^s |u(\xi)|^2 \, d\xi < \infty$. One can approach any function in $\mathcal{F}H^s$ by functions with compact support by truncation, defining u_n by $u_n(\xi) = u(\xi)$ if $|\xi| \leq n$ and $u_n(\xi) = 0$ if $|\xi| > n$, and by the Lebesgue dominated convergence theorem, u_n converges to u in $\mathcal{F}H^s$. Any $u \in \mathcal{F}H^s$ having compact support can be approached by functions in $C_c^\infty(R^N)$ because for a smoothing sequence ϱ_m one has $\varrho_m \star u \to u$ in $L^2(R^N)$, and because the supports stay in a bounded set one has $\varrho_m \star u \to u$ in $\mathcal{F}H^s$. Therefore $C_c^\infty(R^N)$ is dense in $\mathcal{F}H^s$, so that $S(R^N)$ is dense in $\mathcal{F}H^s$. Using the Fourier transform one deduces that $S(R^N)$ is dense in $H^s(R^N)$. Let $m \geq s$ be a nonnegative integer then one can approach any function $v \in S(R^N)$ by a sequence in $C_c^\infty(R^N)$, the convergence being in $H^m(R^N)$ strong, and therefore also in $H^s(R^N)$ strong, and this is done by approaching v by $v(x)\theta(\frac{x}{n})$ with $\theta \in C_c^\infty(R^N)$ with $\theta(x) = 1$ for $|x| \leq 1$. $\qquad\square$

In the characterization of the traces of functions from $H^s(R^N)$ (for $s > \frac{1}{2}$), one will use the following result.

Lemma 15.11. *If $u \in S(R^N)$ and $v \in S(R^{N-1})$ is the restriction of u on the hyperplane $x_N = 0$, i.e.,*

$$v(x') = u(x', 0) \text{ for } x' \in R^{N-1}, \tag{15.19}$$

then one has

$$\mathcal{F}v(\xi') = \int_R \mathcal{F}u(\xi', \xi_N) \, d\xi_N \text{ for } \xi' \in R^{N-1}. \tag{15.20}$$

Proof: Because $\mathcal{F}\delta_0 = 1$, one has (15.14), i.e., $\varphi(0) = \int_R \mathcal{F}\varphi(\xi)\,d\xi$. One uses this relation for the function $x_N \mapsto u(x', x_N)$, and then one takes the Fourier transform in x' of both sides of the equality. $\square$

Notice that in (15.20), the same symbol $\mathcal{F}$ is used for the Fourier transform in N or $N-1$ variables, and even 1 variable in the proof.

[Taught on Friday February 18, 2000.]

Traces of $H^s(R^N)$

Lemma 16.1. *For $s > \frac{1}{2}$, functions of $H^s(R^N)$ have a trace on the hyperplane $x_N = 0$, belonging to $H^{s-\frac{1}{2}}(R^{N-1})$. The mapping γ_0 is surjective from $H^s(R^N)$ onto $H^{s-\frac{1}{2}}(R^{N-1})$.*

Proof: To prove the first part, it is enough to show that there exists C such that for all $u \in \mathcal{S}(R^N)$ and v defined by $v(x') = u(x', 0)$ for $x' \in R^{N-1}$ one has $||v||_{H^{s-(1/2)}(R^{N-1})} \leq C\,||u||_{H^s(R^N)}$. By (15.20) one has $\mathcal{F}v(\xi') = \int_R \mathcal{F}u(\xi', \xi_N)\,d\xi_N$, and *Cauchy–(Bunyakovsky)–Schwarz inequality*[1,2] implies

$$|\mathcal{F}v(\xi')|^2 \leq \left(\int_R (1 + |\xi|^2)^s |\mathcal{F}u(\xi', \xi_N)|^2 \, d\xi_N \right) \left(\int_R \frac{d\xi_N}{(1 + |\xi|^2)^s} \right) \qquad (16.1)$$

Using the change of variable $\xi_N = t\sqrt{1 + |\xi'|^2}$, one has

$$\int_R \frac{d\xi_N}{(1 + |\xi|^2)^s} = \left(\sqrt{1 + |\xi'|^2} \right)^{1-2s} \int_R \frac{dt}{(1 + t^2)^s} = C(s)\left(1 + |\xi'|^2 \right)^{(1/2)-s},$$
$$\qquad (16.2)$$

giving

$$\left(1 + |\xi'|^2 \right)^{s-(1/2)} |\mathcal{F}v(\xi')|^2 \leq C(s) \int_R (1 + |\xi|^2)^s |\mathcal{F}u(\xi', \xi_N)|^2 \, d\xi_N, \qquad (16.3)$$

which gives the desired result by integrating in ξ', because $C(s) < \infty$ if and only if $s > \frac{1}{2}$.

---

[1] Viktor Yakovlevich BUNYAKOVSKY, Ukrainian-born mathematician, 1804–1889. He worked in St Petersburg, Russia. He studied with CAUCHY in Paris (1825), and he proved the "Cauchy–Schwarz" inequality in 1859, 25 years before SCHWARZ.

[2] Hermann Amandus SCHWARZ, German mathematician, 1843–1921. He worked at ETH (Eidgenössische Technische Hochschule), Zürich, Switzerland, and in Berlin, Germany.

In order to prove the surjectivity, one shows that for $v \in H^{s-(1/2)}(R^{N-1})$ the function u defined by

$$\mathcal{F}u(\xi', \xi_N) = \mathcal{F}v(\xi')\varphi\left(\frac{\xi_N}{\sqrt{1+|\xi'|^2}}\right)\frac{1}{\sqrt{1+|\xi'|^2}}, \tag{16.4}$$

$$\text{with } \varphi \in C_c^\infty(R) \text{ and } \int_R \varphi(t)\, dt = 1,$$

is such that $u \in H^s(R^N)$; because (16.4) implies $\int_R \mathcal{F}u(\xi', \xi_N)\, d\xi_N = \mathcal{F}v(\xi')$, it implies $v = \gamma_0 u$ because of (15.20). Indeed, for all $\xi' \in R^{N-1}$, one has

$$\int_R (1+|\xi|^2)^s |\mathcal{F}u(\xi', \xi_N)|^2\, d\xi_N = \int_R (1+|\xi|^2)^s \left|\varphi\left(\frac{\xi_N}{\sqrt{1+|\xi'|^2}}\right)\right|^2 \frac{1}{1+|\xi'|^2}\, d\xi_N =$$
$$C(1+|\xi'|^2)^{s-(1/2)}, \tag{16.5}$$

where $C = \int_R (1+t^2)^s \varphi(t)^2\, dt$, and one has used the change of variable $\xi_N = t\sqrt{1+|\xi'|^2}$. One obtains the desired result by integrating (16.5) in ξ'.   □

There cannot be a good notion of trace (i.e., different from 0) in the case $s \le \frac{1}{2}$ because in that case the subspace of functions of the form $\psi = \varphi_+ + \varphi_-$ with $\varphi_\pm \in C_c^\infty(R_\pm^N)$ (i.e., $\psi \in C_c^\infty(R^N)$ equal to 0 in a neighbourhood of $x_N = 0$) is dense in $H^s(R^N)$.

The condition $s > \frac{1}{2}$ has not appeared in proving the surjectivity, but one should notice that the function u constructed is not only in $H^s(R^N)$ but has other properties, so it belongs to a smaller subspace, where traces may exist and belong to $H^{s-(1/2)}(R^{N-1})$.

For a function $u \in H^2(R^N)$ for example, one can define the trace $\gamma_0 u \in H^{3/2}(R^{N-1})$ but also the normal derivative $\gamma_1 u = \gamma_0 \frac{\partial u}{\partial x_N} \in H^{1/2}(R^{N-1})$ (in general one takes $\gamma_1 u$ to be the normal derivative, with the normal pointing to the outside), and a more precise result is that $u \mapsto (\gamma_0 u, \gamma_1 u)$ is surjective from $H^2(R^N)$ onto $H^{3/2}(R^{N-1}) \times H^{1/2}(R^{N-1})$, and more generally one has the following surjectivity result.

Lemma 16.2. *If $m + \frac{1}{2} < s < m + 1 + \frac{1}{2}$, and for $k = 0, \ldots, m$, one writes*

$$\gamma_k u = \gamma_0 \left(\frac{\partial^k u}{\partial x_N^k}\right), \tag{16.6}$$

then $u \mapsto (\gamma_0 u, \ldots, \gamma_m u)$ is surjective from $H^s(R^N)$ onto $H^{s-(1/2)}(R^{N-1}) \times \ldots \times H^{s-m-(1/2)}(R^{N-1})$.

Proof: If $u \in \mathcal{S}(R^N)$ and $v_k = \gamma_k u$, then $\mathcal{F}v_k(\xi') = \int_R (2i\pi \xi_N)^k \mathcal{F}u(\xi', \xi_N)\, d\xi_N$, so if $v_k \in H^{s-k-(1/2)}(R^{N-1})$, then one defines u_k by

$$\mathcal{F}u_k(\xi', \xi_N) = \mathcal{F}v_k(\xi')\varphi_k\left(\frac{\xi_N}{\sqrt{1+|\xi'|^2}}\right)(1+|\xi'|^2)^{-(k+1)/2}, \tag{16.7}$$

with

$$\varphi_k \in C_c^\infty(R) \text{ and } \int_R (2i\,\pi\,t)^j \varphi_k(t)\,dt = \delta_{jk} \text{ for } j = 0, \ldots, m, \qquad (16.8)$$

so that one has

$$u_k \in H^s(R^N) \text{ with } \gamma_k u_k = v_k \text{ and } \gamma_j u_k = 0 \text{ for } j = 0, \ldots, k-1, k+1, \ldots, m, \qquad (16.9)$$

and surjectivity is proven by using $u = \sum_{k=0}^m u_k$. □

In order to define $H^s(\Omega)$ for $\Omega \neq R^N$, it is useful to find a characterization of $H^s(R^N)$ which does not use the Fourier transform explicitly.

Lemma 16.3. *For $0 < s < 1$, $u \in H^s(R^N)$ is equivalent to*

$$u \in L^2(R^N) \text{ and } \int\int_{R^N \times R^N} \frac{|u(x) - u(y)|^2}{|x-y|^{N+2s}}\,dx\,dy < \infty. \qquad (16.10)$$

Proof: For $h \in R^N$, and $\tau_h u(x) = u(x-h)$, one has

$$\mathcal{F}\tau_h u(\xi) = e^{2i\,\pi(h.\xi)}\mathcal{F}u(\xi), \qquad (16.11)$$

so that

$$\int_{R^N} |\tau_h u - u|^2\,dx = \int_{R^N} |1 - e^{2i\,\pi(h.\xi)}|^2 |\mathcal{F}u(\xi)|^2\,d\xi = \int_{R^N} 4\sin^2\pi(h.\xi)|\mathcal{F}u(\xi)|^2\,d\xi, \qquad (16.12)$$

which implies

$$\int\int_{R^N \times R^N} \frac{|u(x)-u(y)|^2}{|x-y|^{N+2s}}\,dx\,dy = \int_{R^N} \frac{1}{|h|^{N+2s}}\left(\int_{R^N}|\tau_h u - u|^2\,dx\right)dh =$$
$$\int_{R^N}|\mathcal{F}u(\xi)|^2\left(\int_{R^N}\frac{4\sin^2\pi(h.\xi)}{|h|^{N+2s}}\,dh\right)d\xi = C\int_{R^N}|\xi|^{2s}|\mathcal{F}u(\xi)|^2\,d\xi, \qquad (16.13)$$

where one has used the change of variable $h = \frac{z}{|\xi|}$, and using invariance by rotation one finds

$$C = \int\int_{R^N} \frac{4\sin^2\pi z_1}{|z|^{N+2s}}\,dz, \qquad (16.14)$$

which is finite because $|\sin \pi z_1| \leq \pi|z|$ for z near 0 and $s < 1$, and $|\sin \pi z_1| \leq 1$ for z near ∞ and $s > 0$. □

For an open set Ω, one could define $H^s(\Omega)$ for $0 < s < 1$ in (at least) three different ways:

(i) one may use $u \in H^s(\Omega)$ for $u \in L^2(\Omega)$ and $\int\int_{\Omega\times\Omega}\frac{|u(x)-u(y)|^2}{|x-y|^{N+2s}}\,dx\,dy < \infty$,

(ii) one may use $u \in H^s(\Omega)$ for $u = U|_\Omega$ with $U \in H^s(R^N)$,

(iii) one may define $H^s(\Omega)$ by interpolation.

For a bounded open set with Lipschitz boundary, the three definitions give the same space with equivalent norms.

If $u \in W^{1,\infty}(R^N) = Lip(R^N)$, then the restriction of u to $x_N = 0$ is a Lipschitz continuous function, and conversely, if v is a Lipschitz continuous function defined on a closed set A with Lipschitz constant M, then one can extend it to R^N into a Lipschitz continuous function u defined on R^N and having the same Lipschitz constant M by

$$u(x) = sup_{a \in A}(v(a) - M\,|a - x|). \tag{16.15}$$

If $u \in H^1(R^N) = W^{1,2}(R^N)$, the space of traces v is $H^{1/2}(R^{N-1})$ (Lemma 16.1), which by Lemma 16.3 means $\int\int_{R^{N-1} \times R^{N-1}} \frac{|v(x)-v(y)|^2}{|x-y|^N}\,dx\,dy < \infty$ and $v \in L^2(R^{N-1})$.

If $u \in W^{1,p}(R^N)$ and $1 < p < \infty$, Emilio GAGLIARDO has shown that the space of traces v is characterized by $\int\int_{R^{N-1} \times R^{N-1}} \frac{|v(x)-v(y)|^p}{|x-y|^{N+p-2}}\,dx\,dy < \infty$ and $v \in L^p(R^{N-1})$.

If $u \in W^{1,1}(R^N)$, Emilio GAGLIARDO has characterized the space of traces as $v \in L^1(R^{N-1})$. However, Jaak PEETRE has shown that there is no linear continuous lifting from $L^1(R^{N-1})$ to $W^{1,1}(R^N)$; I have not yet read the detail of the proof.

[Taught on Monday February 21, 2000.]

17

Proving that a Point is too Small

Lemma 17.1. *If $u \in H^1(R^N)$ and $N \geq 3$, then $\frac{u}{r} \in L^2(R^N)$ and*

$$\left\|\frac{u}{r}\right\|_2 \leq \frac{2}{N-2} \|grad(u)\|_2 \tag{17.1}$$

Proof: One proves (17.1) for all $u \in C_c^\infty(R^N)$, and then it extends to $H^1(R^N)$ by density. For any $\alpha \in R$ one has

$$0 \leq \int_{R^N} \sum_{j=1}^N \left|\frac{\partial u}{\partial x_j} + \frac{\alpha x_j u}{r^2}\right|^2 dx = \int_{R^N} |grad(u)|^2 dx + \sum_{j=1}^N \int_{R^N} 2u \frac{\partial u}{\partial x_j} \frac{\alpha x_j}{r^2}$$
$$+ \int_{R^N} \frac{\alpha^2 u^2}{r^2} dx = \int_{R^N} |grad(u)|^2 dx - ((N-2)\alpha - \alpha^2) \int_{R^N} \frac{u^2}{r^2} dx, \tag{17.2}$$

where one has performed an integration by parts, allowed because $\frac{x_j}{r^2} \in W_{loc}^{1,1}(R^N)$ for all j (for $N \geq 3$), and $\sum_{j=1}^N \frac{\partial}{\partial x_j}\left(\frac{x_j}{r^2}\right) = \frac{N-2}{r^2}$; one then chooses the best value of α, i.e., $\alpha = \frac{N-2}{2}$. □

The result is obviously false for $N = 2$, even for u smooth with $u(0) \neq 0$ as $\frac{1}{r} \notin L_{loc}^2(R^2)$.

Lemma 17.2. *The space of functions in $C_c^\infty(R^N)$ which are 0 in a small ball around 0 is dense in $H^1(R^N)$ for $N \geq 3$.*

Proof: As $C_c^\infty(R^N)$ is dense in $H^1(R^N)$ for all N, one must approach any $u \in C_c^\infty(R^N)$ by a sequence of functions which are 0 in a small ball around 0, and this is done by taking

$$u_n(x) = u(x)\theta(n x) \text{ with } \theta \in C^\infty(R^N), \theta(r) = 0 \text{ for } r \leq 1, \theta(r) = 1 \text{ for } r \geq 2. \tag{17.3}$$

One has $u_n \to u$ in $L^2(R^N)$ by the Lebesgue dominated convergence theorem, and similarly $\frac{\partial u}{\partial x_j}\theta(n x) \to \frac{\partial u}{\partial x_j}$ in $L^2(R^N)$, and in order to show that $u n \frac{\partial \theta}{\partial x_j}(n x) \to 0$ in $L^2(R^N)$, one also applies the Lebesgue dominated convergence theorem to $\frac{u}{r} f(n x)$ with $f(x) = r \frac{\partial \theta}{\partial x_j}$. □

Lemma 17.2 is actually also true for $N = 2$, and one may use a variant for Lemma 17.1 given at Lemma 17.4, and then a variant for the truncation formula (17.3), but Lemma 17.3 gives a different proof.

Lemma 17.3. *The space of functions in $C_c^\infty(R^2)$ which are 0 in a small ball around 0 is dense in $H^1(R^2)$.*

Proof: One uses the Hahn–Banach theorem, and one shows that if $T \in H^{-1}(R^2) = \left(H^1(R^2)\right)'$, and $\langle T, \varphi \rangle = 0$ for all $\varphi \in C_c^\infty(R^2)$ which are 0 in a small ball around 0, then $T = 0$. Because $\langle T, \varphi \rangle = 0$ for all $\varphi \in C_c^\infty(\omega)$ for any open set ω such that $0 \notin \overline{\omega}$, one finds that the support of T can only be $\{0\}$ (if T is not 0, in which case the support of T is empty). As will be seen in Lemma 18.2, if a distribution T has support $\{0\}$, then $T = \sum_\alpha c_\alpha D^\alpha \delta_0$ (finite sum), but if some $c_\alpha \neq 0$ then $T \notin H^{-1}(R^2)$, because $\mathcal{F}T = \sum_\alpha c_\alpha (2i\,\pi\,\xi)^\alpha$ and no nonzero polynomial P satisfies $\int_{R^2} \frac{|P(\xi)|^2}{1+|\xi|^2}\,d\xi < \infty$. □

One deduces that if $\Omega = R^N \setminus F$, where F is a finite number of points and $N \geq 2$ then $H_0^1(\Omega) = H^1(\Omega) = H^1(R^N)$. This is not true for $N = 1$ as the functions in $H^1(R)$ are continuous. With some technical changes the same proofs adapt to $W^{1,p}(R^N)$ if $1 < p \leq N$. Similar ideas show that one can approach every function of $H^1(R^N)$ for $N \geq 3$ by functions in $C_c^\infty(R^N)$ which vanish in a neighbourhood of a given segment, but that is not true for $N = 2$ as the functions in $H^1(R^2)$ have traces on the segment.

Like for the limiting case of Sobolev's embedding theorem, there are norms which scale in the same way, but which are not comparable; for example if $N = 2$ then $||grad(u)||_2$ and $||u||_\infty$ scale in the same way but functions in $H^1(R^2)$ are not necessarily bounded, and $\left|\left|\frac{u}{r}\right|\right|_2$ also scales in the same way but $u \in H^1(R^2)$ does not imply $\frac{u}{r} \in L^2(R^2)$. However, one has the following result.

Lemma 17.4. *If $\Omega \subset B(0, R_0) \subset R^2$, then there exists C such that*

$$\left|\left|\frac{u}{r\log(r/R_0)}\right|\right|_2 \leq C\,||grad(u)||_2 \text{ for all } u \in H_0^1(\Omega). \tag{17.4}$$

Proof: One proves the inequality for $u \in C_c^\infty(B(0, R_0))$ so that it is true for $u \in C_c^\infty(\Omega)$ and it then extends to $H_0^1(\Omega)$. For f smooth, one develops

$$\int_{B(0,R_0)} \sum_{j=1}^N \left| \frac{\partial u}{\partial x_j} + x_j f(r) u \right|^2 dx \geq 0, \tag{17.5}$$

and one uses the integration by parts

$$\int_{B(0,R_0)} 2u\,\frac{\partial u}{\partial x_j} x_j f(r)\,dx = -\int_{B(0,R_0)} |u|^2 \left(f(r) + x_j^2 \frac{f'(r)}{r} \right) dx, \tag{17.6}$$

which is valid if $r\,f(r)$ and $r^2 f'(r)$ belong to $L^1(0, R_0 - \varepsilon)$ for every $\varepsilon > 0$, and one deduces

$$\int_{B(0,R_0)} |grad(u)|^2 \, dx \geq \int_{B(0,R_0)} |u|^2 (2f + r\,f' - r^2 f^2) \, dx. \qquad (17.7)$$

If one takes $f = \frac{g}{r^2}$, one has $2f + r\,f' - r^2 f^2 = \frac{g'}{r} - \frac{g^2}{r^2}$, and one then approaches $g = \frac{-1}{2\log(r/R_0)}$, which corresponds to multiplying $|u|^2$ by $\frac{1}{4r^2(\log(r/R_0))^2}$. $\square$

Because the logarithm vanishes for $r = R_0$, it is important to have Ω bounded; however, the same argument works if Ω is unbounded and is outside a ball $B(0, R_0)$, but there is a problem with the entire space; actually, Jacques-Louis LIONS and Jacques DENY have shown that the completion of $C_c^\infty(R^2)$ for the norm $||grad(u)||_2$ is not a space of distributions on R^2.

[Taught on Wednesday February 23, 2000.]

Compact Embeddings

In order to identify distributions with support at a point, one uses the following result.

Lemma 18.1. *Let $T \in \mathcal{D}'(\Omega)$ have its support in a compact $K_0 \subset \Omega$, and assume that T is a distribution of order m. Then if $\varphi_0 \in C_c^\infty(\Omega)$ satisfies $D^\alpha \varphi_0(x) = 0$ for all $x \in K_0$ and all multi-indices α such that $|\alpha| \leq m$, one has $\langle T, \varphi_0 \rangle = 0$.*

Proof: As T is assumed to be of order m, for every compact $K \subset \Omega$ there exists $C(K)$ such that $|\langle T, \varphi \rangle| \leq C(K) \sup_{x \in K, |\alpha| \leq m} |D^\alpha \varphi(x)|$ for all $\varphi \in C_c^\infty(\Omega)$ such that $support(\varphi) \subset K$. Let $\varepsilon_0 > 0$ be such that $\{x \in R^N \mid d(x, K_0) \leq \varepsilon_0\} \subset \Omega$. For $0 < \varepsilon < \varepsilon_0$, let $K_\varepsilon = \{x \in R^N \mid d(x, K_0) \leq \varepsilon\}$, and let χ_ε be the characteristic function of K_ε. Let $\varrho_1 \in C_c^\infty(R^N)$ with $support(\varrho_1) \subset \overline{B(0,1)}$ and $\int_{R^N} \varrho_1(x)\,dx = 1$, and as usual $\varrho_\delta(x) = \frac{1}{\delta^N} \varrho_1\left(\frac{x}{\delta}\right)$ for $\delta > 0$.

If $3\delta < \varepsilon_0$, let $\theta_\delta = \chi_{2\delta} \star \varrho_\delta$, so that $\theta_\delta \in C_c^\infty(\Omega)$, $\theta_\delta(x) = 1$ if $x \in K_\delta$ and $support(\theta_\delta) \subset K_{3\delta} \subset K_{\varepsilon_0}$. One has $\langle T, \varphi_0 \rangle = \langle T, \theta_\delta \varphi_0 \rangle$, because the difference is $\langle T, (1 - \theta_\delta)\varphi_0 \rangle$ and the support of $1 - \theta_\delta$ is included in $\Omega \setminus K_0$, i.e., the largest open set where T is 0. One proves that $\langle T, \theta_\delta \varphi_0 \rangle \to 0$ as $\delta \to 0$ by showing that for any multi-index α such that $|\alpha| \leq m$ one has $\sup_{x \in K_{\varepsilon_0}} |D^\alpha(\theta_\delta \varphi_0)(x)| \to 0$ as $\delta \to 0$.

One has

$$|D^\beta \theta_\delta(x)| \leq C\,\delta^{-|\beta|} \text{ for all } x \text{ and } |\beta| \leq m. \tag{18.1}$$

Because of Taylor's expansion formula and the vanishing of the derivatives of φ_0 on K up to order m, one has

$$|D^\gamma \varphi_0(x)| \leq d(x, K_0)^{m-|\gamma|} \eta\big(d(x, K_0)\big) \text{ for } |\gamma| \leq m \text{ and } \eta(t) \to 0 \text{ as } t \to 0. \tag{18.2}$$

By Leibniz's formula, using $d(x, K_0) \leq 3\delta$ for $x \in support(\theta_\delta)$, one deduces $|D^\alpha(\theta_\delta \varphi_0)(x)| \leq C\,\eta\big(d(x, K_0)\big)$. $\qquad\square$

Lemma 18.2. *If T has support at a point $a \in \Omega$, then T is a finite combination of derivatives of the Dirac mass at a.*

Proof: If $K = \overline{B(a,r)} \subset \Omega$, then T has finite order m on K, and by the preceding result, $D^\alpha \varphi_0(a) = 0$ for all $|\alpha| \le m$ implies $\langle T, \varphi_0 \rangle = 0$. A result of linear algebra says that on any vector space if for linear forms $L_0, \dots, L_p$ every u satisfying $L_1 u = \dots = L_p u = 0$ also satisfies $L_0 u = 0$ then there are scalars $\lambda_1, \dots, \lambda_p$ such that $L_0 = \sum_{j=1}^p \lambda_j L_j$. Therefore there are scalars λ_α for $|\alpha| \le m$ such that

$$\langle T, \varphi \rangle = \sum_{|\alpha| \le m} \lambda_\alpha D^\alpha \varphi(a) \text{ for all } \varphi \in C_c^\infty(\Omega) \text{ with } support(\varphi) \subset K, \quad (18.3)$$

i.e., $T = \sum_{|\alpha| \le m} (-1)^{|\alpha|} \lambda_\alpha D^\alpha \delta_a$.   □

Most compactness results rely on the theorems of ARZELÁ and ASCOLI, and the basic result of interest here is that if one works on a compact set K of R^N and if one has a sequence $u_n \in C(K)$ of functions which have the same modulus of uniform continuity, i.e., $|u_n(x) - u_n(y)| \le \omega(|x - y|)$ for all $x, y \in K$ and all n, with $\omega(t) \to 0$ as $t \to 0$, then there exists a subsequence u_m which converges uniformly on K, to $u_\infty \in C(K)$ (using a diagonal argument one extracts a subsequence which converges on a countable dense set of K, and the subsequence also converges at the other points by equicontinuity, and the limit is continuous for the same reason).

To prove compactness in $L^p(\Omega)$ for $1 \le p < \infty$, one extends the functions by 0 and one applies a compactness result in $L^p(R^N)$, usually attributed to KOLMOGOROV, but I have seen the names FRÉCHET and M. RIESZ mentioned too.

Lemma 18.3. *If a sequence u_n is bounded in $L^p(R^N)$ and satisfies:*
(i) For every $\varepsilon > 0$, there exists $R(\varepsilon)$ such that $\int_{|x| \ge R(\varepsilon)} |u_n|^p \, dx \le \varepsilon$ for all n.
(ii) For every $\varepsilon > 0$ there exists $\delta > 0$ such that if $|h| \le \delta$ then $\int_{R^N} |u_n(x + h) - u_n(x)|^p \, dx \le \varepsilon$.
Then there exists a subsequence u_m which converges strongly to $u_\infty \in L^p(R^N)$.

Proof: It is enough to show that for every $\alpha > 0$ one can write $u_n = v_n + w_n$ such that $\|w_n\|_p \le \alpha$ and that from v_n one can extract a converging subsequence v_m; for the subsequence u_m one then has $\limsup_{m, m' \to \infty} \|u_m - u_{m'}\|_p \le 2\alpha$. Starting from the selected subsequence, one then repeats the argument with α replaced by $\frac{\alpha}{2}$, and so on, and a diagonal subsequence is a Cauchy sequence.

Using $\theta \in C_c^\infty(R^N)$ such that $0 \le \theta \le 1$ and $\theta(x) = 1$ for $|x| \le R(\varepsilon)$, one chooses $v_n = \theta u_n$ and $w_n = (1 - \theta) u_n$, and one notices that $\|w_n\|_p \le \varepsilon$ by (i), and because $\tau_h v_n - v_n = (\tau_h \theta - \theta) \tau_h u_n + \theta(\tau_h u_n - u_n)$ and $\|\tau_h \theta - \theta\|_\infty \le M |h|$, one finds that v_n is bounded in $L^p(R^N)$, has its support in a fixed bounded set, and satisfies (ii).

Assuming that the functions u_n satisfy (ii) and have their support in a fixed bounded set, one uses $v_n = u_n \star \varrho_\delta$ for a special smoothing sequence ϱ_δ, and $w_n = u_n - u_n \star \varrho_\delta$; one can apply the Arzelá–Ascoli theorem to the sequence v_n, as they form a bounded sequence of Lipschitz continuous functions having their support in a fixed compact set; because $w_n(x) = \int_{R^N} \varrho_\delta(y)(u_n(x) - u_n(x-y))\,dy$, one has $||w_n||_p \le \int_{R^N} \varrho_\delta(y)||u_n - \tau_y u_n||_p\, dy \le \varepsilon$. $\square$

If Ω is a bounded open with Lipschitz boundary, then there exists a continuous extension P from $W^{1,p}(\Omega)$ to $W^{1,p}(R^N)$ (Lemma 12.4), and this was used in Lemma 14.5 for showing that the injection of $W^{1,p}(\Omega)$ into $L^p(\Omega)$ is compact. The continuous extension may not exist if the boundary is not Lipschitz, but Lemma 18.4 shows that the compactness property does not require as much smoothness of the boundary.

Lemma 18.4. *If Ω is a bounded open set with a continuous boundary, then the injection of $W^{1,p}(\Omega)$ into $L^p(\Omega)$ is compact.*

Proof: The preceding argument does not apply, and one must find a different proof. Using a partition of unity one has to consider the case of Ω_F with F uniformly continuous, for a subsequence having support in a bounded set. One notices that in dimension 1 one has $W^{1,p}(0,\infty) \subset C_b(0,\infty)$, so that one has

$$\int_{F(x')}^{F(x')+3\varepsilon} |u(x',x_N)|^p\, dx \le C\varepsilon \int_{F(x')}^{\infty} \left(|u(x',x_N)|^p + \left|\frac{\partial u}{\partial x_N}(x',x_N)\right|^p\right) dx_N,$$
(18.4)

which one may then integrate in x'. Using the uniform continuity of F one can construct $\theta \in C^\infty(R^N)$ with $0 \le \theta \le 1$, $\theta(x) = 0$ if $x_N < F(x') + \varepsilon$ and $\theta(x) = 1$ if $x_N > F(x') + 2\varepsilon$. One uses then $v_n = \theta\, u_n$ and $w_n = (1-\theta)u_n$. $\square$

It is often useful to know different proofs of the same result. If Ω is an open set with finite measure, then for $p \in [1,\infty]$ Poincaré's inequality holds for $W_0^{1,p}(\Omega)$, as was shown in Lemma 10.2. The use of the Fourier transform is restricted to the case $p = 2$, so Lemma 18.5 is more restrictive than Lemma 10.2 concerning Poincaré's inequality, but it also shows that the injection of $H_0^1(\Omega)$ into $L^2(\Omega)$ is compact (which by the equivalence lemma 11.1 implies Poincaré's inequality).

Lemma 18.5. *If $\Omega \subset R^N$ is an open set with finite measure, then Poincaré's inequality holds for $H_0^1(\Omega)$, and the injection of $H_0^1(\Omega)$ into $L^2(\Omega)$ is compact.*

Proof: Let $u \in H_0^1(\Omega)$, extended by 0 outside Ω, and let $A = ||u||_2$ and $B = ||grad(u)||_2$. Because $\mathcal{F}u(\xi) = \int_\Omega u(x)e^{-2i\pi(x.\xi)}\,dx$ one has $|\mathcal{F}u(\xi)| \le A\, meas(\Omega)^{1/2}$ for all $\xi \in R^N$, so that $\int_{|\xi|\le\varrho} |\mathcal{F}u(\xi)|^2\, d\xi \le A^2 meas(\Omega)\omega_N\varrho^N$, where ω_N is the volume of the unit ball. Because $\int_{|\xi|\ge\varrho} |\mathcal{F}u(\xi)|^2\, d\xi$ is bounded by $\int_{|\xi|\ge\varrho} \frac{4\pi^2|\xi|^2}{4\pi^2\varrho^2}|\mathcal{F}u(\xi)|^2\, d\xi \le \frac{B^2}{4\pi^2\varrho^2}$, one deduces by adding these two inequalities that $A^2 = \int_{R^N} |\mathcal{F}u(\xi)|^2\, d\xi \le A^2 meas(\Omega)\omega_N\varrho^N + \frac{B^2}{4\pi^2\varrho^2}$ for every $\varrho > 0$, which explains the choice $meas(\Omega)\omega_N \varrho^N = \frac{1}{2}$ gives $A \leq c_N meas(\Omega)^{1/N} B$ for a universal constant c_N (i.e., independent of the open set).

In order to prove that the injection of $H_0^1(\Omega)$ into $L^2(\Omega)$ is compact, one assumes that $u_n \rightharpoonup 0$ in $H_0^1(\Omega)$ weak, and one wants to prove that $u_n \to 0$ in $L^2(\Omega)$ strong. Indeed one may take $||u_n||_2 \leq A$ and $||grad(u_n)||_2 \leq B$, and because $\mathcal{F}u_n(\xi)$ is the L^2 scalar product of u_n by a fixed function in $L^2(\Omega)$, one has $\mathcal{F}u_n(\xi) \to 0$ for every $\xi \in R^N$. Because $|\mathcal{F}u(\xi)| \leq A\, meas(\Omega)^{1/2}$ for all $\xi \in R^N$ one deduces by the Lebesgue dominated convergence theorem that $\int_{|\xi| \leq \varrho} |\mathcal{F}u_n(\xi)|^2\, d\xi \to 0$ for any $\varrho > 0$. Because $\int_{|\xi| \geq \varrho} |\mathcal{F}u_n(\xi)|^2\, d\xi \leq \frac{B^2}{4\pi^2 \varrho^2}$, one deduces that $\limsup_{n \to \infty} ||u_n||_2 \leq \frac{B}{2\pi \varrho}$, and then letting $\varrho \to \infty$ one has $||u_n||_2 \to 0$. $\square$

[Taught on Friday February 25, 2000.]

19

Lax–Milgram Lemma

The main reason why Sobolev spaces are important is that they are the natural functional spaces for solving the boundary value problems of continuum mechanics and physics (at least up to now); they may be elliptic equations like

$$\Delta u = f, \tag{19.1}$$

for which one invokes the names LAPLACE or POISSON, parabolic equations like the *heat equation*

$$\frac{\partial u}{\partial t} - \kappa \, \Delta u = f, \tag{19.2}$$

for which one invokes the name FOURIER, or hyperbolic equations like the *wave equation*

$$\frac{\partial^2 u}{\partial t^2} - c^2 \, \Delta u = f, \tag{19.3}$$

for which one invokes the names D'ALEMBERT[1] or D. BERNOULLI.[2]

The Sobolev space $H^1(\Omega)$ is adapted to problems of the form

$$-\sum_{i,j} \frac{\partial}{\partial x_i}\left(A_{ij}\frac{\partial u}{\partial x_j}\right) = f, \text{ written } -div\big(A\,grad(u)\big) = f, \tag{19.4}$$

when the matrix A (which is usually symmetric in applications) has bounded measurable coefficients and satisfies the *ellipticity condition* that there exists $\alpha > 0$ such that

$$\sum_{ij} A_{ij}(x)\xi_i\xi_j \geq \alpha \, |\xi|^2 \text{ for all } \xi \in R^N, \text{ for a.e. } x \in \Omega. \tag{19.5}$$

---

[1] Jean LE ROND, known as D'ALEMBERT, French mathematician, 1717–1783. He worked in Paris, France.

[2] Daniel BERNOULLI, Swiss mathematician, 1700–1782. He worked in St Petersburg, Russia, and in Basel, Switzerland.

There are various *physical interpretations* possible.

(i) One may consider the *stationary heat equation*, so that u is the *temperature* and $A \, grad(u)$ is the *heat flux*.

(ii) One may consider *electrostatics*, which is a *simplification of Maxwell's equation* where there is no magnetic field and no time dependence, so that u is the *electrostatic potential*, $E = -grad(u)$ is the *electric field*, $D = A \, grad(u)$ is the *polarization field*, and f is the *density of electric charge* (usually denoted by ϱ); A is called the *permittivity* tensor in this case. The *density of electric energy* is $e = \frac{1}{2}(E.D)$, so using the Sobolev space $H^1(\Omega)$ for u corresponds to having a *finite electric energy stored in Ω.*

(iii) One may consider a *different simplification of Maxwell's equation* with no magnetic field and no time dependence either, but where one considers that the *electric current j* (which must satisfy the equation of *conservation of electric charge*

$$\frac{\partial \varrho}{\partial t} + div(j) = 0, \tag{19.6}$$

is related to the electric field by *Ohm's law*[3] $j = \sigma E$ (so A is the *conductivity* tensor σ in this case, whose inverse is the *resistivity* tensor).

Whatever the physical intuition is, it only gives *hints* about the properties of the solution of the equation, and one must prove its existence by precise mathematical arguments. Garrett BIRKHOFF[4] mentioned in [3] that there is a shocking statement by POINCARÉ, who had written in an article that the solution of an equation existed because it was a physical problem; however, he added that POINCARÉ had immediately corrected his mistake and had given a mathematical proof of existence in his next article. Laurence YOUNG mentioned in [19] that even HILBERT had overlooked the question of existence for problems of minimization; he justly pointed out that if one proves that every minimizer of a functional must satisfy some necessary conditions of optimality, it is not enough to check among the functions satisfying all these already known necessary conditions of optimality, which of these gives the lowest value to the functional that one seeks to minimize, unless one has proven first that there exists a minimizer, and that remark is of course valid in the classical situations where one had found only one function satisfying a condition of optimality. Nonexistence may come from the fact that minimizing sequences are unbounded and that the minimum is "attained" at infinity,[5] but

---

[3] Georg Simon OHM, German mathematician, 1789–1854. He worked in München (Munich), Germany.

[4] Garrett BIRKHOFF, American mathematician, 1911–1996. He worked at Harvard University, Cambridge MA.

[5] In a national undergraduate mathematical contest in the late 1960s, students were asked to characterize the range of values of polynomials with real coefficients in two real variables, and some students found a case that the organizers of the competition had overlooked, that of polynomials whose range is $(0, +\infty)$ and do not attain their minimum because minimizing sequences tend to ∞, an example of such a polynomial being $(x\,y - 1)^2 + x^2$.

in infinite-dimensional situations the minimizing sequences may stay bounded and show oscillations (and converge only in a weak topology). It was for such problems without solutions that Laurence YOUNG had introduced what we now call Young measures, whose use I pioneered in the late 1970s in partial differential equations (motivated by questions of continuum mechanics and physics), although I called them parametrized measures, which is the name that I had heard in seminars of control theory, as I only learnt later that they had been introduced by Laurence YOUNG, whom I had first met in 1971.[6]

Equation (19.4), completed with adequate boundary conditions, is dealt in a mathematical way by using the *Lax–Milgram lemma* 19.1, or some variant, and (19.5) plays a crucial role.

Lemma 19.1. *(Lax–Milgram lemma) Let V be a real Hilbert space, with norm* $||\cdot||$, *and let V' be its dual, with dual norm*[7] $||\cdot||_*$; *let* $\mathcal{A} \in \mathcal{L}(V, V')$ *be V-elliptic, i.e.,*

$$\text{there exists } \alpha > 0 \text{ such that } \langle \mathcal{A} v, v \rangle \geq \alpha \, ||v||^2 \text{ for all } v \in V. \qquad (19.7)$$

Then $\mathcal{A}$ *is an isomorphism from V onto V'.*

Proof: From $|\langle \mathcal{A} v, v \rangle| \leq ||\mathcal{A}v||_* ||v||$ one deduces

$$||\mathcal{A}v||_* \geq \alpha \, ||v|| \text{ for all } v \in V, \qquad (19.8)$$

which is called an *a priori inequality*. (19.8) implies that all solutions of $\mathcal{A}u = f$ must satisfy $||u|| \leq \frac{1}{\alpha}||f||_*$, but by itself this inequality does not prove the existence of a solution, as it only says that $\mathcal{A}$ is injective and has closed range.[8] Because the transposed operator $\mathcal{A}^T$ also satisfies (19.7), one deduces that V-ellipticity also implies

$$||\mathcal{A}^T v||_* \geq \alpha \, ||v|| \text{ for all } v \in V, \qquad (19.9)$$

---

[6] It was my first visit to United States, and my command of English was very poor, so I was glad to meet people who spoke French, and Laurence YOUNG spoke it without even a Swiss accent, for he had been in school in Lausanne, Switzerland, when his father taught there.

[7] Although every Hilbert space is *isometric* to its dual, one should be careful about identifying V and V' because that identification may be incompatible with another identification which has already been done for considering functions as distributions, i.e., identifying a locally integrable function f with the Radon measure $f\,dx$, and that corresponds to having identified $L^2(\Omega)$ with its dual. DIRAC's famous notation used in quantum mechanics shows that he had understood the importance of distinguishing elements of L^2 from elements of its dual: a "bra" $\langle a|$ is an element of H' and a "ket" $|b\rangle$ is an element of H, so that the bracket $\langle a|b\rangle$ is a complex number, while $|b\rangle\langle a|$ is a linear operator, i.e., an element of $\mathcal{L}(H, H)$; one should be aware that mathematicians have a different notation $a(b)$ or the tensor product $b \otimes a$, and use the notations (b_1, b_2) for two elements of H so that the scalar product is linear in b_1 and antilinear in b_2 while for physicists $\langle a|b\rangle$ is linear in b and antilinear in a.

[8] If E is a Banach space and F is a normed space, and there exists $\beta > 0$ such that $M \in \mathcal{L}(E, F)$ satisfies $||e||_E \leq \beta \, ||M e||_F$ for all $e \in E$, then M is injective

so that $\mathcal{A}^T$ is injective, and that is equivalent to $\mathcal{A}$ having a dense range, so that the range of $\mathcal{A}$ being closed and dense must coincide with V'; of course, a bound for the norm of $\mathcal{A}^{-1}$ is $\frac{1}{\alpha}$. $\qquad\square$

In practice, one has a real Hilbert space V and $\mathcal{A} \in \mathcal{L}(V, V')$ is defined implicitly by a bilinear continuous form a on $V \times V$ by

$$a(u, v) = \langle \mathcal{A}u, v \rangle \text{ for all } u, v \in V, \tag{19.10}$$

and $f \in V'$ is given by a linear continuous form L on V, so that one does not need to characterize what V' is, and solving $\mathcal{A}u = f$ is equivalent to the *variational formulation*

$$\text{find } u \in V \text{ such that } a(u, v) = L(v) \text{ for all } v \in V. \tag{19.11}$$

For (19.4), the Hilbert space V is a closed subset of $H^1(\Omega)$ containing $H_0^1(\Omega)$, adapted to the boundary conditions that must be added to (19.4); the bilinear continuous form a is the restriction to $V \times V$ of

$$a(u, v) = \int_\Omega \left(A\, grad(u), grad(v) \right) dx \text{ for } u, v \in H^1(\Omega); \tag{19.12}$$

the linear continuous form L is the restriction to V of

$$L(v) = \int_\Omega \left(f_0 v + \sum_{j=1}^N f_j \frac{\partial v}{\partial x_j} \right) dx + \langle G, \gamma_0 v \rangle \text{ for } v \in H^1(\Omega), \tag{19.13}$$

where $f_0, \ldots, f_N \in L^2(\Omega)$ and G belongs to the dual of the space of traces on $\partial\Omega$ of functions in $H^1(\Omega)$ (a simple example is $\langle G, \gamma_0 v \rangle = \int_{\partial\Omega} g\, \gamma_0 v\, dH^{N-1}$ for a suitable function g defined on $\partial\Omega$, but if Ω is bounded with Lipschitz boundary then the space of traces is $H^{1/2}(\partial\Omega)$ and G must be taken in the dual $H^{-1/2}(\partial\Omega)$).

From (19.5), one deduces that

$$\langle Av, v \rangle = a(v, v) \geq \alpha \int_\Omega |grad(v)|^2 dx \text{ for all } v \in H^1(\Omega), \tag{19.14}$$

so that V-ellipticity holds if (and only if) Poincaré's inequality holds for V.

The case of a homogeneous Dirichlet condition corresponds to $V = H_0^1(\Omega)$, defined as the closure in $H^1(\Omega)$ of $C_c^\infty(\Omega)$ (and characterized by $\gamma_0 u = 0$ on

---

(as $M e = 0$ implies $||e|| \leq 0$) and if $f_n \to f_\infty$ with $f_n \in Range(M)$ for all n, then $f_n = M e_n$ and $||e_n - e_m||_E \leq \beta\, ||f_n - f_m||_F$ shows that e_n is a Cauchy sequence, which converges then to $e_\infty \in E$, and $f_\infty = M e_\infty \in Range(M)$, showing that the range of M is closed. Conversely, if F is also a Banach space and $M \in \mathcal{L}(E, F)$ is injective with closed range, then there must exist $\beta > 0$ such that the previous inequality holds, because $Range(M)$ is a Banach space and the *closed graph theorem* implies that M is an isomorphism from E onto $Range(M)$, and β can be chosen the norm of the inverse.

$\partial\Omega$ if Ω is bounded with Lipschitz boundary); Poincaré's inequality certainly holds for $H_0^1(\Omega)$ when Ω has finite measure or when Ω is included in a strip with finite width; if Poincaré's inequality holds, then for every $f \in H^{-1}(\Omega)$ there exists a unique solution $u \in V$ of $a(u,v) = L(v)$ for every $v \in V$, or equivalently for every $v \in C_c^\infty(\Omega)$ by density, and that is exactly (19.4).

For the case of a nonhomogeneous Dirichlet condition, i.e., $\gamma_0 u = g$ on $\partial\Omega$, one asks that g belongs to $\gamma_0 H^1(\Omega)$ (which is $H^{1/2}(\partial\Omega)$ if Ω is bounded with Lipschitz boundary), so that there exists $u_1 \in H^1(\Omega)$ with $\gamma_0 u_1 = g$; one then looks for a solution $u = u_1 + U$ with $U \in H_0^1(\Omega)$ satisfying the equation with f replaced by $f + div\big(A\,grad(u_1)\big) \in H^{-1}(\Omega)$, so that there exists a unique solution for $f \in H^{-1}(\Omega)$ and $g \in \gamma_0 H^1(\Omega)$.

If $V = H^1(\Omega)$ then Poincaré's inequality never holds.[9] If there is a solution $u \in H^1(\Omega)$ of $a(u,v) = L(v)$ for every $v \in H^1(\Omega)$ and L has the simple form $L(v) = \int_\Omega f\,v\,dx + \int_{\partial\Omega} g\,\gamma_0 v\,dH^{N-1}$ with $f \in L^2(\Omega)$ and $g \in L^2(\partial\Omega)$, it is useful to characterize what a solution can be in this case. Taking all $v \in C_c^\infty(\Omega)$ gives the equation (19.4) in Ω. Then assuming that the coefficients a_{ij} are Lipschitz continuous and that the boundary of Ω is smooth, one can show that $u \in H^2(\Omega)$ and then an integration by parts shows that one has the *Neumann condition*[10]

$$(A\,\gamma_0 grad(u).n) = g \text{ on } \partial\Omega. \tag{19.15}$$

If the boundary is not smooth enough the solution may not belong to $H^2(\Omega)$, and the coefficients may not be smooth either, and an interpretation of the boundary condition will be studied later, but there is another important point to discuss, concerning existence.

If Ω has finite measure, $1 \in H^1(\Omega)$ and, because $a(u,1) = 0$ for every $u \in V = H^1(\Omega)$, a necessary condition for the existence of a solution is that $L(1) = 0$. With the physical interpretation of a stationary heat equation it means that the total amount of heat is 0, adding the source of heat inside

---

[9] When $1 \in V$, i.e., if $meas(\Omega) < \infty$, Poincaré's inequality cannot hold; if $\Omega = R^N$ or simply if Ω contains balls of arbitrarily large radius, a scaling argument shows that Poincaré's inequality does not hold. For an arbitrary (nonempty) connected open set Ω, one may assume that $meas(\Omega) = \infty$ and that Ω contains the origin; let $\gamma_0 = meas(\{x \in \Omega \mid |x| < 1\}) > 0$, and for $n \geq 1$ let $\gamma_n = meas(\{x \in \Omega \mid 2^{n-1} < |x| < 2^n\}) > 0$; one assumes that there exists $\kappa > 0$ such that $||v||_2 \leq \kappa\,||grad(v)||_2$ for all $v \in H^1(\Omega)$, and one chooses $v = h(r)$ with $h(r) = 1$ for $r < 2^n$, $h(r) = 0$ for $r > 2^{n+1}$ and h affine on $[2^n, 2^{n+1}]$, i.e., $h(r) = 2 - 2^{-n}r$ for $2^n \leq r \leq 2^{n+1}$; one has $||grad(v)||_2^2 = \gamma_{n+1} 2^{-2n}$ and $||v||_2^2 \geq \gamma_0 + \ldots + \gamma_n$, so that one obtains in particular $\gamma_n \leq 2^{-2n}\kappa^2 \gamma_{n+1}$ for all $n \geq 1$; if there exists λ such that $\gamma_n \leq C\,2^{\lambda n}$ for all $n \geq 1$ one deduces that $\gamma_n \leq C\,2^\lambda \kappa^2 2^{(\lambda-2)n}$ for all $n \geq 1$, and by repeating the argument m times one has $\gamma_n \leq C_m 2^{(\lambda-2m)n}$ for all $n \geq 1$, which implies $meas(\Omega) = \sum_{n\geq 0} \gamma_n < \infty$, a contradiction; that one λ exists follows from the upper bound of the measure of the ball, $\gamma_n \leq C\,2^{N\,n}$.

[10] Franz Ernst NEUMANN, German mathematician, 1798–1895. He worked in Königsberg, then in Germany, now Kaliningrad, Russia.

Ω which is $\int_\Omega f\, dx$ and the heat flux imposed on the boundary $\partial\Omega$, which is $\int_{\partial\Omega} g\, dH^{N-1}$; if this condition is not satisfied then the solution of the evolution heat equation will not converge to a limit, and it actually tends to infinity (with a sign depending upon the sign of the total heat imposed; of course, when the temperature becomes too large, the modeling by a linear equation is not very good, and in a real problem the absolute temperature cannot become negative anyway).

If the necessary condition $L(1) = 0$ is satisfied and if the injection of $H^1(\Omega)$ into $L^2(\Omega)$ is compact then a solution exists, but it is not unique as one may add an arbitrary constant to the solution (an example of a *Fredholm alternative*). If one denotes by u_Ω the average of u on Ω, and by $u_{\partial\Omega}$ the average of $\gamma_0 u$ on $\partial\Omega$, then the compactness assumption implies that Poincaré's inequality $||u||_2 \leq C||grad(u)||_2$ holds for all $u \in H^1(\Omega)$ satisfying $u_\Omega = 0$, and that it also holds for all $u \in H^1(\Omega)$ satisfying $u_{\partial\Omega} = 0$. Even if the compactness condition does not hold but one of these Poincaré's inequalities is true, then there exists a solution.

Using Poincaré's inequality for all $u \in H^1(\Omega)$ satisfying $u_\Omega = 0$, one changes V to

$$V = \{u \in H^1(\Omega) \mid u_\Omega = 0\}, \tag{19.16}$$

and the bilinear form is then V-elliptic and a solution exists. One only has $a(u,v) = L(v)$ for $v \in C_c^\infty(\Omega)$ satisfying $\int_\Omega v\, dx = 0$ (and $C_c^\infty(\Omega) \not\subset V$), and there exists a *Lagrange multiplier*[11] λ such that

$$\begin{aligned} a(u,v) &= L(v) + \lambda \int_\Omega v\, dx \text{ for all } v \in C_c^\infty(\Omega), \text{ i.e.,} \\ -div\big(A\, grad(u)\big) &= f + \lambda \text{ in } \Omega, \end{aligned} \tag{19.17}$$

and then one obtains the boundary condition and λ is such that the necessary condition must hold; if the necessary condition holds from the start, then $\lambda = 0$ and one has found a solution of the problem.

Using Poincaré's inequality for all $u \in H^1(\Omega)$ satisfying $u_{\partial\Omega} = 0$, one changes V to

$$V = \{u \in H^1(\Omega) \mid u_{\partial\Omega} = 0\}, \tag{19.18}$$

and the bilinear form is then V-elliptic and a solution exists. One now has $C_c^\infty(\Omega) \subset V$, so that (19.4) holds; then one obtains the boundary condition with a Lagrange multiplier μ appearing in the boundary condition, i.e.,

$$a(u,v) = L(v) + \mu \int_{\partial\Omega} \gamma_0 v\, dH^{N-1} \text{ for all } v \in H^1(\Omega), \tag{19.19}$$

and μ is such that the necessary condition must hold; if the necessary condition holds from the start, then $\mu = 0$ and one has found a solution of the problem.

[Taught on Monday February 28, 2000.]

---

[11] Giuseppe Lodovico LAGRANGIA (Joseph Louis LAGRANGE), Italian-born mathematician, 1736–1813. He worked in Torino (Turin) Italy, in Berlin, Germany, and in Paris, France. He was made a count in 1808 by NAPOLÉON.

The Space $H(div; \Omega)$

Let Ω be a bounded open set of R^N with Lipschitz boundary, and for $f \in L^2(\Omega)$ and $g \in L^2(\partial\Omega)$, let $u \in H^1(\Omega)$ satisfy

$a(u, v) = L(v)$ for all $v \in H^1(\Omega)$, where

$$a(\varphi, \psi) = \int_\Omega \left(\sum_{i,j} A_{ij} \frac{\partial\varphi}{\partial x_j} \frac{\partial\psi}{\partial x_i} \right) dx \text{ for all } \varphi, \psi \in H^1(\Omega) \qquad (20.1)$$
$$L(\psi) = \int_\Omega f\,\psi\,dx + \int_{\partial\Omega} g\,\gamma_0\psi\,dH^{N-1} \text{ for all } \psi \in H^1(\Omega).$$

Using all $v \in C_c^\infty(\Omega)$ one deduces that

$$-\sum_{i,j} \frac{\partial}{\partial x_i} \left(A_{ij} \frac{\partial u}{\partial x_j} \right) = f \text{ in } \Omega, \qquad (20.2)$$

and the question is to understand what boundary condition u satisfies.

If one assumes that

$$A_{ij} \in W^{1,\infty}(\Omega) \text{ for } i, j = 1, \dots, N, \text{ and } u \in H^2(\Omega), \qquad (20.3)$$

then $A_{ij} \frac{\partial u}{\partial x_j} \in H^1(\Omega)$ for all $i, j = 1, \dots, N$, and an integration by parts gives

$$\int_\Omega f\,v\,dx = a(u, v) - \int_{\partial\Omega} \left(\sum_{i,j} \gamma_0 \left(A_{ij} \frac{\partial u}{\partial x_j} \right) \gamma_0 v\,\nu_i \right) dH^{N-1} \text{ for all } v \in H^1(\Omega),$$
$$(20.4)$$

so that one has

$$\sum_{i,j} \gamma_0 \left(A_{ij} \frac{\partial u}{\partial x_j} \right) \gamma_0 v\,\nu_i = g \text{ on } \partial\Omega, \qquad (20.5)$$

which can only happen if $g \in H^{1/2}(\partial\Omega)$, of course.

In applications, one does not always have Lipschitz coefficients A_{ij}, because one often mixes different materials and there are interfaces of discontinuity for the coefficients. In applications, one does not always have smooth boundaries, and corners in the boundary put a limit on the regularity of the solution. For a convex domain Ω, $u \in H_0^1(\Omega)$ and $\Delta u \in L^2(\Omega)$ imply $u \in H^2(\Omega)$; the following example shows that it is not true for plane domains with corners with angles $> \pi$.

Let Ω be the sector $0 < \theta < \theta_0$ with $\pi < \theta_0 < 2\pi$, and let $u = r^\alpha \cos(\alpha\,\theta)\varphi$ with $\varphi \in C_c^\infty(R^2)$ with $\varphi = 1$ near the origin. Because $u_0 = r^\alpha \cos(\alpha\,\theta)$ is harmonic,[1] i.e., satisfies $\Delta u_0 = 0$, one sees that Δu is 0 near the origin; the normal derivative of u on the side $\theta = 0$ is 0, and it is also 0 on the side $\theta = \theta_0$ if $\alpha\,\theta_0 = \pi$, which gives $\frac{1}{2} < \alpha < 1$, so that one does not have $u \in H^2(\Omega)$, which requires $\alpha > 1$.

Another way to treat the problem of giving a meaning to the Neumann condition, without having $u \in H^2(\Omega)$, is the following argument of Jacques-Louis LIONS.[2]

Definition 20.1. $H(div;\Omega) = \{u \in L^2(\Omega;R^N) \mid div(u) \in L^2(\Omega)\}$. $\square$

Of course, $H(div;\Omega)$ is a Hilbert space.

One localizes by multiplying all the components of u by the same function θ, noticing that if $v_j = \theta\,u_j$ for $j = 1,\dots,N$, then one has $div(v) = \theta\,div(u) + (u.grad(\theta))$.

If P is an invertible matrix and $\Omega' = P\,\Omega$, one transports a scalar function φ defined on Ω to the scalar function ψ defined on Ω' by

$$\psi(P\,x) = \varphi(x) \text{ for } x \in \Omega, \tag{20.6}$$

and one wants to transport $u \in H(div;\Omega)$ to $v \in H(div;\Omega')$ in such a way that one has

$$\int_\Omega (u.grad(\varphi))\,dx = \int_{\Omega'} (v.grad(\psi))\,dx', \tag{20.7}$$

but $(grad(\psi)(P\,x).P\,y)=(grad(\varphi)(x).y)$ gives $grad(\psi)(P\,x)=P^{-T}grad(\varphi)(x)$ (writing $P^{-T} = (P^T)^{-1}$), and one asks that

$$\begin{aligned}(u(x).grad(\varphi)(x)) = (v(P\,x).grad(\psi)(P\,x))|det(P)|, \text{ i.e.,}\\ v(P\,x) = |det(P)|^{-1}P\,u(x).\end{aligned} \tag{20.8}$$

If P is an orthogonal[3] matrix then $v(P\,x) = P\,u(x)$.

Once one works on Ω_F for a Lipschitz continuous function F, one proves easily that $\left(\mathcal{D}(\overline{\Omega_F})\right)^N$ is dense in $H(div;\Omega_F)$.

---

[1] It is the real part of z^α, and every holomorphic function has its real part and its imaginary part harmonic.

[2] I believe that he proved it while I was a student, because in the first courses that I followed he used the argument with the $H^2(\Omega)$ hypothesis, and later he started teaching the new argument with the space $H(div;\Omega)$.

[3] This is the case used most of the time, but it is misleadingly simple. The general linear case considered is of course just a simplified version for the case of diffeomorphisms, and is related to the fact that elements of $H(div;\Omega)$ should be considered as $(N-1)$-differential forms.

All this analysis serves for proving that $\left(\mathcal{D}(\overline{\Omega})\right)^N$ is dense in $H(div; \Omega)$. The next step is to prove that one can define the normal trace $(u.\nu)$; for smooth functions it means $\sum_j \gamma_0 u_j \nu_j$, but for $H(div; \Omega)$ the definition uses a completion argument.

Lemma 20.2. *The mapping $u \mapsto (u.\nu) = \sum_j \gamma_0 u_j \nu_j$, defined from $\left(\mathcal{D}(\overline{\Omega})\right)^N$ into $L^\infty(\partial\Omega)$, extends into a linear continuous map from $H(div; \Omega)$ into $\left(\gamma_0 H^1(\Omega)\right)'$, the dual of the space of traces of functions of $H^1(\Omega)$, i.e., $H^{-1/2}(\partial\Omega)$ (as $\partial\Omega$ has no boundary, $H_0^{1/2}(\partial\Omega) = H^{1/2}(\partial\Omega)$). Moreover, the mapping is surjective.*

Proof: For $u \in \left(\mathcal{D}(\overline{\Omega})\right)^N$ and $v \in H^1(\Omega)$ one has

$$\int_\Omega \left(\sum_j u_j \frac{\partial v}{\partial x_j} + div(u)\, v\right) dx = \int_{\partial\Omega} \left(\sum_j \gamma_0 u_j\, \nu_j\right) \gamma_0 v\, dH^{N-1}, \quad (20.9)$$

and as the left side of the identity is continuous on $H(div; \Omega) \times H^1(\Omega)$, so is the right side, on which one writes $\langle (u.\nu), \gamma_0 v \rangle$ as a linear continuous form on $\gamma_0 H^1(\Omega)$; notice that if one starts from an element of $\gamma_0 H^1(\Omega)$ it does not matter which v one chooses which has this element as its trace, as the left side will give the same value whatever the choice is.

In order to show surjectivity, one takes $g \in \left(\gamma_0 H^1(\Omega)\right)'$ and one solves $\int_\Omega (grad(u_*).grad(v))\,dx + \int_\Omega u_* v\, dx = \langle g, \gamma_0 v\rangle$ for all $v \in H^1(\Omega)$, which has a unique solution $u_* \in H^1(\Omega)$, which satisfies $-\Delta u_* + u_* = 0$ in Ω and therefore $\xi_* = grad(u_*)$ belongs to $H(div; \Omega)$ and satisfies $div(\xi_*) = u_*$, and the precise variational formulation says that $(\xi_*.\nu) = g$.   □

The example of R^2 with $u_1 = f_1(x_1) f_2(x_2)$ and $u_2 = g_1(x_1) g_2(x_2)$ shows that one has $u \in H(div; R^2)$ if $f_1, g_2 \in H^1(R)$ and $f_2, g_1 \in L^2(R)$, so that u_1 can be discontinuous along the line $x_2 = 0$ while u_2 must be continuous, and $(u.\nu) = -u_2$ if $\Omega = R_+^2$.

In a problem of electrostatics, the potential u is in $H^1(\Omega)$ and has a trace on the boundary; more generally, on any interface u takes the same value on both sides of the interface. The polarization field D satisfies $div(D) = \varrho$, so that $D \in H(div; \Omega)$ if $\varrho \in L^2(\Omega)$, and the normal component of D is continuous at any interface (if it does not support a nonzero charge). For the electric field E, it is the tangential component of E which is continuous, and its value is the tangential derivative of the trace of u; one can actually define the space $H(curl; \Omega)$ by

$$H(curl; \Omega) = \left\{ E \in \left(L^2(\Omega)\right)^N \mid \frac{\partial E_i}{\partial x_j} - \frac{\partial E_j}{\partial x_i} \in L^2(\Omega) \text{ for all } i, j = 1, \ldots, N \right\},$$

$$(20.10)$$

and prove an analogous theorem, that the tangential trace is defined.

[Taught on Wednesday March 1, 2000.]

Background on Interpolation; the Complex Method

Although the term interpolation space only appeared much later, the subject has its origin in questions studied by M. RIESZ, and then by THORIN[1], and also by MARCINKIEWICZ; they might have been motivated by studying the properties of the Hilbert transform.

A *holomorphic function* in an open set of the complex plane is a complex-valued function which has a derivative in the complex sense, i.e.,

$$\frac{f(z) - f(z_0)}{z - z_0} \to f'(z_0) \text{ as } z \to z_0, \tag{21.1}$$

and the *Cauchy–Riemann equation* is satisfied, i.e.,

$$f(x + iy) = P(x, y) + iQ(x, y) \text{ implies}$$
$$\frac{\partial P}{\partial x} = \frac{\partial Q}{\partial y}$$
$$\frac{\partial P}{\partial y} = -\frac{\partial Q}{\partial x}, \tag{21.2}$$

so that both P and Q are *harmonic*, i.e., satisfy

$$\Delta P = \Delta Q = 0, \tag{21.3}$$

where the Laplacian is[2]

$$\Delta = \frac{\partial^2}{\partial x^2} + \frac{\partial^2}{\partial y^2}. \tag{21.4}$$

If one works in the upper half plane $y > 0$, and one imposes the real part of f on the boundary, then P is determined (if the given trace is nice enough) and then the partial derivatives of Q are known, so that Q is defined up to

[1] G. Olof THORIN, Swedish mathematician. He was a student of Marcel RIESZ in Lund, Sweden.
[2] Geometers have a different notation, and their Laplacian is what analysts write as $-\Delta$.

addition of an arbitrary real constant. In this way one is led to study the following transform named after HILBERT,

$$H\,u = \frac{1}{\pi}pv\frac{1}{x} \star u, \text{ i.e., } H\,u(x) = \lim_{\varepsilon \to 0}\frac{1}{\pi}\int_{|y-x|>\varepsilon}\frac{u(y)}{x-y}\,dy, \qquad (21.5)$$

which relates the real part to the imaginary part on the boundary. Using the Fourier transform one can show that

$$H \text{ is a surjective isometry of } L^2(R) \text{ onto itself, and } H^2 = -I, \qquad (21.6)$$

and more precisely, one has the following result.

Lemma 21.1. *For $u \in L^2(R)$, one has*

$$\mathcal{F}(H\,u)(\xi) = -i\,sign(\xi)\mathcal{F}u(\xi) \text{ a.e. } \xi \in R. \qquad (21.7)$$

Proof: One may use Laurent SCHWARTZ's extension of the Fourier transform to tempered distributions, because $pv\frac{1}{x} \in \mathcal{S}'(R)$, as it is the sum of a distribution with compact support and a bounded function.

$$x\,pv\frac{1}{x} = 1 \text{ implies } \frac{d}{d\xi}\left(\mathcal{F}\left(pv\frac{1}{x}\right)\right) = -2i\pi\,\delta_0$$
$$\text{i.e., } \mathcal{F}\left(pv\frac{1}{x}\right)(\xi) = -i\pi\,sign(\xi) + C, \text{ a.e. } \xi \in R, \qquad (21.8)$$

and one deduces $C = 0$ from the fact that $pv\frac{1}{x}$ is real and *odd* so that its Fourier transform must be odd (of course Laurent SCHWARTZ had defined in a natural way what it means to be even or odd for a distribution). Then one has $\mathcal{F}(H\,u)(\xi) = -i\,sign(\xi)\mathcal{F}u(\xi)$ a.e. $\xi \in R$, so that $||H\,u||_2 = ||u||_2$ and $H^2 = -I$. □

M. RIESZ proved that the Hilbert transform is continuous from $L^p(R)$ into itself for $1 < p < \infty$, but the result is not true for $p = 1$ or for $p = \infty$.[3] I suppose that it was in relation to the properties of the Hilbert transform that M. RIESZ proved the following "interpolation" result in 1926, in the case $p_\theta \leq q_\theta$; this restriction was removed by THORIN, in 1938; it is often called the *convexity theorem*.

Theorem 21.2. *If $1 \leq p_0, p_1, q_0, q_1 \leq \infty$, and a linear map A is continuous from $L^{p_0}(\Omega)$ into $L^{q_0}(\Omega')$ and from $L^{p_1}(\Omega)$ into $L^{q_1}(\Omega')$ then for $0 < \theta < 1$ it is continuous from $L^{p_\theta}(\Omega)$ into $L^{q_\theta}(\Omega')$, where*

$$\frac{1}{p_\theta} = \frac{1-\theta}{p_0} + \frac{\theta}{p_1}; \; \frac{1}{q_\theta} = \frac{1-\theta}{q_0} + \frac{\theta}{q_1}, \qquad (21.9)$$

---

[3] The Hilbert transform maps the Hardy space $\mathcal{H}^1(R)$ into $L^1(R)$, but that is a tautology, as it is defined as $\{u \in L^1(R) \mid H\,u \in L^1(R)\}$. It is a deeper result that the Hilbert transform actually maps $\mathcal{H}^1(R)$ into itself, a deep result that $BMO(R)$ is the dual of $\mathcal{H}^1(R)$, so that the Hilbert transform maps $BMO(R)$ into itself and therefore maps $L^\infty(R)$ into $BMO(R)$, and a deep result that $BMO(R) = \{u = f + H\,g \mid, f, g \in L^\infty(R)\}$.

and one has

$$||A||_{\mathcal{L}(L^{r_\theta}(\Omega);L^{q_\theta}(\Omega'))} \leq ||A||^{1-\theta}_{\mathcal{L}(L^{r_0}(\Omega);L^{q_0}(\Omega'))}||A||^{\theta}_{\mathcal{L}(L^{r_1}(\Omega);L^{q_1}(\Omega'))}. \quad \Box \quad (21.10)$$

If the Hilbert transform was mapping $L^1(R)$ into itself, then by this interpolation result it would map $L^p(R)$ into itself for $1 < p < 2$, and by transposition for $2 < p < \infty$, but it does not map $L^1(R)$ into $L^1(R)$. However, there exists a constant C such that if $u \in L^1(R)$ one has

$$meas\{x \mid |H\,u(x)| > t\} \leq \frac{C\,||u||_1}{t} \text{ for all } t > 0, \qquad (21.11)$$

and from that result, the continuity in $L^2(R)$ and the symmetry of the Hilbert transform one can deduce that it maps $L^p(R)$ into itself for $1 < p < \infty$.

THORIN's proof used a property of the modulus of holomorphic functions, the *three lines theorem* (a variant of Hadamard's *three circles theorem*), stating that if $f(z)$ is holomorphic in the strip $0 < \Re z = x < 1$, continuous on the closed strip $0 \leq x \leq 1$ and such that $|f(i\,y)| \leq M_0$ and $|f(1 + i\,y)| \leq M_1$ for all $y \in R$, then one has $|f(\theta + i\,y)| \leq M_0^{1-\theta}M_1^{\theta}$ for all $\theta \in (0,1)$ and all $y \in R$.

After the idea of THORIN was used again by STEIN, a general *complex interpolation method* was developed, by Alberto CALDERÓN, by Jacques-Louis LIONS, and by M. KREIN.

If $f \in L^p(\Omega)$, then Hölder's inequality gives $\int_E |f|\,dx \leq ||f||_p meas(E)^{1/p'}$ for all measurable subsets E of Ω, and MARCINKIEWICZ introduced a space sometimes called weak L^p (which one should not confuse with L^p equipped with the weak topology), and denoted by $L^{p,\infty}$ in the scale of Lorentz spaces, which is the space of (equivalence classes of) measurable functions g for which there exists C such that

$$\int_E |g|\,dx \leq C\,meas(E)^{1/p'} \text{ for all measurable subsets } E \subset \Omega. \qquad (21.12)$$

It contains $L^p(\Omega)$ but if $\Omega \subset R^N$ and $1 \leq p < \infty$ it also contains functions like $\frac{1}{|x|^{N/p}}$. In 1939, MARCINKIEWICZ published the following result, as a note without proof, and proofs were written later by Mischa COTLAR[4] and by Antoni ZYGMUND.

Lemma 21.3. *If* $1 \leq p_0, p_1, q_0, q_1 \leq \infty$, *and a linear map* A *is continuous from* $L^{p_0}(\Omega)$ *into* $L^{q_0,\infty}(\Omega')$ *and from* $L^{p_1}(\Omega)$ *into* $L^{q_1,\infty}(\Omega')$ *then for* $0 < \theta < 1$ *it is continuous from* $L^{p_\theta}(\Omega)$ *into* $L^{q_\theta}(\Omega')$ *under the condition that* $p_\theta \leq q_\theta$, *where* p_θ *and* q_θ *are given by (21.9).* $\Box$

The results of M. RIESZ, THORIN, and MARCINKIEWICZ, were generalized as theories of *interpolation*, and the main contributors were Nachman

---

[4] Mischa COTLAR, Ukrainian-born mathematician, born in 1913. He worked in Buenos Aires and in La Plata, Argentina, at Rutgers University, Piscataway, NJ, and in Caracas, Venezuela.

ARONSZAJN, Alberto CALDERÓN, Emilio GAGLIARDO, KREIN, Jacques-Louis LIONS and Jaak PEETRE, but similar techniques have been used by specialists of harmonic analysis, like STEIN. I suppose that a part of the motivation of Jacques-Louis LIONS was the question of identifying traces of Sobolev spaces and their variants, following the characterization of traces of $W^{1,p}(R^N)$ by Emilio GAGLIARDO.

Definition 21.4. *Let E_0 and E_1 be normed spaces, continuously embedded into a topological vector space $\mathcal{E}$ so that $E_0 \cap E_1$ and $E_0 + E_1$ are defined.*

An intermediate space *between E_0 and E_1 is any normed space E such that $E_0 \cap E_1 \subset E \subset E_0 + E_1$ (with continuous embeddings).*

An interpolation space *between E_0 and E_1 is any intermediate space E such that every linear mapping from $E_0 + E_1$ into itself which is continuous from E_0 into itself and from E_1 into itself is automatically continuous from E into itself. It is said to be of* exponent θ *(with $0 < \theta < 1$), if there exists a constant C such that one has*

$$||A||_{\mathcal{L}(E;E)} \leq C\,||A||_{\mathcal{L}(E_0;E_0)}^{1-\theta}||A||_{\mathcal{L}(E_1;E_1)}^{\theta} \text{ for all } A \in \mathcal{L}(E_0;E_0) \cap \mathcal{L}(E_1;E_1). \quad \square$$
(21.13)

One is interested in general methods (or *functors*) which construct interpolation spaces from two arbitrary normed spaces (or Banach spaces, or Hilbert spaces).

Definition 21.5. *For two Banach spaces E_0, E_1, the* complex method *consists in looking at the space of real analytic functions f with values in E_0+E_1, defined on the open strip $0 < x < 1$, continuous on the closed strip $0 \leq x \leq 1$, and such that*

$f(i\,y)$ *is bounded in E_0 and $f(1 + i\,y)$ is bounded in E_1*

equipped with the norm $||f|| = \max\{\sup_y ||f(i\,y)||_0, \sup_y ||f(1 + i\,y)||_1\}$,
(21.14)

and for $0 < \theta < 1$, one defines

$$[E_0, E_1]_\theta = \{a \in E_0+E_1 \mid a = f(\theta)\}, \text{ with the norm } ||a||_{[E_0,E_1]_\theta} = \inf_{f(\theta)=a} ||f||. \quad \square$$
(21.15)

Of course such a space contains $E_0 \cap E_1$, as one can take f to be a constant function taking its value in $E_0 \cap E_1$.

Lemma 21.6. *The interpolation property holds for the spaces of Definition 21.5.*

Proof: If $A \in \mathcal{L}(E_0; F_0) \cap \mathcal{L}(E_1; F_1)$, then $g(x + i\,y) = A\,f(x + i\,y)$ satisfies a property similar to f with the spaces F_0 and F_1, so that one has

$$||A\,a||_{[F_0,F_1]_\theta} \leq \max\{||A||_{\mathcal{L}(E_0;F_0)}, ||A||_{\mathcal{L}(E_1;F_1)}\}||a||_{[E_0,E_1]_\theta}. \quad (21.16)$$

One may replace $\max\{||A||_{\mathcal{L}(E_0;F_0)}, ||A||_{\mathcal{L}(E_1;F_1)}\}$ by $||A||^{1-\theta}_{\mathcal{L}(E_0;F_0)}||A||^{\theta}_{\mathcal{L}(E_1;F_1)}$, by considering instead $g(x + i\,y) = e^{-s\,\theta+s(x+i\,y)}A\,f(x + i\,y)$, which makes the quantity $\max\{e^{-s\,\theta}||A||_{\mathcal{L}(E_0;F_0)}, e^{s\,(1-\theta)}||A||_{\mathcal{L}(E_1;F_1)}\}$ appear, and then one minimizes in s by taking $e^s = \frac{||A||_{\mathcal{L}(E_0;F_0)}}{||A||_{\mathcal{L}(E_1;F_1)}}$. $\qquad\qquad\qquad\square$

At least for the case of Jacques-Louis LIONS, one motivation for introducing interpolation spaces was the question of traces for variants of Sobolev spaces. For example, if $\Omega = R^N_+ = \{x \in R^N \mid x_N > 0\}$, and one wants to describe the trace on the boundary of a function $u \in W^{1,p}(\Omega)$, one notices that

$u \in W^{1,p}(R^N_+)$ is equivalent to

$$u \in L^p\big(R_+; W^{1,p}(R^{N-1})\big) \text{ and } \frac{du}{dx_N} \in L^p\big(R_+; L^p(R^{N-1})\big), \qquad (21.17)$$

and he introduced a more general framework, which Jaak PEETRE also did independently so that it gave a joint article, where they considered (strongly measurable) functions defined on $(0, \infty)$ with values in $E_0 + E_1$ and such that

$$t^{\alpha_0} u \in L^{p_0}(R_+; E_0) \text{ and } t^{\alpha_1}\frac{du}{dt} \in L^{p_1}(R_+; E_1), \qquad (21.18)$$

and looked for the space spanned by $u(0)$ for a special range of parameters where $u(0)$ is automatically defined. It seems that this is a four-parameter family of spaces, but changing t into t^{β} shows that three parameters are enough; it was Jaak PEETRE who finally proved that the family actually depends only upon two parameters, and after simplification it led to the K-method and the J-method that we are going to study.

[Taught on Wednesday March 8, 2000.]

Real Interpolation; K-Method

Definition 22.1. *Let E_0 and E_1 be two normed spaces, continuously embedded into a topological vector space $\mathcal{E}$ so that*

$E_0 \cap E_1$ is equipped with the norm $\|a\|_{E_0 \cap E_1} = \max\{\|a\|_0, \|a\|_1\}$
$E_0 + E_1$ is equipped with the norm $\|a\|_{E_0 + E_1} = \inf_{a=a_0+a_1}\big(\|a_0\|_0 + \|a_1\|_1\big).$

$$(22.1)$$

Following Jaak PEETRE, for $a \in E_0 + E_1$ and $t > 0$ one defines

$$K(t; a) = \inf_{a=a_0+a_1}\big(\|a_0\|_0 + t\,\|a_1\|_1\big), \qquad (22.2)$$

and for $0 < \theta < 1$ and $1 \le p \le \infty$ (or for $\theta = 0, 1$ with $p = \infty$), one writes

$$(E_0, E_1)_{\theta,p} = \big\{a \in E_0 + E_1 \mid t^{-\theta}K(t; a) \in L^p\big(R_+; \tfrac{dt}{t}\big)\big\}$$
$$\text{with the norm } \|a\|_{(E_0,E_1)_{\theta,p}} = \|t^{-\theta}K(t; a)\|_{L^p(0,\infty;dt/t)}. \quad \square \qquad (22.3)$$

An idea of Emilio GAGLIARDO is to consider a plane with coordinates x_0, x_1 and to associate to each $a \in E_0 + E_1$ a set

$$G(a) = \{(x_0, x_1) \mid \text{ there exists a decomposition } a = a_0 + a_1 \\ \text{with } \|a_0\|_0 \le x_0, \|a_1\|_1 \le x_1\}. \qquad (22.4)$$

Each *Gagliardo set* $G(a)$ is convex because if $a = b_0 + b_1$ with $\|b_0\|_0 \le y_0$ and $\|b_1\|_1 \le y_1$, then for $0 < \eta < 1$ one has $a = c_0 + c_1$ with $c_0 = (1-\eta)a_0 + \eta\, b_0$ and $c_1 = (1-\eta)a_1 + \eta\, b_1$ and the triangle inequality gives $\|c_0\|_0 \le (1-\eta)x_0 + \eta\, y_0$ and $\|c_1\|_1 \le (1 - \eta)x_1 + \eta\, y_1$. Using the function $t \mapsto K(t; a)$ is one way of describing the boundary of this convex set.

For $t > 0$, $a \mapsto K(t; a)$ is a norm equivalent to the norm on $E_0 + E_1$ given in (22.1). $K(t; a)$ is nondecreasing in t and $\frac{K(t;a)}{t}$ is nonincreasing in t, and moreover $K(t; a)$ is concave in t, as an infimum of affine functions, so that it is continuous. One can give a definition of the space involving a sum instead of an integral: on an interval $e^n \le t \le e^{n+1}$ one has $K(e^n; a) \le K(t; a) \le e\, K(e^n; a)$

for $n \in Z$, and as the measure of (e^n, e^{n+1}) for the measure $\frac{dt}{t}$ is 1, one sees that

$$a \in (E_0, E_1)_{\theta,p} \text{ is equivalent to } e^{-n\theta}K(e^n; a) \in l^p(Z), \text{ and}$$
$$||e^{-n\theta}K(e^n; a)||_{l^p(Z)} \text{ is an equivalent norm on } (E_0, E_1)_{\theta,p}. \tag{22.5}$$

Lemma 22.2. *If* $0 < \theta < 1$ *and* $1 \leq p \leq q \leq \infty$, *one has* $(E_0, E_1)_{\theta,p} \subset (E_0, E_1)_{\theta,q}$ *(with continuous embedding).*

Proof: Using the equivalent definition (22.5), one notices that l^p is increasing with p.

Another way to prove the same result is to notice that if $1 \leq p < \infty$, and $t_0 > 0$ one has $K(t; a) \geq K(t_0; a)$ for $t > t_0$, so that

$$||a||^p_{(E_0,E_1)_{\theta,p}} \geq K(t_0; a)^p \int_{t_0}^{\infty} t^{-\theta p} \frac{dt}{t} = K(t_0; a)^p \frac{t_0^{-\theta p}}{\theta p}$$
$$\text{implying } t_0^{-\theta} K(t_0; a) \leq C \, ||a||_{(E_0,E_1)_{\theta,p}}, \text{ i.e.,} \tag{22.6}$$
$$||t^{-\theta} K(t; a)||_{L^{\infty}(0,\infty,dt/t)} \leq C \, ||a||_{(E_0,E_1)_{\theta,p}},$$

and by Hölder's inequality one obtains

$$||a||_{(E_0,E_1)_{\theta,q}} = ||t^{-\theta} K(t; a)||_{L^q(0,\infty,dt/t)} \leq C' \, ||a||_{(E_0,E_1)_{\theta,p}} \text{ for } p \leq q \leq \infty. \quad \square \tag{22.7}$$

Because for $a \in E_0 + E_1$ one has $K(t; a) \geq \min\{1, t\}||a||_{E_0+E_1}$, one sees that if $(E_0, E_1)_{\theta,p}$ is not reduced to 0 one must have $t^{-\theta} \min\{1, t\} \in L^p(R_+; \frac{dt}{t})$, so that the space $(E_0, E_1)_{\theta,p}$ is reduced to 0 if $\theta < 0$ or if $\theta > 1$, and also in the cases $\theta = 0$ or $\theta = 1$, if $p < \infty$.

Because for $a \in E_0 \cap E_1$ one has the decompositions $a = a + 0$ and $a = 0 + a$, one finds that $K(t; a) \leq \min\{1, t\}||a||_{E_0 \cap E_1}$, so that for all the pairs (θ, p) which are considered one has $E_0 \cap E_1 \subset (E_0, E_1)_{\theta,p}$ (with continuous embedding).

It will be important to characterize as much as possible what these interpolation spaces are in each context, but the interpolation property comes automatically.

Lemma 22.3. *If* A *is linear from* $E_0 + E_1$ *into* $F_0 + F_1$ *and maps* E_0 *into* F_0 *with* $||A\,x||_{F_0} \leq M_0||x||_{E_0}$ *for all* $x \in E_0$, *and maps* E_1 *into* F_1 *with* $||A\,x||_{F_1} \leq M_1||x||_{E_1}$ *for all* $x \in E_1$, *then* A *is linear continuous from* $(E_0, E_1)_{\theta,p}$ *into* $(F_0, F_1)_{\theta,p}$ *for all* θ, p, *and for* $0 < \theta < 1$ *one has*

$$||A\,a||_{(F_0,F_1)_{\theta,p}} \leq M_0^{1-\theta} M_1^{\theta} ||a||_{(E_0,E_1)_{\theta,p}} \text{ for all } a \in (E_0, E_1)_{\theta,p}. \tag{22.8}$$

Proof: For each decomposition $a = a_0 + a_1$ with $a_0 \in E_0$ and $a_1 \in E_1$, one has $A\,a = A\,a_0 + A\,a_1$, and

$$K(t; A\,a) \leq ||A\,a_0||_{F_0} + t\,||A\,a_1||_{F_1} \leq M_0||a_0||_{E_0} + t\,M_1||a_1||_{E_1} =$$
$$M_0\left(||a_0||_{E_0} + \frac{t\,M_1}{M_0}||a_1||_{E_1}\right). \tag{22.9}$$

Taking the infimum on all decompositions of a, (22.9) implies

$$K(t; A a) \leq M_0 K\left(\frac{t M_1}{M_0}; a\right). \tag{22.10}$$

Using $s = \frac{t M_1}{M_0}$ one deduces that $t^{-\theta} K(t; A a) \leq M_0^{1-\theta} M_1^{\theta} s^{-\theta} K(s; a)$ and as $\frac{dt}{t} = \frac{ds}{s}$ one finds that

$$||A a||_{(F_0, F_1)_{\theta, p}} \leq M_0^{1-\theta} M_1^{\theta} ||s^{-\theta} K(s; a)||_{L^p(0, \infty, dt/t)} = M_0^{1-\theta} M_1^{\theta} ||a||_{(E_0, E_1)_{\theta, p}}$$
for all $a \in (E_0, E_1)_{\theta, p}$,

$$\tag{22.11}$$

which is (22.8).   □

   An important example is the case $E_0 = L^1(\Omega)$, $E_1 = L^\infty(\Omega)$, for which the corresponding interpolation spaces are the *Lorentz spaces*:[1] for $1 < p < \infty$ and $1 \leq q \leq \infty$ one writes

$$L^{p,q}(\Omega) = \left(L^1(\Omega), L^\infty(\Omega)\right)_{1/p', q}, \tag{22.12}$$

and one will find that $L^{p,p}(\Omega) = L^p(\Omega)$ (with equivalent norms). For a function $f \in L^1(\Omega) + L^\infty(\Omega)$ one can calculate explicitly $K(t; f)$, and the formula makes use of the *nonincreasing rearrangement* of f, a notion introduced by HARDY and LITTLEWOOD. For a measurable scalar function f on Ω such that

$$\text{for every } \lambda > 0, meas\{x \in \Omega \mid |f(x)| > \lambda\} < \infty, \tag{22.13}$$

one can define the nonincreasing rearrangement of f, denoted by f^*. It is the only (real) nonincreasing function defined on $\left(0, meas(\Omega)\right)$ which is equimeasurable to $|f|$, and it can be defined by

$$\lambda \in \left[f^*(t+), f^*(t-)\right] \text{ if and only if}$$
$$meas\{x \in \Omega \mid |f(x)| > \lambda\} \leq t \leq meas\{x \in \Omega \mid |f(x)| \geq \lambda\}. \tag{22.14}$$

If necessary, one extends $f^*(t)$ to be 0 for $t > meas(\Omega)$. One basic property is that for any (piecewise) continuous function Φ defined on $[0, \infty)$ one has

$$\int_\Omega \Phi(|f(x)|)\, dx = \int_0^{meas(\Omega)} \Phi\left(f^*(t)\right) dt. \tag{22.15}$$

Lemma 22.4. *If $E_0 = L^1(\Omega)$ and $E_1 = L^\infty(\Omega)$ then for any function $f \in L^1(\Omega) + L^\infty(\Omega)$ one has*

$$K(t; f) = \int_0^t f^*(s)\, ds \text{ for all } t > 0, \text{ if one extends } f^* \text{ by 0 for } t > meas(\Omega). \tag{22.16}$$

---

[1] LORENTZ had introduced these spaces before the interpolation theories were developed.

Proof: If one decomposes $f = f_0 + f_1$ with $f_0 \in L^1(\Omega)$ and $||f_1||_{L^\infty(\Omega)} \le \lambda$ (and $\lambda > 0$), then the infimum of $||f_0||_{L^1(\Omega)}$ is obtained by taking $f_1(x) = f(x)$ whenever $|f(x)| \le \lambda$, and $f_1(x) = \lambda \frac{f(x)}{|f(x)|}$ whenever $|f(x)| > \lambda$, and this shows that

$$K(t; f) = \inf_{\lambda > 0} \left(t\lambda + \int_{|f(x)| > \lambda} (|f(x)| - \lambda)\, dx \right) = \inf_{\lambda > 0} \left(t\lambda + \int_{f^*(s) > \lambda} (f^*(s) - \lambda)\, ds \right).$$
(22.17)

The infimum is attained for any λ in the interval $[f^*(t+), f^*(t-)]$ and is $\int_0^t f^*(s)\, ds$ (one extends f by 0 outside Ω and f^* by 0 for $t > meas(\Omega)$). Indeed let τ be such that $\lambda \in [f^*(\tau+), f^*(\tau-)]$, then $\int_{f^*(s) > \lambda} (f^*(s) - \lambda)\, ds + t\lambda = \int_0^\tau f^*(s)\, ds + \lambda(t - \tau)$, and it is enough to check that $\int_t^\tau f^*(s)\, ds + \lambda(t - \tau) \ge 0$ for all $\tau > 0$; this is a consequence of $f^*(s) \ge \lambda$ for $s < t$ and $f^*(s) \le \lambda$ for $s > t$.   □

In order to compare two definitions of Lorentz spaces, we shall use the general form of *Hardy's inequality*, which extends (13.3).

Lemma 22.5. *Let* $1 \le q \le \infty$ *and* $\alpha < 1$, *then*

$$t^\alpha \varphi \in L^q\left(R_+; \frac{dt}{t}\right) \text{ and } \psi(t) = \frac{1}{t} \int_0^t \varphi(s)\, ds \text{ imply}$$
$$t^\alpha \psi \in L^q\left(R_+; \frac{dt}{t}\right) \text{ and } ||t^\alpha \psi||_{L^q(0,\infty;dt/t)} \le \frac{1}{1-\alpha} ||t^\alpha \varphi||_{L^q(0,\infty;dt/t)}.$$
(22.18)

Proof: The case $q = \infty$ is obvious, because $|\varphi(t)| \le M\, t^{-\alpha}$ for all $t > 0$ implies $|\psi(t)| \le \frac{M\, t^{-\alpha}}{1-\alpha}$ for all $t > 0$.

For $1 \le q < \infty$, one uses the fact that $C_c(R_+)$ is dense in the space of φ such that $t^\alpha \varphi \in L^q\left(R_+; \frac{dt}{t}\right)$, so that one may assume that $\varphi \in C_c(R_+)$, in which case ψ vanishes near 0 and behaves as $\frac{C}{t}$ for t large. As ψ is of class C^1 and

$$t\, \psi'(t) + \psi(t) = \varphi(t),$$
(22.19)

one multiplies by $t^{\alpha q} |\psi|^{q-2} \psi$ and integrates against $\frac{dt}{t}$; one finds

$$\int_0^\infty t\, \psi' t^{\alpha q} |\psi|^{q-2} \psi \frac{dt}{t} = \frac{1}{q} \int_0^\infty t^{\alpha q} d|\psi|^q = -\alpha \int_0^\infty t^{\alpha q} |\psi|^q \frac{dt}{t},$$
(22.20)

because $t^\alpha \psi(t)$ tends to 0 at ∞. This shows that

$$(1-\alpha) \int_0^\infty |t^\alpha \psi|^q \frac{dt}{t} = \int_0^\infty |t^\alpha \psi|^{q-2} t^\alpha \psi\, t^\alpha \varphi \frac{dt}{t} \le ||t^\alpha \psi||_q^{q-1} ||t^\alpha \varphi||_q, \quad (22.21)$$

by Hölder's inequality, where $|| \cdot ||_q$ denotes the norm of $L^q\left(R_+; \frac{dt}{t}\right)$.   □

Lemma 22.6. *For* $1 < p < \infty$ *and* $1 \leq q \leq \infty$ *one has*

$$L^{p,q}(\Omega) = \left(L^1(\Omega), L^\infty(\Omega)\right)_{1/p',q} =$$
$$\left\{f \in L^1(\Omega) + L^\infty(\Omega) \mid t^{1/p} f^*(t) \in L^q\left(R_+; \tfrac{dt}{t}\right)\right\} \ and \qquad (22.22)$$
$$\|t^{1/p} f^*\|_{L^q(0,\infty,dt/t)} \ is \ an \ equivalent \ norm,$$

showing that $L^{p,p}(\Omega) = L^p(\Omega)$ *with an equivalent norm.* $L^{p,\infty}(\Omega)$ *is the weak* L^p *space of MARCINKIEWICZ, with an equivalent norm.*

Proof: The definition of the interpolation space would have $t^{-\theta} K(t; f) \in L^q\left(R_+; \tfrac{dt}{t}\right)$, with $\theta = \tfrac{1}{p'}$, and as $K(t; f) = \int_0^t f^*(s)\, ds \geq t f^*(t)$ because f^* is nonincreasing, it implies $t^{-\theta} K(t; f) \geq t^{1-\theta} f^*(t) = t^{1/p} f^*(t)$. Conversely, if $t^{1/p} f^* \in L^q\left(R_+; \tfrac{dt}{t}\right)$, then Hardy's inequality (13.3) implies $t^{1/p} \tfrac{1}{t} \int_0^t f^*(s)\, ds \in L^q\left(R_+; \tfrac{dt}{t}\right)$, because $\alpha = \tfrac{1}{p} < 1$, and $t^{1/p} \tfrac{1}{t} \int_0^t f^*(s)\, ds = t^{-\theta} K(t; f)$.

The definition of the weak L^p space of MARCINKIEWICZ is that there exists M such that for every measurable subset ω of Ω one has $\int_\omega |f|\, dx \leq M \, meas(\omega)^{1/p'}$. The statement is then the consequence of the fact that for $t > 0$ one has $\sup_{meas(\omega)=t} \int_\omega |f|\, dx = \int_0^t f^*(s)\, ds$, and this is seen by choosing $\lambda \in [f^*(t+), f^*(t-)]$ and defining $\omega_0 = \{x \mid |f(x)| > \lambda\}$ and $\omega_1 = \{x \mid |f(x)| \geq \lambda\}$, so that $meas(\omega_0) \leq t \leq meas(\omega_1)$ (and $|f(x)| = t$ on $\omega_1 \setminus \omega_0$). If ω is not a subset of ω_1, one increases the integral of $|f|$ by replacing the part of ω which is not in ω_1 by a part of the same measure in $\omega_1 \setminus \omega$; if ω is a subset of ω_1 but does not contain ω_0, one increases the integral of $|f|$ by replacing a part of ω which is not in ω_0 by a corresponding part of the same measure in $\omega_0 \setminus \omega$, so that finally the subsets of measure t for which the integral of $|f|$ is maximum must contain ω_0 and be contained in ω_1.

[Taught on Friday March 10, 2000.]

23

Interpolation of L^2 Spaces with Weights

We want to consider now Sobolev spaces $H^s(\Omega)$ when s is not an integer, which were defined for $\Omega = R^N$ using the Fourier transform at Definition 15.7, and discussed for more general open sets $\Omega \subset R^N$ after Lemma 16.3. They are indeed interpolation spaces with the particular choice $p = 2$, and for $p \neq 2$ the interpolation spaces belong to the larger family of spaces named after Oleg BESOV[1], the *Besov spaces*.

For $s \in R$, if

$$F_s = \{v \in L^2_{loc}(R^N) \mid (1 + 4\pi^2|\xi|^2)^{s/2}v \in L^2(R^N)\}, \qquad (23.1)$$

then the Fourier transform $\mathcal{F}$ is an isometry from $H^s(R^N)$ onto F_s and the inverse Fourier transform $\overline{\mathcal{F}}$ is an isometry from F_s onto $H^s(R^N)$. The interpolation property implies then that $\mathcal{F}$ maps continuously $\left(H^\alpha(R^N), H^\beta(R^N)\right)_{\theta,p}$ into $(F_\alpha, F_\beta)_{\theta,p}$ and $\overline{\mathcal{F}}$ maps $(F_\alpha, F_\beta)_{\theta,p}$ into $\left(H^\alpha(R^N), H^\beta(R^N)\right)_{\theta,p}$, so that $\left(H^\alpha(R^N), H^\beta(R^N)\right)_{\theta,p}$ coincides with the tempered distributions whose Fourier transform belongs to $(F_\alpha, F_\beta)_{\theta,p}$ (and one deduces in the same way that it is an isometry if one uses the corresponding norms).

Identifying interpolation spaces between Sobolev spaces $H^s(R^N)$ is then the same question as interpolating between some L^2 spaces with weights, and this new question can be settled easily in a more general setting.

Lemma 23.1. *For a (measurable) positive function w on Ω, let*

$$E(w) = \left\{u \mid \int_\Omega |u(x)|^2 w(x)\, dx < \infty\right\} \text{ with } ||u||_w = \left(\int_\Omega |u(x)|^2 w(x)\, dx\right)^{1/2} \qquad (23.2)$$

If w_0, w_1 are two such functions, then for $0 < \theta < 1$ one has

$$\left(E(w_0), E(w_1)\right)_{\theta,2} = E(w_\theta) \text{ with equivalent norms, where } w_\theta = w_0^{1-\theta}w_1^\theta. \qquad (23.3)$$

---

[1] Oleg Vladimirovich BESOV, Russian mathematician, born in 1933. He works at the Steklov Institute of Mathematics, Moscow, Russia.

Proof: One uses a variant of the K functional adapted to L^2 spaces, namely

$$K_2(t;a) = \inf_{a=a_0+a_1} \left(||a_0||_0^2 + t^2||a_1||_1^2 \right)^{1/2}, \qquad (23.4)$$

and one checks immediately that $K_2(t;a) \leq K(t;a) \leq \sqrt{2}\,K_2(t;a)$ for all $a \in E_0 + E_1$, whatever the normed spaces E_0, E_1 of the abstract theory are.

For $E_0 = E(w_0)$ and $E_1 = E(w_1)$, for any $a \in E_0 + E_1$ and $t > 0$ one can calculate explicitly $K_2(t;a)$:

$$K_2(t;a)^2 = \inf_{a=a_0+a_1} \left(\int_\Omega (|a_0(x)|^2 w_0(x) + t^2 |a_1(x)|^2 w_1(x))\, dx \right), \qquad (23.5)$$

and one is led to choose for $a_0(x)$ the value λ which minimizes $|\lambda|^2 w_0(x) + t^2 |a(x)-\lambda|^2 w_1(x)$, and as λ is characterized by $\lambda\, w_0(x) - t^2(a(x)-\lambda)w_1(x) = 0$, one finds

$$a_0(x) = \frac{t^2 w_1(x)}{w_0(x)+t^2 w_1(x)}\, a(x)$$

$$a_1(x) = \frac{w_0(x)}{w_0(x)+t^2 w_1(x)}\, a(x) \qquad (23.6)$$

(which are measurable), and this optimal choice gives

$$|a_0(x)|^2 w_0(x) + t^2 |a_1(x)|^2 w_1(x) = \frac{t^2 w_0(x) w_1(x)}{w_0(x) + t^2 w_1(x)}\, |a(x)|^2, \qquad (23.7)$$

so that

$$K_2(t;a) = \left(\int_\Omega \frac{t^2 w_0(x) w_1(x)}{w_0(x) + t^2 w_1(x)}\, |a(x)|^2\, dx \right)^{1/2}. \qquad (23.8)$$

For $0 < \theta < 1$ one has

$$||t^{-\theta} K_2(t;a)||^2_{L^2(0,\infty,dt/t)} = \int_0^\infty \int_\Omega t^{-2\theta} \frac{t^2 w_0(x) w_1(x)}{w_0(x) + t^2 w_1(x)}\, |a(x)|^2\, dx\, \frac{dt}{t}, \qquad (23.9)$$

which one computes by integrating in t first, by Fubini's theorem, and after making the change of variable $t = s\sqrt{\frac{w_0(x)}{w_1(x)}}$, so that $\frac{dt}{t} = \frac{ds}{s}$, one finds

$$\int_0^\infty t^{-2\theta} \frac{t^2 w_0(x) w_1(x)}{w_0(x)+t^2 w_1(x)}\, \frac{dt}{t} = w_0(x)^{1-\theta} w_1(x)^\theta \int_0^\infty \frac{t^{1-2\theta}}{1+t^2}\, dt, \qquad (23.10)$$

which gives

$$||t^{-\theta} K_2(t;a)||_{L^2(0,\infty,dt/t)} = C \left(\int_\Omega |a(x)|^2 w_\theta(x)\, dx \right)^2 \text{ with}$$
$$C^2 = \int_0^\infty \frac{t^{1-2\theta}}{1+t^2}\, dt = \frac{\pi}{2\sin(\pi\,\theta)}. \qquad \square \qquad (23.11)$$

Using the theory of interpolation, one can improve the *Hausdorff–Young inequality*,[2] which asserts that the Fourier transform maps $L^p(R^N)$ into

---

[2] YOUNG had proven the result when p' is an even integer, and HAUSDORFF had proven the general case.

$L^{p'}(R^N)$ if $1 \leq p \leq 2$, and this improvement uses Lorentz spaces. Indeed $\mathcal{F}f(\xi) = \int_{R^N} f(x)e^{-2i\pi(x.\xi)}\,dx$ gives immediately $||\mathcal{F}f||_\infty \leq ||f||_1$, where $||\cdot||_p$ denotes the $L^p(R^N)$ norm; on the other hand $||\mathcal{F}f||_2 = ||f||_2$, so that the interpolation property asserts that

Fourier transform maps $\left(L^1(R^N), L^2(R^N)\right)_{\theta,p}$ into $\left(L^\infty(R^N), L^2(R^N)\right)_{\theta,p}$.
$$(23.12)$$

The important reiteration theorem 26.3 (of Jacques-Louis LIONS and Jaak PEETRE) will show that these spaces are in the family of Lorentz spaces, and the result will then be that for $1 < p < 2$ and $1 \leq q \leq \infty$

Fourier transform maps $L^{p,q}(R^N)$ into $L^{p',q}(R^N)$, for $1<p<2, 1 \leq q \leq \infty$,
$$(23.13)$$

and in particular, because $p < p'$, it maps $L^p(R^N)$ into $L^{p',p}(R^N) \subset L^{p'}(R^N)$.

Results concerning convolution can also be improved using the theory of interpolation and Lorentz spaces, and in particular the Sobolev's embedding theorem can be improved, as noticed by Jaak PEETRE. The classical result is that

for $1 \leq p < N$ one has $W^{1,p}(R^N) \subset L^{p^*}(R^N)$, with
$p^* = \frac{Np}{N-p}$, i.e., $\frac{1}{p^*} = \frac{1}{p} - \frac{1}{N}$,
$$(23.14)$$

and this will be improved into

for $1 \leq p < N$ one has $W^{1,p}(R^N) \subset L^{p^*,p}(R^N) \subset L^{p^*}(R^N)$. (23.15)

In his original proof, Sergei SOBOLEV used a convolution formula

$$u = \sum_j \frac{\partial u}{\partial x_j} \star \frac{\partial E}{\partial x_j} \text{ for an elementary solution } E \text{ of } \Delta, \qquad (23.16)$$

and

$$\begin{cases} E = \frac{C_N}{|x|^{N-2}} \text{ for } N \geq 3 \\ E = C_2 \log(|x|) \text{ for } N = 2 \end{cases} \text{ implies } \frac{\partial E}{\partial x_j} \in L^{N',\infty}(R^N). \qquad (23.17)$$

Together with the reiteration theorem 26.3 and a duality theorem of Jacques-Louis LIONS and Jaak PEETRE, which asserts[3] that $L^{N',\infty}(R^N)$ is the dual of $L^{N,1}(R^N)$, one finds that for $1 < p < N$ and $1 \leq q \leq \infty$, convolution of $L^{p,q}(R^N)$ by $L^{N',\infty}(R^N)$ gives a result in $L^{p^*,q}(R^N)$.

However, this argument does not give Sobolev's embedding theorem for $p = 1$, which was proven by Louis NIRENBERG by a different method (also introduced by Emilio GAGLIARDO), or the improvement that $W^{1,1}(R^N)$ is

---

[3] The result of Jacques-Louis LIONS and Jaak PEETRE is valid for general Banach spaces, and the particular result for Lorentz spaces may have been known before.

continuously embedded in $L^{1^*,1}(R^N)$, which I proved. The reason is that convolution of $L^1(R^N)$ by any Lorentz space $L^{a,b}(R^N)$ gives a result in $L^{a,b}(R^N)$ and not better, because one can approach the Dirac mass δ_0 by a bounded sequence in $L^1(R^N)$.

This is something that one should be aware of, that different ways of using the theory of interpolation may lead to results in different interpolation spaces, usually differing only in the second parameter.

The usual scaling arguments, for example, are insensitive to the second parameter for the Lorentz spaces, and cannot be used to check that a given result is optimal. For example, if $u \in L^1(R^N) + L^\infty(R^N)$ and for $\lambda \neq 0$ let U be defined by $U(x) = u(\lambda x)$ for $x \in R^N$, then any decomposition of $u = a_0 + a_1$ with $a_0 \in L^1(R^N)$ and $a_1 \in L^\infty(R^N)$ gives a decomposition $U = A_0 + A_1$ with $A_j(x) = a_j(\lambda x)$ for $x \in R^N$ and $j = 1, 2$. Then one has $||A_0||_{L^1(R^N)} = |\lambda|^{-N}||a_0||_{L^1(R^N)}$ and $||A_1||_{L^\infty(R^N)} = ||a_1||_{L^\infty(R^N)}$, so that

$$U(x) = u(\lambda x) \text{ for } x \in R^N \text{ implies } K(t; U) = |\lambda|^{-N} K(t\,|\lambda|^N; u), \text{ so}$$
$$||U||_{L^{p,q}(R^N)} = |\lambda|^{-N/p}||u||_{L^{p,q}(R^N)}, \tag{23.18}$$

and the parameter q does not appear in the way the norm changes.

[Taught on Monday March 13, 2000.]

Real Interpolation; J-Method

Jaak PEETRE developed another interpolation method, the J-method, which is a dual method compared to the K-method.

The K-method is the natural result of investigations which originated in questions of traces: if $u \in L^{p_0}(R_+; E_0)$ and $u' \in L^{p_1}(R_+; E_1)$ with $1 \le p_0, p_1 \le \infty$, then $u \in C^0([0,1]; E_0 + E_1)$, so that $u(0)$ exists and the question is to characterize the space of such values at 0 (traces). As one can change $u(t)$ in $u(t^\lambda)$ with $\lambda > 0$ and not change $u(0)$, one then discovers naturally that one can consider spaces of functions such that $t^{\alpha_0} u \in L^{p_0}(R_+; E_0)$ and $t^{\alpha_1} u' \in L^{p_1}(R_+; E_1)$, and for some set of parameters $u(0)$ exists. These ideas may have started with Emilio GAGLIARDO, and I do not know if he had first identified the traces of functions from $W^{1,p}(R^N)$ on an hyperplane before or after thinking of the general framework, but certainly Jacques-Louis LIONS and Jaak PEETRE perfected the framework, and the K-method is Jaak PEETRE's further simplification, which shows that the family of interpolations spaces that they had introduced only depends upon two parameters.

If one wants to characterize the duals of the spaces obtained, then one finds easily that these dual spaces are naturally defined as integrals, and one considers then questions like that of identifying which are the elements $a \in E_0 + E_1$ which can be written as $\int_0^\infty v(t)\,dt$ where $t^{\beta_0}v \in L^{q_0}(R_+; E_0)$ and $t^{\beta_1}v \in L^{q_1}(R_+; E_1)$, for the range of parameters where the integral is defined. Again, looking at $v(t^\lambda)$ shows that there are not really four parameters, but one important observation is that these spaces are (almost) the same as the ones defined by traces, and I do not know if Emilio GAGLIARDO had investigated such questions before the basic work of Jacques-Louis LIONS and Jaak PEETRE. The J-method is then the simplification by Jaak PEETRE of the preceding framework.

Definition 24.1. *For $v \in E_0 \cap E_1$ and $t > 0$, one writes*

$$J(t; v) = \max\{||v||_0, t||v||_1\}. \quad \Box$$

(24.1)

The case $t = 1$ corresponds to the usual norm on $E_0 \cap E_1$, which makes both injections into E_0 or E_1 continuous and with norms at most 1; $J(t; v)$ then gives a family of equivalent norms on $E_0 \cap E_1$.

Definition 24.2. *For $0 < \theta < 1$ and $1 \leq p \leq \infty$, or for $\theta = 0, 1$, and $p = 1$, one defines*

$$(E_0, E_1)_{\theta,p;J} = \left\{ a = \int_0^\infty v(t) \frac{dt}{t} \in E_0 + E_1 \mid v(t) \in E_0 \cap E_1 \text{ a.e. } t > 0, \right.$$
$$\left. \text{and } t^{-\theta} J(t; v(t)) \in L^p\left(R_+; \frac{dt}{t}\right) \right\} \text{ with}$$
$$||a||_{\theta,p;J} = \inf_v ||t^{-\theta} J(t; v)||_{L^p(0,\infty;dt/t)}, \text{ the infimum being taken on all}$$
$$v \text{ with } \int_0^\infty v(t) \frac{dt}{t} = a. \quad \square$$
$$(24.2)$$

As every $a \in E_0 \cap E_1$ can be written as $a = \int_0^\infty \varphi(t) a \frac{dt}{t}$ with φ having compact support in R_+ and satisfying $\int_0^\infty \varphi(t) \frac{dt}{t} = 1$, one could consider other values of θ, p, but the infimum of $||t^{-\theta} J(t; v)||_{L^p(0,\infty;dt/t)}$ would be 0 in these cases. Indeed one may replace φ by $\varphi(\lambda t)$ and let λ tend to ∞, and the infimum tends to 0 if $\theta < 0$ or if $\theta = 0$ and $p > 1$; similarly letting λ tend to 0 the infimum tends to 0 if $\theta > 1$ or if $\theta = 1$ and $p > 1$.

As noted by Jaak PEETRE, one can avoid every question of measurability by using a discrete description, based on defining

$$a_n = \int_{2^n}^{2^{n+1}} v(t) \frac{dt}{t} \in E_0 \cap E_1 \text{ and } a = \sum_{n \in Z} a_n. \quad (24.3)$$

The important property of this family of spaces is the following *equivalence* result, which says that apart from the extreme cases $\theta = 0, 1$, where the two methods use different values of p anyway, the J-method gives the same spaces as the K-method.

Lemma 24.3. *For $0 < \theta < 1$ and $1 \leq p \leq \infty$, the J-method gives the same spaces as the K-method, with equivalent norms.*

Proof: Let $a \in (E_0, E_1)_{\theta,p;J}$, so that $a = \int_0^\infty u(s) \frac{ds}{s}$ with $s^{-\theta} J(s; u(s)) \in L^p(0, \infty, \frac{ds}{s})$. Using the decompositions $u = u + 0 = 0 + u$, one deduces that

$$\text{for } u \in E_0 \cap E_1 \text{ one has } K(t; u) \leq \min\{||u||_0, t\,||u||_1\}, \quad (24.4)$$

and because $a \mapsto K(t; a)$ is a norm, one deduces that

$$K(t; a) \leq \int_0^\infty K(t; u(s)) \frac{ds}{s} \leq \int_0^\infty \min\{||u(s)||_0, t\,||u(s)||_1\} \frac{ds}{s}, \quad (24.5)$$

and because for $u \in E_0 \cap E_1$ one has $||u||_0 \leq J(s; u)$ and $||u||_1 \leq \frac{1}{s} J(s; u)$, one deduces that

$$\min\{||u(s)||_0, t\,||u(s)||_1\} \leq \min\left\{1, \frac{t}{s}\right\} J(s; u(s)), \quad (24.6)$$

so that

$$t^{-\theta}K(t;a) \le \int_0^\infty \min\left\{\left(\frac{t}{s}\right)^{-\theta}, \left(\frac{t}{s}\right)^{1-\theta}\right\} s^{-\theta}J(s;u(s))\,\frac{ds}{s}. \tag{24.7}$$

This is a convolution product for the multiplicative group R_+ with Haar measure $\frac{dt}{t}$, of the function $t^{-\theta}J(t;u(t))$ which belongs to $L^p(R_+;\frac{dt}{t})$ and the function $\min\{t^{-\theta},t^{1-\theta}\}$, which belongs to $L^1(R_+;\frac{dt}{t})$, so that one has

$$||t^{-\theta}K(t;a)||_{L^p(0,\infty;dt/t)} \le C\,||t^{-\theta}J(t;u(t))||_{L^p(0,\infty;dt/t)}, \tag{24.8}$$

proving that $(E_0,E_1)_{\theta,p;J} \subset (E_0,E_1)_{\theta,p}$ with continuous embedding.

In order to prove the opposite continuous embedding, one must start from $a \in (E_0,E_1)_{\theta,p}$ and construct $u(t) \in E_0 \cap E_1$ such that $a = \int_0^\infty u(t)\,\frac{dt}{t}$ and $t^{-\theta}J(t;u(t)) \in L^p(R_+;\frac{dt}{t})$, and for that it is enough to ensure that one can construct such a u satisfying $J(t;u(t)) \le C\,K(t;a)$ for all $t > 0$. This fact is true in a slightly more general context, one chooses

$$u(t) = u_n \text{ for } e^n < t < e^{n+1}, \text{ so that } a = \int_0^\infty u(t)\,\frac{dt}{t} \text{ means } a = \sum_{-\infty}^{+\infty} u_n, \tag{24.9}$$

and the basic construction is shown in Lemma 24.4. Because $a \in (E_0,E_1)_{\theta,p} \subset (E_0,E_1)_{\theta,\infty}$, one has $K(t;a) \le C\,t^\theta$ and the hypothesis of Lemma 24.4 is indeed satisfied. $\square$

Lemma 24.4. *If $a \in E_0 + E_1$ satisfies*

$$K(t;a) \to 0 \text{ as } t \to 0 \text{ and } \frac{K(t;a)}{t} \to 0 \text{ as } t \to \infty, \tag{24.10}$$

then, for a universal constant C,

> *for $n \in Z$ there exists $u_n \in E_0 \cap E_1$ so that if*
> *u is defined by $u(t) = u_n$ for $e^n < t < e^{n+1}$, then* $\qquad$ (24.11)
> *$a = \int_0^\infty u(t)\,\frac{dt}{t}$ and $J(t;u(t)) \le C\,K(t;a)$ for all $t > 0$.*

Proof: Let $C_0 > 1$, and for each $n \in Z$ let $a = a_{0,n} + a_{1,n}$ with $a_{0,n} \in E_0$, $a_{1,n} \in E_1$ and $||a_{0,n}||_0 + e^n||a_{1,n}||_1 \le C_0 K(e^n;a)$. In particular $||a_{0,n}||_0 \to 0$ as $n \to -\infty$ and $||a_{1,n}||_1 \to 0$ as $n \to +\infty$, and one chooses

$$u_n = a_{0,n+1} - a_{0,n} = a_{1,n} - a_{1,n+1} \in E_0 \cap E_1, \tag{24.12}$$

and for $i < j$ one has $u_i + \ldots + u_j = a_{0,j+1} - a_{0,i} = a - a_{1,j+1} - a_{0,i}$, which converges to a in $E_0 + E_1$ as $i \to -\infty$ and $j \to +\infty$. Because $K(t;a)$ is nondecreasing in t and $\frac{K(t;a)}{t}$ is nonincreasing in t, one has

$K(e^n; a) \leq K(t; a) \leq K(e^{n+1}; a)$ and $\frac{t}{e^{n+1}} K(e^{n+1}; a) \leq K(t; a) \leq \frac{t}{e^n} K(e^n; a)$ for $e^n < t < e^{n+1}$, and

$$\|u_n\|_0 \leq \|a_{0,n+1}\|_0 + \|a_{0,n}\|_0 \leq C_0 K(e^{n+1}; a) + C_0 K(e^n; a) \leq$$
$$C_0(1 + e) K(t; a)$$
$$t\,\|u_n\|_1 \leq t\,\|a_{1,n+1}\|_1 + t\,\|a_{1,n}\|_1 \leq C_0 \frac{t}{e^{n+1}} K(e^{n+1}; a) + C_0 \frac{t}{e^n} K(e^n; a) \leq$$
$$C_0(1 + e) K(t; a),$$

$$(24.13)$$

so that $J\big(t; u(t)\big) \leq C_0(1 + e) K(t; a)$. $\qquad\qquad\square$

[Taught on Wednesday March 15, 2000.]

25

Interpolation Inequalities, the Spaces $(E_0, E_1)_{\theta,1}$

Definition 25.1. *Let E be a normed space; for $0 \leq \theta \leq 1$, one says that*

E *is of class* $\mathcal{K}(\theta)$ *if* $E_0 \cap E_1 \subset E \subset (E_0, E_1)_{\theta,\infty;K}$

E *is of class* $\mathcal{J}(\theta)$ *if* $(E_0, E_1)_{\theta,1;J} \subset E \subset E_0 + E_1$ (25.1)

E *is of class* $\mathcal{H}(\theta)$ *if* $(E_0, E_1)_{\theta,1;J} \subset E \subset (E_0, E_1)_{\theta,\infty;K}$. $\square$

Of course, for $0 < \theta < 1$ the indices J and K may be dropped as the two interpolation methods give the same spaces, but for the extreme values $\theta = 0, 1$ the only interpolation spaces that we use correspond to $p = 1$ for the J-method or $p = \infty$ for the K-method.

The *reiteration theorem* 26.3 will state that if F_0 is of class $\mathcal{H}(\theta_0)$ and F_1 is of class $\mathcal{H}(\theta_1)$ with $\theta_0 \neq \theta_1$, then for $0 < \theta < 1$ and $1 \leq p \leq \infty$ one has $(F_0, F_1)_{\theta,p} = (E_0, E_1)_{\eta,p}$ with $\eta = (1 - \theta)\theta_0 + \eta\theta_1$. Therefore, if $F_0 = (E_0, E_1)_{\theta_0,p_0}$ and $F_1 = (E_0, E_1)_{\theta_1,p_1}$, the interpolation space $(F_0, F_1)_{\theta,p}$ is the same, whatever the precise values p_0 and p_1 are, if $\theta_0 \neq \theta_1$. However, the interpolation spaces do depend upon p_0 and p_1 in the case $\theta_0 = \theta_1$, and in that case they may be new spaces, i.e., not included in the family indexed by θ, p.

Lemma 25.2. *(i) For a normed space E,*

E *is of class* $\mathcal{K}(\theta)$ *if and only if* $E_0 \cap E_1 \subset E$ *and there exists C such that* $K(t; a) \leq C t^\theta ||a||_E$ *for all $t > 0$ and all $a \in E$.*

(25.2)

(ii) For a Banach space E,

E *is of class* $\mathcal{J}(\theta)$ *if and only if* $E \subset E_0 + E_1$ *and there exists C such that* $||a||_E \leq C t^{-\theta} J(t; a)$ *for all $t > 0$ and all $a \in E_0 \cap E_1$*
    or if and only if
      there exists C such that $||a||_E \leq C ||a||_0^{1-\theta}||a||_1^\theta$ *for all $a \in E_0 \cap E_1$.*

(25.3)

Proof: (i) It means $||a||_{(E_0,E_1)_{\theta,\infty;K}} \leq C\,||a||_E$ for all $a \in E$. Because $||a||_{(E_0,E_1)_{\theta,\infty;K}} = ||t^{-\theta}K(t;a)||_{L^\infty(0,\infty)}$, the condition is $K(t;a) \leq C\,t^\theta||a||_E$ for all $t > 0$.

(ii) It means $||a||_E \leq C\,||a||_{(E_0,E_1)_{\theta,1;J}}$ for all $a \in E$, and the necessary condition follows from the fact that

for $a \in E_0 \cap E_1$ one has $||a||_{(E_0,E_1)_{\theta,1;J}} \leq t_0^{-\theta}J(t_0;a)$ for all $t_0 > 0$.   (25.4)

Indeed, if $\varphi \in L^1\left(R_+; \frac{dt}{t}\right)$ and $\int_0^\infty \varphi(t)\,\frac{dt}{t} = 1$, then every $a \in E_0 \cap E_1$ can be written as $a = \int_0^\infty u(t)\,\frac{dt}{t}$ with $u(t) = \varphi(t)a$ for $t > 0$, and one has

$$||a||_{(E_0,E_1)_{\theta,1;J}} \leq \int_0^\infty t^{-\theta}|\varphi(t)|J(t;a)\,\frac{dt}{t}; \qquad (25.5)$$

then for $t_0 > 0$, one takes a sequence φ_n converging to δ_{t_0} (for example $\varphi_n(t) = n$ if $t_0 < t < \frac{n+1}{n}t_0$ and $\varphi_n(t) = 0$ otherwise), and one obtains (25.4) by letting $n \to \infty$. Having shown that $||a||_E \leq C\,t^{-\theta}J(t;a) = C\,\max\{||a||_0, t\,||a||_1\}$ for all $a \in E_0 \cap E_1$ and all $t > 0$, one takes the minimum in t, which is attained for $t = \frac{||a||_0}{||a||_1}$, so that $\min_{t>0} t^{-\theta}J(t;a) = C\,||a||_0^{1-\theta}||a||_1^\theta$ for all $a \in E_0 \cap E_1$, and the condition $||a||_E \leq C\,t^{-\theta}J(t;a)$ for all $a \in E_0 \cap E_1$ and all $t > 0$ is then equivalent to $||a||_E \leq C\,||a||_0^{1-\theta}||a||_1^\theta$ for all $a \in E_0 \cap E_1$. Conversely, assume that there exists a constant C such that $||a||_E \leq C\,t^{-\theta}J(t;a)$ for all $a \in E_0 \cap E_1$ and all $t > 0$; for $b \in (E_0,E_1)_{\theta,1;J}$ and a decomposition $b = \int_0^\infty u(t)\,\frac{dt}{t}$ with $u(t) \in E_0 \cap E_1$ for $t > 0$ and $\int_0^\infty t^{-\theta}J(t;u(t))\,\frac{dt}{t} < \infty$, one has

$$||b||_E \leq \int_0^\infty ||u(t)||_E\,\frac{dt}{t} \leq \int_0^\infty C\,t^{-\theta}J(t;u(t))\,\frac{dt}{t}, \qquad (25.6)$$

the integral converging in E because it is a Banach space; taking the infimum on all decompositions of b gives $||b||_E \leq C||b||_{(E_0,E_1)_{\theta,1;J}}$. $\qquad\square$

Actually, Jacques-Louis LIONS and Jaak PEETRE had observed something slightly more general, which is very useful, and often used with the reiteration theorem 26.3.

Lemma 25.3. *For a Banach space F, a linear mapping L from $E_0 \cap E_1$ into F extends into a linear continuous mapping from $(E_0,E_1)_{\theta,1;J}$ into F if and only if*

there exists C such that $||L\,a||_F \leq C\,||a||_0^{1-\theta}||a||_1^\theta$ for all $a \in E_0 \cap E_1$. (25.7)

Proof: The continuity from $(E_0,E_1)_{\theta,1;J}$ into F is $||L\,a||_F \leq C\,||a||_{(E_0,E_1)_{\theta,1;J}}$ for all $a \in (E_0,E_1)_{\theta,1}$, or just for all $a \in E_0 \cap E_1$, which is dense in $(E_0,E_1)_{\theta,1;J}$; then one uses (25.3) with $E = (E_0,E_1)_{\theta,1;J}$. Conversely, for $b \in (E_0,E_1)_{\theta,1;J}$, $b = \int_0^\infty u(t)\,\frac{dt}{t}$ with $u(t) \in E_0 \cap E_1$ for $t > 0$ and $\int_0^\infty t^{-\theta}J(t;u(t))\,\frac{dt}{t} < \infty$, one has

$$||L\,b||_F \leq \int_0^\infty ||L\,u(t)||_F\,\frac{dt}{t} \leq \int_0^\infty C\,t^{-\theta}J(t;u(t))\,\frac{dt}{t}, \qquad (25.8)$$

the integral converging in F because it is a Banach space; taking the infimum on all decompositions of b gives $||L\,b||_F \leq C||b||_{(E_0, E_1)_{\theta,1;J}}$.   □

For example, Sobolev space $H^{1/2}(R)$ is not embedded in $L^\infty(R)$, but for the slightly smaller interpolation space $\left(H^1(R), L^2(R)\right)_{1/2,1}$ one has

$$\left(H^1(R), L^2(R)\right)_{1/2,1} \subset C_0(R), \tag{25.9}$$

the space of continuous functions tending to 0 at ∞, because of the fact that $H^1(R) \subset C_0(R)$ with the precise estimate

$$||u||_{L^\infty(R)} \leq ||u||_{L^2(R)}^{1/2}||u'||_{L^2(R)}^{1/2} \text{ for all } u \in H^1(R); \tag{25.10}$$

then by using the reiteration theorem 26.3, one finds that

$$\text{for } 0 < s < \frac{1}{2}, H^s(R) \subset L^{p(s),2}, \text{ with } \frac{1}{p(s)} = \frac{1}{2} - s. \tag{25.11}$$

Therefore, one should be aware that some results which are not true for limiting cases, like Sobolev's embedding theorems, may be obtained by the theory of interpolation because the limiting case is actually true if one uses a slightly different space, and the difference does not really matter, because of the reiteration theorem 26.3.

Exchanging the two spaces is a special case of the reiteration theorem 26.3, but it is seen easily directly.

Lemma 25.4. *One has $(E_1, E_0)_{\theta,p} = (E_0, E_1)_{1-\theta,p}$ for $0 < \theta < 1$ and $1 \leq p \leq \infty$; the same result holds for $\theta = 0$ or 1, and $p = 1$ or $p = \infty$.*

Proof: One uses $F_0 = E_1$ and $F_1 = E_0$, and denoting by $K_F(t; a)$ the K functional using the spaces F_0, F_1, for any decomposition $a = a_0 + a_1$ with $a_0 \in E_0$ and $a_1 \in E_1$, one has

$$K_F(t; a) = \inf(||a_1||_1 + t\,||a_0||_0) = t \inf\left(||a_0||_0 + \frac{1}{t}||a_1||_1\right) = t\,K\left(\frac{1}{t}; a\right), \tag{25.12}$$

so that the change of variable $t = \frac{1}{s}$ gives $||t^{-\theta}K_F(t; a)||_{L^p(0,\infty;dt/t)} = ||s^{1-\theta}K(s; a)||_{L^p(0,\infty;dt/t)}$. For the limiting case $\theta = 0$ or 1, and $p = 1$, one uses the J-method, and for $a \in E_0 \cap E_1$ one observes that $J_F(t; a) = \min\{||a||_{E_1}, t\,||a||_{E_0}\} = t \min\{||a||_{E_0}, \frac{1}{t}||a||_{E_1}\} = t\,J(\frac{1}{t}; a)$; with the change of variable $t = \frac{1}{s}$ and $v(s) = u(t)$, one has $||t^{-\theta}J_F\left(t; u(t)\right)||_{L^p(0,\infty;dt/t)} = ||s^{1-\theta}J\left(s; v(s)\right)||_{L^p(0,\infty;dt/t)}$.   □

[Taught on Friday March 17, 2000.]

The Lions–Peetre Reiteration Theorem

The reiteration theorem 26.3 is proven in two steps.

Lemma 26.1. *If $G_0 \subset (E_0, E_1)_{\theta_0, \infty}$ and $G_1 \subset (E_0, E_1)_{\theta_1, \infty}$ with $0 \leq \theta_0 \neq \theta_1 \leq 1$ (and continuous embeddings), then for $0 < \theta < 1$ and $1 \leq p \leq \infty$, or for $\theta = 0$ or 1 with $p = \infty$, one has $(G_0, G_1)_{\theta, p} \subset (E_0, E_1)_{\eta, p}$ with $\eta = (1 - \theta)\theta_0 + \theta\,\theta_1$.*

Proof: One uses the K-method and the continuous embeddings mean that

there exists C_0 with $K(t; g_0) \leq C_0 t^{\theta_0} \|g_0\|_{G_0}$ for all $g_0 \in G_0$ and all $t > 0$,

there exists C_1 with $K(t; g_1) \leq C_1 t^{\theta_1} \|g_1\|_{G_1}$ for all $g_1 \in G_1$ and all $t > 0$.
$$\tag{26.1}$$

For $a \in G_0 + G_1$, let

$$K_G(t; a) = \inf_{a = g_0 + g_1} \left(\|g_0\|_{G_0} + t \|g_1\|_{G_1} \right), \tag{26.2}$$

then

$$K(t; a) \leq K(t; g_0) + K(t; g_1) \leq C_0 t^{\theta_0} \|g_0\|_{G_0} + C_1 t^{\theta_1} \|g_1\|_{G_1}, \tag{26.3}$$

and minimizing among all decompositions of a one deduces that

$$t^{-\eta} K(t; a) \leq C_0 t^{\theta_0 - \eta} K_G \left(\frac{C_1}{C_0} t^{\theta_1 - \theta_0}; a \right). \tag{26.4}$$

Because $\theta_1 \neq \theta_0$, one may use the change of variable

$$s = t^{\theta_1 - \theta_0}, \tag{26.5}$$

and as $s^{-\theta} = t^{-\theta(\theta_1 - \theta_0)} = t^{\theta_0 - \eta}$ one finds

$$t^{-\eta} K(t; a) \leq s^{-\theta} K_G \left(\frac{C_1}{C_0} s; a \right), \tag{26.6}$$

and using $\frac{ds}{s} = |\theta_1 - \theta_0| \frac{dt}{t}$), one deduces that $a \in (G_0, G_1)_{\theta, p}$ implies $a \in (E_0, E_1)_{\eta, p}$. $\qquad\square$

Lemma 26.2. *If $(E_0, E_1)_{\theta_0,1} \subset H_0$ and $(E_0, E_1)_{\theta_1,1} \subset H_1$ with $0 \leq \theta_0 \neq \theta_1 \leq 1$ (and continuous embeddings), then for $0 < \theta < 1$ and $1 \leq p \leq \infty$, or for $\theta = 0$ or 1 with $p = 1$, one has $(E_0, E_1)_{\eta,p} \subset (H_0, H_1)_{\theta,p}$ with $\eta = (1 - \theta)\theta_0 + \theta\,\theta_1$.*

Proof: One uses the J-method and the fact that for $t > 0$ one has

$$\|u\|_{H_0} \leq C_0 \|u\|_{(E_0,E_1)_{\theta_0,1}} \leq C_0' \, t^{-\theta_0} J(t; u) \text{ for all } u \in E_0 \cap E_1$$
$$\|u\|_{H_1} \leq C_1 \|u\|_{(E_0,E_1)_{\theta_1,1}} \leq C_1' \, t^{-\theta_1} J(t; u) \text{ for all } u \in E_0 \cap E_1. \tag{26.7}$$

For $a \in (E_0, E_1)_{\eta,p}$ one has $a = \int_0^\infty u(t) \frac{dt}{t}$ with $u(t) \in E_0 \cap E_1$ and $t^{-\eta} J(t; u(t)) \in L^p(R_+; \frac{dt}{t})$. One chooses

$$\lambda = \theta_1 - \theta_0 \neq 0, \tag{26.8}$$

and because $u(t) \in E_0 \cap E_1 \subset H_0 \cap H_1$, one can estimate

$$J_H(t^\lambda; u(t)) = \max\{\|u(t)\|_{H_0}, t^\lambda \|u(t)\|_{H_1}\}, \text{ and}$$
$$\|u(t)\|_{H_0} \leq C_0' t^{-\theta_0} J(t; u(t)) \text{ and } \|u(t)\|_{H_1} \leq C_1' t^{-\theta_1} J(t; u(t)), \text{ so that}$$
$$\max\{\|u(t)\|_{H_0}, t^\lambda \|u(t)\|_{H_1}\} \leq C t^{-\theta_0} J(t; u(t)), \text{ with } C = \max\{C_0', C_1'\}. \tag{26.9}$$

One has $-\lambda\theta - \theta_0 = -(\theta_1 - \theta_0)\theta - \theta_0 = -\eta$, so that

$$(t^\lambda)^{-\theta} J_H(t^\lambda; u(t)) \leq C t^{-\eta} J(t; u(t)), \tag{26.10}$$

and then

$$v(t^\lambda) = u(t) \text{ implies } a = \int_0^\infty u(t) \frac{dt}{t} = \lambda \int_0^\infty v(t) \frac{dt}{t}$$
$$\text{and } t^{-\theta} J_H(t; v(t)) \in L^p(R_+; \frac{dt}{t}), \tag{26.11}$$

showing that $a \in (H_0, H_1)_{\theta,p}$. $\qquad\square$

Then Lemma 26.1 and Lemma 26.2 imply the reiteration theorem of Jacques-Louis LIONS and Jaak PEETRE.

Theorem 26.3. *(reiteration theorem) If $0 \leq \theta_1 \neq \theta_0 \leq 1$, $(E_0, E_1)_{\theta_0,1} \subset F_0 \subset (E_0, E_1)_{\theta_0,\infty}$ and $(E_0, E_1)_{\theta_1,1} \subset F_1 \subset (E_0, E_1)_{\theta_1,\infty}$, then for $0 < \theta < 1$ and $1 \leq p \leq \infty$ one has $(F_0, F_1)_{\theta,p} = (E_0, E_1)_{\eta,p}$ with $\eta = (1 - \theta)\theta_0 + \theta\,\theta_1$, with equivalent norms.*

Proof: Both Lemma 26.1 and Lemma 26.2 apply, showing that $(F_0, F_1)_{\theta,p}$ is both included in and contains $(E_0, E_1)_{\eta,p}$. $\qquad\square$

As an application, let us consider the limiting case of Sobolev's embedding theorem in R^2, where $H^1(R^2)$ is not embedded in $L^\infty(R^2)$ but nevertheless for $0 < s < 1$ the space $H^s(R^2) = (H^1(R^2), L^2(R^2))_{1-s,2}$ is actually embedded into $(L^\infty(R^2), L^2(R^2))_{1-s,2}$, which is $L^{a(s),2}(R^2)$ with $\frac{1}{a(s)} = \frac{1-s}{2}$ by the reiteration theorem 26.3. The result follows from the fact that

$$X = (H^2(R^2), L^2(R^2))_{1/2,1} \subset \mathcal{F}L^1(R^2) \subset C_0(R^2) \subset L^\infty(R^2). \tag{26.12}$$

As both X and $H^1(R^2)$ are of class $\mathcal{H}(1/2)$ for $E_0 = H^2(R^2)$ and $E_1 = L^2(R^2)$, the reiteration theorem 26.3 implies

$$
\begin{aligned}
H^s(R^2) = \big(H^1(R^2), L^2(R^2)\big)_{1-s,2} = \big(X, L^2(R^2)\big)_{1-s,2} \subset \\
\big(L^\infty(R^2), L^2(R^2)\big)_{1-s,2} = L^{a(s),2}(R^2).
\end{aligned}
\tag{26.13}
$$

In proving (26.12), one notices that

$$
\text{for } s > \frac{N}{2} \text{ one has } H^s(R^N) \subset \mathcal{F}L^1(R^N) \subset C_0(R^N),
\tag{26.14}
$$

because $u \in H^s(R^N)$ implies $(1 + |\xi|)^s \mathcal{F}u \in L^2(R^N)$ and as $(1 + |\xi|)^{-s} \in L^2(R^N)$, one deduces that $\mathcal{F}u \in L^1(R^N)$. Because $H^2(R^2) \subset L^\infty(R^2)$, one deduces that $||\mathcal{F}u||_{L^1(R^2)} \leq C\,||D^2u||_{L^2(R^2)} + C\,||u||_{L^2(R^2)}$ for all $u \in H^2(R^2)$, and by rescaling, i.e., applying the inequality to u_λ defined by $u_\lambda(x) = v(\lambda\,x)$ for all $x \in R^N$, one deduces that

$$
||\mathcal{F}u||_{L^1(R^2)} \leq C\,|\lambda|\,||D^2u||_{L^2(R^2)} + \frac{C}{|\lambda|}||u||_{L^2(R^2)} \text{ for all } \lambda > 0,
\tag{26.15}
$$

because $||\mathcal{F}u_\lambda||_{L^1(R^2)} = ||\mathcal{F}u||_{L^1(R^2)}$ for all $\lambda > 0$ (so that $||\mathcal{F}u||_{L^1(R^2)}$ scales as $||u||_{L^\infty(R^2)}$); taking the best λ gives $||u||_{\mathcal{F}L^1(R^2)} \leq C'\,||D^2u||_{L^2(R^2)}^{1/2}||u||_{L^2(R^2)}^{1/2}$; this implies

$$
\begin{aligned}
||u||_{\mathcal{F}L^1(R^2)} \leq C'\,||u||_{H^2(R^2)}^{1/2}||u||_{L^2(R^2)}^{1/2}, \text{ which is equivalent to} \\
\big(H^2(R^2), L^2(R^2)\big)_{1/2,1} \subset \mathcal{F}L^1(R^2)
\end{aligned}
\tag{26.16}
$$

by Lemma 25.2; of course, the same scaling argument works directly with $L^\infty(R^2)$ in place of $\mathcal{F}L^1(R^2)$.

[Taught on Monday March 20, 2000.]

Maximal Functions

Before a general theory of interpolation had been developed, for which the interpolation property is proven for linear continuous mappings, some non-linear interpolation method had already been used, for example for proving that the maximal function maps $L^p(R^N)$ into itself for $1 < p \leq \infty$. Probably because this classical proof is well known to specialists of harmonic analysis, who are experts in the theory of singular integrals, they rarely mention the theory of interpolation when they use this type of argument.

Definition 27.1. *For $f \in L^1_{loc}(R^N)$, the* maximal function $M f$ *is defined by*

$$M f(x) = \sup_{r>0} \frac{\int_{B(x,r)} |f(y)|\, dy}{|B(x,r)|}, \tag{27.1}$$

where $B(x,r)$ is the ball centered at x and with radius $r > 0$, and $|B(x,r)|$ is its volume. □

The concept was introduced by HARDY and LITTLEWOOD, who proved the following result in dimension 1, the general case being due to WIENER.

Lemma 27.2. *If $1 < p \leq \infty$, then $f \in L^p(R^N)$ implies $M f \in L^p(R^N)$, with*

$$||M f||_p \leq C(p)||f||_p \text{ for all } f \in L^p(R^N), \text{ and } C(p) \to \infty \text{ as } p \to 1. \square$$
$$\tag{27.2}$$

This will be proven below, but the fact that the result is not true for $p = 1$ and that $C(p)$ must tend to ∞ as p tends to 1 is seen easily by considering for f the characteristic function of the unit ball, for which one has $M f(x) \geq \frac{1}{(1+r)^N}$ (with $r = |x|$, as usual), because for $s = 1 + r$ one has $\int_{B(x,s)} |f(y)|\, dy = |B(0,1)|$, as $B(x,s)$ contains $B(0,1)$, and $|B(x,s)| = s^N |B(0,1)|$. Therefore $M f$ does not belong to $L^1(R^N)$, and as $||M f||_p \to \infty$ as $p \to 1$, one must have $C(p) \to \infty$. The same argument shows that

$$\text{if } f \in L^1(R^N) \text{ and } f \neq 0, \text{ then } M f \notin L^1(R^N), \tag{27.3}$$

as it is bounded below by $\frac{C}{r^N}$ for r large, and in this case one has a result involving the weak L^1 space, which has a definition analogous to that of the Marcinkiewicz spaces for $p > 1$, which coincides with the Lorentz spaces $L^{p,\infty}$ for $p > 1$; however, the weak L^1 space is not included in the family of interpolation spaces between $L^1(R^N)$ and $L^\infty(R^N)$, as it is not a subset of $L^1(R^N) + L^\infty(R^N)$ (but it may have been included in the original definition of Lorentz spaces). The proof of the result for $p = 1$ uses the classical covering lemma 27.3, which was probably known to either VITALI[1] or BESICOVITCH,[2] who have proven more refined covering results.

Lemma 27.3. *Let A be a measurable subset of R^N, covered by a family of (closed) balls $B_i = B(C_i, r_i), i \in I$, whose radii satisfy $0 < r_i \leq R_0 < \infty$ for all $i \in I$. Then for each $\varepsilon > 0$ there exists a subfamily $J \subset I$ such that the balls B_j are disjoint for $j \in J$, and $|A| \leq (3 + \varepsilon)^N \sum_{j \in J} |B_j|$.*

Proof: For $0 < \alpha < 1$ one chooses a ball B_{j_1} with radius $r_{j_1} \geq (1-\alpha) \sup_{i \in I} r_i$, and one discards all the balls which intersect B_{j_1}, and one repeats the process as long as there are any balls left. In that way, one has selected a finite or infinite subfamily J such that the balls B_j are disjoint for $j \in J$ by construction. If $\sum_{j \in J} |B_j| = +\infty$ the result is proven. If $\sum_{j \in J} |B_j| < \infty$ and if the family is infinite one has $|B_{j_n}| \to 0$, so that $r_{j_n} \to 0$ as $n \to \infty$, and therefore all the balls have been discarded at some time; indeed, if one has $(1 - \alpha)r_i > r_{j_n}$ then the ball B_i must have been discarded before step n, or the ball B_{j_n} could not have been selected at step n; if the family is finite then all the balls have been either selected or discarded after a finite number of steps. Any ball B_i has been discarded because it intersects a selected ball B_{j_m}, so that one has $r_{j_m} \geq (1 - \alpha)r_i$, which implies that $B_i \subset B(C_{j_m}, k\,r_{j_m})$ with $k > 1 + \frac{2}{1-\alpha}$; therefore, $A \subset \bigcup_{j \in J} B(C_j, k\,r_j)$, so that $|A| \leq k^N \sum_{j \in J} |B_j|$ and taking α small one can choose $k \leq 3 + \varepsilon$. $\qquad \square$

Lemma 27.4. *For $f \in L^1(R^N)$ one has*

$$meas\{x \in R^N \mid |M\,f(x)| \geq t\} \leq \frac{3^N \|f\|_1}{t} \text{ for every } t > 0. \qquad (27.4)$$

Proof: Let

$$\Omega_s = \{x \in R^N \mid |M\,f(x)| > s\}, \qquad (27.5)$$

---

[1] Giuseppe VITALI, Italian mathematician, 1875–1932. After teaching in high school in Genova (Genoa), he worked in Modena, in Padova (Padua), and in Bologna, Italy. The department of pure and applied mathematics of Università degli Studi di Modena e Reggio Emilia is named after him.

[2] Abram Samoilovitch BESICOVITCH, Russian-born mathematician, 1891–1970. He worked in Petrograd, Russia, in Liverpool and in Cambridge, England, where he held the Rouse Ball professorship (1950–1958).

so that for every $x \in \Omega_s$ there exists $r(x) > 0$ with

$$\frac{\int_{B(x,r(x))} |f(y)| \, dy}{|B(x,r(x))|} > s. \tag{27.6}$$

One uses the lemma for the covering of Ω_s by all the balls $B(x, r(x))$ with $x \in \Omega_s$, and the radii $r(x)$ are bounded because

$$s|B(0,1)|r(x)^N = s|B(x,r(x))| < \int_{B(x,r(x))} |f(y)| \, dy \le ||f||_1. \tag{27.7}$$

One finds a disjoint family of balls with centers $x \in X \subset \Omega_s$ such that

$$|\Omega_s| \le (3+\varepsilon)^N \sum_{x \in X} |B(x,r(x))| \le (3+\varepsilon)^N \sum_{x \in X} \frac{\int_{B(x,r(x))} |f(y)| \, dy}{s} \le$$
$$\frac{(3+\varepsilon)^N}{s} ||f||_1, \tag{27.8}$$

and by letting ε tend to 0 gives $|\Omega_s| \le \frac{3^N}{s} ||f||_1$ and choosing $s = t - \eta$ for $\eta > 0$ and letting η tend to 0 gives the desired bound. $\qquad\square$

Definition 27.5. $||f||_{*1}$ *denotes*

the smallest constant $C \ge 0$ such that $\text{meas}\{x \mid |f(x)| \ge t\} \le \dfrac{C}{t}$ *for all $t > 0$,* $\tag{27.9}$

which is not a norm. $\qquad\square$

Because for $f \in L^1(\Omega)$ one has $t \, \text{meas}\{x \mid |f(x)| \ge t\} \le ||f||_1$, one deduces that $||f||_1 \le ||f||_{*1}$ for all $f \in L^1(\Omega)$. Despite the notation, $||\cdot||_{*1}$ is not a norm; one sees easily that

$$||\lambda f||_{*1} = |\lambda| \, ||f||_{*1} \text{ for all scalars } \lambda, \tag{27.10}$$

but the triangle inequality does not always hold. For example, if for $\Omega = (0,1)$ one takes $f(t) = \frac{1}{t}$ and $g(t) = \frac{1}{1-t}$, and $h = f + g$, so that $||f||_{*1} = ||g||_{*1} = 1$, one has $h(t) = \frac{1}{t(1-t)}$ and by symmetry its nonincreasing rearrangement is $h^*(t) = h(\frac{t}{2})$ for $0 < t < 1$, and the supremum of $t \, h(\frac{t}{2}) = \frac{4}{2-t}$ is $||h||_{*1} = 4$.

Actually, one always has

$$\sqrt{||f_1 + f_2||_{*1}} \le \sqrt{||f_1||_{*1}} + \sqrt{||f_2||_{*1}} \text{ for all } f_1, f_2, \tag{27.11}$$

because for $0 < s < t < \infty$ one has

$$\{x \mid |f_1(x) + f_2(x)| \ge t\} \subset \{x \mid |f_1(x)| \ge s\} \cup \{x \mid |f_2(x)| \ge t - s\}, \tag{27.12}$$

so that

$$\text{meas}(\{x \mid |f_1(x) + f_2(x)| \ge t\}) \le \frac{||f_1||_{*1}}{s} + \frac{||f_2||_{*1}}{t-s}; \tag{27.13}$$

the minimum in s of the right side is attained with $s = \gamma\sqrt{||f_1||_{*1}}$ and $t - s = \gamma\sqrt{||f_2||_{*1}}$, so that $t = \gamma(\sqrt{||f_1||_{*1}} + \sqrt{||f_2||_{*1}})$, i.e., $\gamma = \dfrac{t}{\sqrt{||f_1||_{*1}} + \sqrt{||f_2||_{*1}}}$, and putting in (27.13) one obtains

$$meas(\{x \mid |f_1(x) + f_2(x)| \geq t\}) \leq \frac{\sqrt{||f_1||_{*1}} + \sqrt{||f_2||_{*1}}}{\gamma} = \frac{(\sqrt{||f_1||_{*1}} + \sqrt{||f_2||_{*1}})^2}{t}. \tag{27.14}$$

Definition 27.6. *If f is a measurable function on Ω for which there exists C such that $meas\{x \in \Omega \mid |f(x)| \geq s\} \leq \frac{C}{s}$ for $s > s_0$ (and $s_0 > 0$), then one defines*

$$K_*(t; f) = \inf_{f = g + h}(||g||_{*1} + t\,||h||_\infty) \text{ for } t > 0. \qquad\Box \tag{27.15}$$

Because $||g||_{*1} \leq ||g||_1$ for every $g \in L^1(\Omega)$, one deduces that

$$K_*(t; f) \leq K(t; f) \text{ for } f \in L^1(\Omega) + L^\infty(\Omega). \tag{27.16}$$

Lemma 27.7. *If there exists C such that $meas\{x \in \Omega \mid |f(x)| \geq s\} \leq \frac{C}{s}$ for $s > s_0$, then one has $t\,f^*(t) \leq K_*(t; a)$ for all $t > 0$.*

Proof: Because $|f_1| \leq |f_2|$ a.e. in Ω implies $||f_1||_{*1} \leq ||f_2||_{*1}$, one deduces that among all the decompositions $f = g + h$ with $||h||_\infty \leq \lambda$, the one for which $||g||_{*1}$ is lowest corresponds to $|g| = (|f| - \lambda)_+$ (and $|h| = \min\{|f|, \lambda\}$). For $\varepsilon > 0$ there exists $\lambda \geq 0$ such that $||(|f| - \lambda)_+||_{*1} + t\lambda \leq (1 + \varepsilon)K_*(t; f)$; if $g = (|f| - \lambda)_+$, then as the nonincreasing rearrangement of g is $(f^* - \lambda)_+$, one has $t\,(f^*(t) - \lambda)_+ \leq ||g||_{*1} \leq (1 + \varepsilon)K_*(t; f) - t\lambda$. If $\lambda \leq f^*(t)$ it means $t\,(f^*(t) - \lambda) \leq ||g||_{*1} \leq (1 + \varepsilon)K_*(t; f) - t\lambda$, while if $\lambda > f^*(t)$ it means $0 \leq ||g||_{*1} \leq (1+\varepsilon)K_*(t; f) - t\lambda$, and in both cases one has $t\,f^*(t) \leq (1+\varepsilon)K_*(t; f)$, and letting ε tend to 0 gives the desired bound. $\qquad\Box$

For $f \in L^1(\Omega) + L^\infty(\Omega)$ one then has $t\,f^*(t) \leq K_*(t; f) \leq K(t; f)$ for all $t > 0$.

Lemma 27.8. *Let $0 < \theta < 1$ and $1 \leq q \leq \infty$. If there exists C such that $meas\{x \in \Omega \mid |f(x)| \geq s\} \leq \frac{C}{s}$ for $s > s_0$ and $t^{-\theta}K_*(t; f) \in L^q(R_+; \frac{dt}{t})$ then $f \in L^1(\Omega) + L^\infty(\Omega)$ and $t^{-\theta}K(t; f) \in L^q(R_+; \frac{dt}{t})$, i.e., $f \in L^{p,q}(\Omega)$ for $p = \frac{1}{1-\theta}$, and $||t^{-\theta}K(t; f)||_{L^q(0,\infty;dt/t)} \leq \frac{1}{\theta}||t^{-\theta}K_*(t; f)||_{L^q(0,\infty;dt/t)}$.*

Proof: As $K_*(t; f)$ is nondecreasing in t, $t^{-\theta}K_*(t; f) \in L^q(R_+; \frac{dt}{t})$ implies $t^{-\theta}K_*(t; f) \in L^\infty(R_+; \frac{dt}{t})$, i.e., $K_*(t; f) \leq C\,t^\theta$ for $t > 0$, so that $f^*(t) \leq C\,t^{\theta-1}$ for $t > 0$, and therefore $f \in L^1(\Omega) + L^\infty(\Omega)$. One has $t^{1-\theta}f^*(t) \in L^q(R_+; \frac{dt}{t})$, and by Hardy's inequality (13.3) one deduces $t^{-\theta}K(t; f) \in L^q(R_+; \frac{dt}{t})$ with the precise estimate shown. $\qquad\Box$

One can now finish the proof of the Hardy–Littlewood/Wiener theorem that the maximal function maps $L^p(R^N)$ into itself for $1 < p \leq \infty$, and obtain the same result for Lorentz spaces.

Lemma 27.9. *For* $1 < p < \infty$ *and* $1 \leq q \leq \infty$, $f \in L^{p,q}(R^N)$ *implies* $M f \in L^{p,q}(R^N)$ *and* $||M f||_{L^{p,q}(R^N)} \leq \frac{3^{N/p}}{p-1}||f||_{L^{p,q}(R^N)}$.

Proof: For $g, h \in L^1(R^N) + L^\infty(R^N)$ one has

$$\int_{B(x,r)} |g(y) + h(y)|\, dy \leq \int_{B(x,r)} |g(y)|\, dy + \int_{B(x,r)} |h(y)|\, dy \leq$$
$$|B(x,r)|(M g(x) + M h(x)) \text{ a.e. } x \in R^N \text{ for all } r > 0, \tag{27.17}$$

so that the maximal function is *subadditive*, i.e.,

$$M(g + h) \leq M g + M h \text{ a.e. in } R^N. \tag{27.18}$$

For each decomposition $f = g + h$ with $g \in L^1(R^N)$ and $h \in L^\infty(R^N)$, one then has $M f \leq M g + M h$, i.e., $M f = g_0 + h_0$ with $0 \leq g_0 \leq M g$ and $0 \leq h_0 \leq M h$, so that

$$t\,(M f)^*(t) \leq K_*(t; M f) \leq ||g_0||_{*1} + t\,||h_0||_\infty \leq ||M g||_{*1} + t\,||M h||_\infty \leq$$
$$3^N||g||_1 + t\,||h||_\infty, \tag{27.19}$$

so that

$$t\,(M f)^*(t) \leq 3^N \inf_{f=g+h}\left(||g||_1 + \frac{t}{3^N}||h||_\infty\right) = 3^N K\left(\frac{t}{3^N}; f\right). \tag{27.20}$$

If $f \in L^{p,q}(R^N)$ then $t^{-\theta}K(t; f) \in L^q(R_+; \frac{dt}{t})$ with $\theta = \frac{1}{p'}$, and one deduces that

$$||t^{1-\theta}(M f)^*||_{L^q(0,\infty;dt/t)} \leq 3^{N(1-\theta)}||t^{-\theta}K(t; f)||_{L^q(0,\infty;dt/t)}, \tag{27.21}$$

and then Hardy's inequality (Lemma 22.5) implies

$$||t^{-\theta}K(t; M f)||_{L^q(0,\infty;dt/t)} \leq \frac{3^{N(1-\theta)}}{\theta}||t^{-\theta}K(t; f)||_{L^q(0,\infty;dt/t)}. \quad \square \tag{27.22}$$

[Taught on Wednesday March 22, 2000.]

Bilinear and Nonlinear Interpolation

Another family of nonlinear interpolation results is based on the method of traces of Jacques-Louis LIONS and Jaak PEETRE. One considers the space of (weakly) differentiable functions from $(0, \infty)$ to $E_0 + E_1$ such that $t^{\alpha_0} u \in L^{p_0}(R_+; E_0)$ and $t^{\alpha_1} u' \in L^{p_1}(R_+; E_1)$ and for suitable values of $\alpha_0, p_0, \alpha_1, p_1$ (namely $\alpha_0 + \frac{1}{p_0} > 0$ and $\alpha_1 + \frac{1}{p_1} < 1$) these functions are automatically continuous on $[0, 1]$ with values in $E_0 + E_1$ and the space spanned by $u(0)$ is an interpolation space. Using the change of function $v(s) = u(s^\gamma)$ with $\gamma > 0$ amounts to replacing α_0, α_1 by β_0, β_1 defined by $\beta_0 = \gamma \alpha_0 + \frac{\gamma - 1}{p_0}$ and $\beta_1 = \gamma \alpha_1 - \frac{\gamma - 1}{p_1}$, or $\beta_0 + \frac{1}{p_0} = \gamma(\alpha_0 + \frac{1}{p_0})$ and $\beta_1 + \frac{1}{p_1} = \gamma(\alpha_1 + \frac{1}{p_1}) + 1 - \gamma$, so that the family of interpolation spaces depends upon at most three parameters. However, Jaak PEETRE proved that the corresponding space is equal to $(E_0, E_1)_{\theta, p}$, and one can choose γ such that $\beta_0 + \frac{1}{p_0} = \theta$ and $\beta_1 + \frac{1}{p_1} = \theta$, and p is defined by $\frac{1}{p} = \frac{1-\theta}{p_0} + \frac{\theta}{p_1}$. This will be shown later, but assuming that the characterization has been obtained, one can deduce a few properties.

The interpolation property for a linear operator $A \in \mathcal{L}(E_0; F_0) \cap \mathcal{L}(E_1; F_1)$ follows immediately, because $v(t) = A u(t)$ gives $t^{\alpha_0} v \in L^{p_0}(R_+; F_0)$ and $t^{\alpha_1} v' \in L^{p_1}(R_+; F_1)$. Actually, as was noticed by Jacques-Louis LIONS, one can deduce a nonlinear interpolation theorem.

Lemma 28.1. *If $E_0 \subset E_1$, $F_0 \subset F_1$, and A is a possibly nonlinear operator from E_1 into F_1 which satisfies*

$$\begin{aligned} &\|A(u) - A(v)\|_1 \le M_1 \|u - v\|_1 \text{ for all } u, v \in E_1 \\ &A \text{ maps } E_0 \text{ into } F_0 \text{ with } \|A(u)\|_0 \le M_0 \|u\|_0 \text{ for all } u \in E_0, \end{aligned} \tag{28.1}$$

then for $0 < \theta < 1$ and $1 \le p \le \infty$,

$$\begin{aligned} &A \text{ maps } (E_0, E_1)_{\theta, p} \text{ into } (F_0, F_1)_{\theta, p}, \text{ and} \\ &\|A(u)\|_{(F_0, F_1)_{\theta, p}} \le C \|u\|_{(E_0, E_1)_{\theta, p}} \text{ for all } u \in (E_0, E_1)_{\theta, p}. \end{aligned} \tag{28.2}$$

Proof: Defining $v(t) = A\big(u(t)\big)$, one has $||v(t + h) - v(t)||_{F_1} = ||A\big(u(t + h)\big) - A\big(u(t)\big)||_{F_1} \leq M_1||u(t + h) - u(t)||_{E_1}$, and dividing by $|h|$ and letting h tend to 0 one deduces that $||v'(t)||_{F_1} \leq M_1||u'(t)||_{E_1}$ for a.e. $t \in (0, \infty)$. Therefore, as for the linear case, one deduces that $t^{\alpha_0}v \in L^{p_0}(R_+; F_0)$ and $t^{\alpha_1}v' \in L^{p_1}(R_+; F_1)$. □

In 1970, Jacques-Louis LIONS had asked me to consider the case where A is only Hölder continuous, where his idea does not work, and I noticed that his result can be proven directly by the K-method in a way which can be extended to the case of Hölder continuous mappings as will be shown later, and this was also noticed by Jaak PEETRE. One just notices that for every decomposition $a = a_0 + a_1$ one has $A(a) = b_0 + b_1$ with $b_0 = A(a_0)$ and $b_1 = A(a) - A(a_0)$ and $||b_0||_0 \leq M_0||a_0||_0$ and $||b_1||_1 = ||A(a) - A(a_0)||_1 \leq M_1||a - a_0||_1 = M_1||a_1||_1$, so that $K\big(t; A(a)\big) \leq M_0 K\big(\frac{tM_1}{M_0}; a\big)$.

There were other interpolation theorems, for example by Jaak PEETRE or by Felix BROWDER[1] but under the assumption that the mapping is Lipschitz continuous from E_0 to F_0 and from E_1 to F_1. An application considered by Jacques Louis LIONS was to interpolate the regularity of the solution of some variational inequalities, as he had done for linear (elliptic or parabolic) equations with Enrico MAGENES, but in his example the mapping considered is not Lipschitz continuous from E_0 to F_0, and I suppose that it was the reason for his particular hypothesis.

The same idea of Jacques-Louis LIONS applies to a bilinear setting (and I have generalized it to a nonlinear setting).

Lemma 28.2. *Let B be bilinear from $(E_0 + E_1) \times (F_0 + F_1)$ into $G_0 + G_1$, satisfying*

B *maps* $E_0 \times F_0$ *into* G_0, *and* $||B(e_0, f_0)||_{G_0} \leq M_0||e_0||_{E_0}||f_0||_{F_0}$
    for all $e_0 \in E_0, f_0 \in F_0$,
B *maps* $E_0 \times F_1$ *into* G_1, *and* $||B(e_0, f_1)||_{G_1} \leq M_1||e_0||_{E_0}||f_1||_{F_1}$
    for all $e_0 \in E_0, f_1 \in F_1$,   (28.3)
B *maps* $E_1 \times F_0$ *into* G_1, *and* $||B(e_1, f_0)||_{G_1} \leq M_1||e_1||_{E_1}||f_0||_{F_0}$
    for all $e_1 \in E_1, f_0 \in F_0$.

Then

for $0 < \theta, \eta < 1$ with $\theta + \eta < 1$ and $1 \leq p, q, r \leq \infty$ with $\frac{1}{r} = \frac{1}{p} + \frac{1}{q}$,
B *maps* $(E_0, E_1)_{\theta,p} \times (F_0, F_1)_{\eta,q}$ *into* $(G_0, G_1)_{\theta+\eta,r}$ *and*
$||B(e, f)||_{(G_0,G_1)_{\theta+\eta,r}} \leq C\,||e||_{(E_0,E_1)_{\theta,p}}||f||_{(F_0,F_1)_{\eta,q}}$ (28.4)
    for all $e \in (E_0, E_1)_{\theta,p}, f \in (F_0, F_1)_{\eta,q}$.

---

[1] Felix Earl BROWDER, American mathematician, born in 1928. He worked at Yale University, at The University of Chicago, Chicago, IL, and at Rutgers University, Piscataway, NJ.

Proof: Any $e \in (E_0, E_1)_{\theta,p}$ can be written as $e = u(0)$ with $t^\theta \|u(t)\|_{E_0} \in L^p(R_+; \frac{dt}{t})$ and $t^\theta \|u'(t)\|_{E_1} \in L^p(R_+; \frac{dt}{t})$, and any $f \in (F_0, F_1)_{\eta,q}$ can be written as $f = v(0)$ with $t^\eta \|v(t)\|_{F_0} \in L^q(R_+; \frac{dt}{t})$ and $t^\eta \|v'(t)\|_{F_1} \in L^q(R_+; \frac{dt}{t})$. One defines $w(t) = B(u(t), v(t))$, and $B(e, f) = w(0)$; because $t^{\theta+\eta} \|w(t)\|_{G_0} \in L^r(R_+; \frac{dt}{t})$ and $w'(t) = B(u(t), v'(t)) + B(u'(t), v(t))$, one has $t^{\theta+\eta} \|w'(t)\|_{G_1} \in L^r(R_+; \frac{dt}{t})$, so that $B(e, f) \in (G_0, G_1)_{\theta+\eta,r}$ with corresponding bounds. □

The same result was essentially obtained by O'NEIL, who derived precise bounds for the convolution product analogous to those for the product.[2]

The product corresponds to the choice $E_0 = F_0 = G_0 = L^\infty(\Omega)$ and $E_1 = F_1 = G_1 = L^1(\Omega)$ (with $M_0 = M_1 = 1$), and in this case the result states that if $1 < p, q, r < \infty$ with $\frac{1}{r} = \frac{1}{p} + \frac{1}{q}$ and $1 \le a, b, c \le \infty$ and $\frac{1}{c} = \frac{1}{a} + \frac{1}{b}$, then the product is continuous from $L^{p,a}(\Omega) \times L^{q,b}(\Omega)$ into $L^{r,c}(\Omega)$ (the limiting cases will be discussed when studying the duals of interpolation spaces); as a particular case, the product is continuous from $L^p(\Omega) \times L^q(\Omega)$ into $L^r(\Omega)$, a simple consequence of Hölder's inequality.

The convolution product corresponds to the choice $E_0 = F_0 = G_0 = L^1(R^N)$ and $E_1 = F_1 = G_1 = L^\infty(R^N)$ (with $M_0 = M_1 = 1$), and in this case the result states that if $1 < p, q, s < \infty$ with $\frac{1}{s} = \frac{1}{p} + \frac{1}{q} - 1$ and $1 \le a, b, c \le \infty$ and $\frac{1}{c} = \frac{1}{a} + \frac{1}{b}$, then the convolution product is continuous from $L^{p,a}(R^N) \times L^{q,b}(R^N)$ into $L^{r,c}(R^N)$. As a particular case, the convolution product is continuous from $L^p(R^N) \times L^q(R^N)$ into $L^{s,1}(R^N)$, an improvement from Young's inequality (2.3); one cannot take $a = p$ and $b = q$, which would give $\frac{1}{2} < c < 1$, but one may choose $a \ge p$ and $b \ge q$ such that $c = 1$ (Jaak PEETRE has actually defined interpolation spaces with $0 < \theta < 1$ and $0 < p \le \infty$, but for $0 < p < 1$ they are only *quasi-normed spaces*).

There is another bilinear interpolation result, due to Jacques-Louis LIONS and Jaak PEETRE, with quite different assumptions.

Lemma 28.3. *Let B be bilinear from $(E_0 + E_1) \times (F_0 + F_1)$ into $G_0 + G_1$, satisfying*

> *B maps $E_0 \times F_0$ into G_0, and $\|B(e_0, f_0)\|_{G_0} \le M_0 \|e_0\|_{E_0} \|f_0\|_{F_0}$*
> *for all $e_0 \in E_0, f_0 \in F_0$,*
> *B maps $E_1 \times F_1$ into G_1, and $\|B(e_1, f_1)\|_{G_1} \le M_1 \|e_1\|_{E_1} \|f_1\|_{F_1}$*   (28.5)
> *for all $e_1 \in E_1, f_1 \in F_1$.*

Then

> *for $0 < \theta < 1$ and $1 \le p, q, r \le \infty$ with $\frac{1}{r} = \frac{1}{p} + \frac{1}{q} - 1$,*
> *B maps $(E_0, E_1)_{\theta,p} \times (F_0, F_1)_{\theta,q}$ into $(G_0, G_1)_{\theta,r}$, and*
> *$\|B(e, f)\|_{(G_0,G_1)_{\theta,r}} \le C \|e\|_{(E_0,E_1)_{\theta,p}} \|f\|_{(F_0,F_1)_{\theta,q}}$*   (28.6)
> *for all $e \in (E_0, E_1)_{\theta,p}, f \in (F_0, F_1)_{\theta,q}$.* □

---

[2] HARDY and LITTLEWOOD have shown that $\int_0^t (f g)^* ds \le \int_0^t f^* g^* ds$ for all $t > 0$.

It should be noticed that there are situations where both theorems can be used but give different results, in the second parameter (the first one is usually the same, compatible with scaling properties). For example, applying this last bilinear theorem to the product with $E_0 = F_0 = G_0 = L^\infty(\Omega)$, and $E_1 = F_1 = L^2(\Omega)$ and $G_1 = L^1(\Omega)$, one only obtains that the product maps $L^{p,a}(\Omega) \times L^{p,b}(\Omega)$ into $L^{p/2,c}(\Omega)$ with $2 < p < \infty$ and $1 \le a, b, c \le \infty$ and $\frac{1}{c} = \frac{1}{a} + \frac{1}{b} - 1$, while the first bilinear theorem gives the result in $L^{p/2,1}(\Omega)$ (but it has used more general information, that the product of a function in $L^1(\Omega)$ and a function in $L^\infty(\Omega)$ is defined).

A similar situation arises for the so-called Riesz–Thorin theorem, which states that if a linear mapping is continuous from $L^{p_0}(\Omega)$ into $L^{q_0}(\Omega')$ and from $L^{p_1}(\Omega)$ into $L^{q_1}(\Omega')$, then for $0 < \theta < 1$ it is continuous from $L^{p_\theta}(\Omega)$ into $L^{q_\theta}(\Omega')$, where $\frac{1}{p_\theta} = \frac{1-\theta}{p_0} + \frac{\theta}{p_1}$ and $\frac{1}{q_\theta} = \frac{1-\theta}{q_0} + \frac{\theta}{q_1}$. M. RIESZ had only proven this result under the additional assumption that $p_\theta \le q_\theta$, and this condition was removed by THORIN. The K-method follows the approach of M. RIESZ and implies that the mapping is continuous from $L^{p_\theta,p}(\Omega)$ into $L^{q_\theta,p}(\Omega')$ for any $p \in [1,\infty]$, but if one chooses $p = p_\theta$, the space $L^{q_\theta,p_\theta}(\Omega')$ is only included in $L^{q_\theta}(\Omega')$ if $p_\theta \le q_\theta$. The complex method is the generalization of THORIN's idea, and an example of STEIN; although it is more precise in this example, for other questions, it has a defect of having only one parameter.

[Taught on Friday March 24, 2000.]

Obtaining L^p by Interpolation, with the Exact Norm

Talking about variants of interpolation methods, it is useful to obtain $L^p(\Omega)$ as an interpolation space between $L^1(\Omega)$ and $L^\infty(\Omega)$, but with the exact L^p norm. I describe below the way I solved this question in 1970, in order to answer a question of Haïm BREZIS.

Definition 29.1. *For $a \in E_0 + E_1$, one defines*

$$L^*(s; a) = \inf\{||a_0||_0 \mid a = a_0 + a_1 \text{ with } a_0 \in E_0, a_1 \in E_1 \text{ and } ||a_1||_1 \leq s\}. \quad \Box$$
(29.1)

In relation to the associated Gagliardo (convex) set $G(a)$ introduced in (22.4), i.e., the set of $(x_0, x_1) \in [0, \infty) \times [0, \infty)$ such that there exists a decomposition $a = a_0 + a_1$ with $a_0 \in E_0, a_1 \in E_1$ and $||a_0||_0 \leq x_0, ||a_1||_1 \leq x_1$, the boundary of this set has the equation $x_0 = L^*(x_1; a)$.

Lemma 29.2. *For $E_0 = L^1(\Omega)$ and $E_1 = L^\infty(\Omega)$, and $1 < p < \infty$ one has*

$$\int_0^\infty p(p-1)s^{p-2}L^*(s; a)\, ds = \int_\Omega |a(x)|^p\, dx \text{ for all } a \in L^1(\Omega) + L^\infty(\Omega).$$
(29.2)

Proof: The optimal decomposition consists in taking $a_1(x) = a(x)$ if $|a(x)| \leq s$ and $a_1(x) = \frac{s\,a(x)}{|a(x)|}$ if $|a(x)| > s$, so that

$$L^*(s; a) = \int_{|a(x)| \geq s} (|a(x)| - s)\, dx. \quad (29.3)$$

Then, using Fubini's theorem, one has

$$\int_0^\infty p(p-1)s^{p-2}L^*(s; a)\, ds = \int_0^\infty p(p-1)s^{p-2}\left(\int_{|a(x)|\geq s}(|a(x)| - s)\, dx\right) ds =$$
$$\int_\Omega \left(\int_0^{|a(x)|} p(p-1)s^{p-2}(|a(x)| - s)\, ds\right) dx = \int_\Omega |a(x)|^p\, dx. \quad \Box$$
(29.4)

The extension to Φ convex of class C^2 on R with $\Phi(0) = \Phi'(0) = 0$ is straightforward and was suggested by Haïm BREZIS when I showed him my construction:

$$\int_0^\infty \Phi''(s)L^*(s;a)\,ds = \int_\Omega \Phi(|a(x)|)\,dx \text{ for all } a \in L^1(\Omega) + L^\infty(\Omega), \quad (29.5)$$

and one uses Taylor's expansion formula with remainder,

$$\Phi(h) = \Phi(0) + \Phi'(0)h + \int_0^h (h-s)\Phi''(s)\,ds. \tag{29.6}$$

If Φ is convex with $\Phi(0) = \Phi'(0) = 0$ then Φ'' is a nonnegative measure and one must use a *Stieltjes integral*.[1]

Therefore, the same approach can deal with *Orlicz spaces*.[2]

Lemma 29.3. *If A is a linear mapping from $L^1(\Omega) + L^\infty(\Omega)$ into $L^1(\Omega') + L^\infty(\Omega')$ which is continuous from $L^1(\Omega)$ into $L^1(\Omega')$ with norm M_1 and which is continuous from $L^\infty(\Omega)$ into $L^\infty(\Omega')$ with norm M_∞, then for $1 < p < \infty$ it is continuous from $L^p(\Omega)$ into $L^p(\Omega')$ with norm $\leq M_1^{1/p} M_\infty^{1/p'}$.*

Proof: For every decomposition $a = a_0 + a_1$ with $||a_1||_1 \leq s$ one has a decomposition $Aa = Aa_0 + Aa_1$ with $||Aa||_1 \leq M_\infty s$ and as $||Aa||_0 \leq M_1||a_0||_0$ one deduces that

$$L^*(M_\infty s; Aa) \leq M_1 L^*(s;a), \tag{29.7}$$

so that

$$\int_{\Omega'} |Aa(x)|^p\,dx = \int_0^\infty p(p-1)s^{p-2}L^*(s;Aa)\,ds =$$
$$M_\infty^{p-1}\int_0^\infty p(p-1)\sigma^{p-2}L^*(M_\infty\sigma;Aa)\,d\sigma \leq$$
$$M_1 M_\infty^{p-1}\int_0^\infty p(p-1)\sigma^{p-2}L^*(\sigma;a)\,d\sigma = M_1 M_\infty^{p-1}\int_\Omega |a(x)|^p\,dx. \ \square \tag{29.8}$$

In order to describe some technical improvements concerning embedding theorems of spaces of Sobolev type into Lorentz spaces, it is useful to derive equivalent ways to check that a function belongs to a Lorentz space $L^{p,q}(\Omega)$, with $1 < p < \infty$ and $1 \leq q \leq \infty$. The definition used is that $f \in L^{p,q}(\Omega)$ means that $t^{-1/p'}K(t;f) \in L^q(R_+; \frac{dt}{t})$, and as $K(t;f)$ can be expressed in terms of the nonincreasing rearrangement f^* of f (defined on $(0, meas(\Omega))$ and extended by 0 in order to have it defined on $(0,\infty)$) by $K(t;f) = \int_0^t f^*(s)\,ds$, $f \in L^{p,q}(\Omega)$ is equivalent to $t^{-1/p'}\left(\int_0^t f^*(s)\,ds\right) \in L^q(R_+; \frac{dt}{t})$, but because $t f^*(t) \leq \int_0^t f^*(s)\,ds$, $f \in L^{p,q}(\Omega)$ implies $t^{1/p}f^*(t) \in L^q(R_+; \frac{dt}{t})$, or

---

[1] Thomas Jan STIELTJES, Dutch-born mathematician, 1856–1894. He worked in Leiden, The Netherlands, and in Toulouse, France.

[2] Władysław Roman ORLICZ, Polish mathematician, 1903–1990. He worked in Lwów (then in Poland, now Lvov, Ukraine) and in Poznan, Poland.

$t^{(1/p)-(1/q)} f^*(t) \in L^q(0,\infty)$, which was the definition used by LORENTZ. It is indeed equivalent if $1 < p < \infty$ by Hardy's inequality (13.3), but this definition can also be used for $p = 1$.

Lemma 29.4. *Assume that*

$$for\ every\ \lambda > 0\ one\ has\ meas\{x \mid |f(x)| > \lambda\} < \infty, \qquad (29.9)$$

and for $n \in Z$,

$$\begin{aligned} one\ chooses\ a_n &\in [f^*(e^n+), f^*(e^n-)],\ so\ that \\ meas\{x \mid |f(x)| > a_n\} &\le e^n \le meas\{x \mid |f(x)| \ge a_n\}. \end{aligned} \qquad (29.10)$$

Then one has

$$f \in L^{p,q}(\Omega)\ if\ and\ only\ if\ e^{n/p} a_n \in l^q(Z). \qquad (29.11)$$

If moreover $a_n \to 0$ *as* $n \to +\infty$, *then*

$$f \in L^{p,q}(\Omega)\ if\ and\ only\ if\ e^{n/p}(a_n - a_{n+1}) \in l^q(Z). \qquad (29.12)$$

Proof: For $e^n < t < e^{n+1}$ one has $f^*(e^{n+1}-) \le f^*(t) \le f^*(e^n+)$, so that $a_{n+1} \le f^*(t) \le a_n$; this implies

$$\alpha \, ||e^{n/p} a_n||_{l^q(Z)} \le ||t^{1/p} f^*(t)||_{L^q(0,\infty;dt/t)} \le \beta \, ||e^{n/p} a_n||_{l^q(Z)}, \qquad (29.13)$$

for two positive constants α, β, which is (29.11). Of course (29.11) implies (29.12), and to prove the converse, one writes $b_n = a_n - a_{n+1}$, so that $a_n = b_n + b_{n+1} + \ldots$, and $e^n a_n = e^n b_n + e^{-1} e^{n+1} b_{n+1} + \ldots$, so that $e^n a_n$ is obtained from $e^n b_n$ by a convolution with c_n defined by $c_n = 0$ for $n > 0$ and $c_n = e^n$ for $n \le 0$, and then an application of Young's inequality (2.3) gives the result. $\square$

[Taught on Monday April 3, 2000.]

My Approach to Sobolev's Embedding Theorem

One can obtain a nonoptimal embedding theorem $W^{1,p}(R^N) \subset L^q(R^N)$ by decomposing $u \in W^{1,p}(R^N)$ as

$$u = (u - u \star \varrho_\varepsilon) + u \star \varrho_\varepsilon, \qquad (30.1)$$

where ϱ_ε is a special smoothing sequence, using the estimates

$$||\tau_h u - u||_p \leq |h| \, ||grad(u)||_p \text{ for all } h,$$
$$u - u \star \varrho_\varepsilon = \int (u - \tau_y u)\varrho_\varepsilon(y) \, dy, \text{ so that} \qquad (30.2)$$
$$||u - u \star \varrho_\varepsilon||_p \leq \int |y| \, |\varrho_\varepsilon(y)| \, dy \, ||grad(u)||_p = C\,\varepsilon \, ||grad(u)||_p,$$

and

$$||u \star \varrho_\varepsilon||_\infty \leq ||u||_p ||\varrho_\varepsilon||_{p'} \leq C \, ||u||_p \varepsilon^{-N/p'}. \qquad (30.3)$$

This means that using $E_0 = L^p(R^N)$ and $E_1 = L^\infty(R^N)$, one has

$$K(t;u) \leq C\,\varepsilon + C\,t\,\varepsilon^{-N/p'} \text{ for all } \varepsilon > 0, \qquad (30.4)$$

and taking the best ε gives

$$K(t;u) \leq C\,t^{p'/(N+p')}, \text{ i.e., } W^{1,p}(R^N) \subset (E_0, E_1)_{\theta,\infty} = L^{q_\theta,\infty}(R^N), \qquad (30.5)$$

with $\theta = \frac{p'}{N+p'} > 0$, so that $q_\theta > p$ and choosing any $q \in (p, q_\theta)$ one has shown that

one has $q > p$ and $||u||_q \leq A \, ||u||_p + B \, ||grad(u)||_p$ for all $u \in W^{1,p}(R^N)$. $\qquad (30.6)$

Notice that (30.5) and (30.6) are not as precise as Sobolev's embedding theorem, so the precise value of q obtained is not important (as long as $q > p$), but I have observed that from a nonoptimal embedding theorem like (30.6), one can derive the best-known embeddings after using two different scaling arguments.

The first step is the usual scaling argument; one should not add $||u||_p$ and $||grad(u)||_p$ which are not measured in the same unit, and one applies the inequality to u_λ defined by $u_\lambda(x) = u(\frac{x}{\lambda})$, and (30.6) becomes

$$|\lambda|^{N/q}||u||_q \le A\,|\lambda|^{N/p}||u||_p + B\,|\lambda|^{(N/p)-1}||grad(u)||_p$$
$$\text{for all } u \in W^{1,p}(R^N) \text{ and all } \lambda > 0. \tag{30.7}$$

Notice that in the case $1 \le p < N$ one must have $q \le p^* = \frac{Np}{N-p}$, or a contradiction would be obtained by letting λ tend to 0. One chooses the best λ in (30.7), and one finds that there exists $\eta \in (0, 1]$ such that

$$||u||_q \le C\,||u||_p^{1-\eta}||grad(u)||_p^{\eta} \text{ for all } u \in W^{1,p}(R^N). \tag{30.8}$$

This inequality is now scale invariant, and that requires that η satisfies

$$\frac{1}{q} = \frac{1-\eta}{p} + \eta\left(\frac{1}{p} - \frac{1}{N}\right) = \frac{1}{p} - \frac{\eta}{N}. \tag{30.9}$$

My second step is a different scaling argument, which consists in applying the inequality to functions $\psi(u)$. When I looked for a good choice of ψ, I was led to consider a sequence of functions $\varphi_n(|u|)$ adapted to u, which involves the levels a_n used in Lemma 29.4, namely

$$\varphi_n \text{ is defined by } \varphi_n(0) = 0, \varphi'(v) = 1 \text{ if } a_n < v < a_{n-1}, \text{ and}$$
$$\varphi'(v) = 0 \text{ if } v < a_n \text{ or } v > a_{n-1}, \tag{30.10}$$

and defining

$$\gamma_n = ||grad(\varphi_n(u))|| \text{ for } n \in Z, \tag{30.11}$$

one notices that

$$|grad(u)| \in L^p(R^N) \text{ is equivalent to } \gamma_n \in l^p(Z). \tag{30.12}$$

For any $r < \infty$ one has

$$\int_{R^N} |\varphi_n(u)|^r\,dx \ge |a_n - a_{n-1}|^r meas\{x \mid |u(x)| \ge a_{n-1}\} \ge |a_n - a_{n-1}|^r e^{n-1}$$
$$\int_{R^N} |\varphi_n(u)|^r\,dx \le |a_n - a_{n-1}|^r meas\{x \mid |u(x)| > a_n\} \le |a_n - a_{n-1}|^r e^n, \tag{30.13}$$

and having comparable lower and upper bounds is the main reason for my choice of the functions φ_n. Using (30.8) for $\varphi_n(u)$ instead of u, one deduces that

$$|a_n - a_{n-1}|e^{(n-1)/q} \le C\left(|a_n - a_{n-1}|e^{n/p}\right)^{1-\eta}\gamma_n^\eta, \text{ i.e.,}$$
$$|a_n - a_{n-1}|e^{n/p^*} \le (C\,e^{1/q})^{1/\eta}\gamma_n, \text{ for all } n \in Z, \tag{30.14}$$

where one has used (30.9), using $\frac{1}{p^*} = \frac{1}{p} - \frac{1}{N}$ even for $p \ge N$. The second part of (30.14) is $|a_n - a_{n-1}|e^{n/p^*} \in l^p(Z)$, and in the case $1 \le p < N$ corresponding to $p^* = \frac{Np}{N-p} < \infty$, Lemma 29.4 gives the improvement by Jaak PEETRE of the original result of Sergei SOBOLEV, namely $W^{1,p}(R^N) \subset L^{p^*,p}(R^N)$.

In the case $p = N$ one has $p^* = \infty$, so that

$$\text{for } p = N \text{ one has } |a_n - a_{n-1}| \in l^N(Z). \tag{30.15}$$

Let $b_n = a_n - a_{n+1} \geq 0$, then as n tends to $-\infty$ one has $a_n = b_n + b_{n+1} + \ldots + b_{m-1} + a_m$, and by Hölder's inequality one has $|a_n - a_m| \leq (|b_n|^N + \ldots + |b_{m-1}|^N)^{1/N}|n - m|^{1/N'}$, and one deduces that

for every $\varepsilon > 0$ there exists $C(\varepsilon)$ such that $|a_n|^{N'} \leq \varepsilon |n| + C(\varepsilon)$ for all $n \leq 0$. (30.16)

For $\kappa > 0$ one chooses $\varepsilon < \frac{1}{\kappa}$, and one computes the integral of $e^{\kappa |u|^{N'}}$ on the set where $|u(x)| \geq a_0$; on the set where $a_{n+1} \leq |u(x)| < a_n$, which has a measure $\leq e^{n+1}$, one has $|u(x)| \leq a_n$, so that $\kappa |u|^{N'} \leq \kappa |a_n|^{N'} \leq \kappa \varepsilon |n| + \kappa C(\varepsilon)$ for all $n \leq 0$, and therefore

$$\int_{|u(x)| \geq a_0} e^{\kappa |u|^{N'}} dx \leq \sum_{n=-\infty}^{0} e^n\, e^{\kappa \varepsilon |n| + \kappa C(\varepsilon)} < \infty. \tag{30.17}$$

Therefore $u \in W^{1,N}(R^N)$ implies $e^{\kappa |u|^{N'}} \in L^1_{loc}(R^N)$ for all $\kappa > 0$, which is the improvement by Neil TRUDINGER of a result of Fritz JOHN and Louis NIRENBERG, who had improved the result of Sergei SOBOLEV for the limiting case $p = N$ by introducing the space $BMO(R^N)$ (containing $W^{1,N}(R^N)$), and they proved that for every function in $BMO(R^N)$ one has $e^{\varepsilon |u|} \in L^1_{loc}(R^N)$ for ε small enough.

For $p > N$ one has $p^* < 0$, so that from $|a_n - a_{n-1}|e^{n/p^*} \in l^p(Z)$, one deduces that $\sum_{n=-\infty}^{0} |a_n - a_{n-1}| < \infty$, and therefore that $|a_n| \leq M$ for all $n \in Z$. Having proven that $W^{1,p}(R^N) \subset L^\infty(R^N)$, the scaling argument shows that one must have

$$||u||_\infty \leq C\, ||u||_p^{1-\theta}||grad(u)||_p^\theta, \text{ and } \frac{1}{p} - \frac{\theta}{N} = 0, \text{ i.e., } \theta = \frac{N}{p}, \tag{30.18}$$

which must have been what Sergei SOBOLEV had proven, and in order to find MORREY's improvement of the Hölder continuity, one applies (30.18) to $v = \tau_h u - u$, for which one has $||v||_p \leq |h|\,||grad(u)||_p$ and $||grad(v)||_p \leq 2||grad(u)||_p$, and one obtains the estimate

$$||\tau_h u - u||_\infty \leq C\, |h|^{1-(N/p)}||grad(u)||_p, \text{ i.e.,}$$
$$W^{1,p}(R^N) \subset C^{0,\alpha}(R^N) \text{ for } \alpha = 1 - \frac{N}{p}. \tag{30.19}$$

[Taught on Wednesday April 5, 2000.]

My Generalization of Sobolev's Embedding Theorem

The original method of proof of Sergei SOBOLEV consisted in writing

$$u = \sum_j \frac{\partial u}{\partial x_j} \star \frac{\partial E}{\partial x_j} \text{ for an elementary solution } E \text{ of } \Delta, \qquad (31.1)$$

and it is not adapted to the case where the derivatives are in different spaces.

A different proof, by Louis NIRENBERG, and also by Emilio GAGLIARDO, can be used for the case where

$$\frac{\partial u}{\partial x_j} \in L^{p_j}(R^N) \text{ for } j = 1, \ldots, N; \qquad (31.2)$$

In the late 1970s, I had heard a talk about this question by Alois KUFNER, then I was told that it had been noticed earlier by TROISI.

The case where the derivatives are in the same Lorentz space $L^{p,q}(R^N)$ with $1 < p < N$ can be treated with the theory of interpolation, as was done by Jaak PEETRE, but the limiting case where

$$\frac{\partial u}{\partial x_j} \in L^{N,p}(R^N) \text{ for } j = 1, \ldots, N, \qquad (31.3)$$

was treated by Haïm BREZIS and Stephen WAINGER by analyzing a formula of O'NEIL about the nonincreasing rearrangement of a convolution product. The case $p = 1$ in (31.3) gives $u \in C_0(R^N)$, by noticing that $C_c(R^N)$ is dense in $L^{N,1}(R^N)$, whose dual is $L^{N',\infty}(R^N)$, which contains the derivatives $\frac{\partial E}{\partial x_j}$. The case $p = \infty$ in (31.3) gives $e^{\varepsilon |u|} \in L^1_{loc}(R^N)$, and u actually belongs to $BMO(R^N)$.

As far as I know, these classical methods do not permit one to treat the case where the derivatives are in different Lorentz spaces; of course, *this question is quite academic*, but serves as a training ground for situations which often occur where one has different information in different directions, for example because some coordinates represent *space* and another one represents *time* (and one has simplified the physical reality so that the model used has a *fake velocity of light* equal to $+\infty$).

First, it is useful to observe that Sobolev's embedding theorem for $p = 1$ is related to the *isoperimetric inequality*. The classical isoperimetric inequality says that among measurable sets A of R^N with a given volume, the $(N - 1)$-dimensional measure of the boundary ∂A is minimum when A is a sphere; equivalently, for a given measure of the boundary, the volume is maximum for a sphere. Analytically it means that

$$meas(A) \leq C_0 \big(meas(\partial A)\big)^{N/(N-1)}, \qquad (31.4)$$

and it tells what the best constant C_0 is, while Sobolev's embedding theorem for $W^{1,1}(R^N)$ gives

$$\int_{R^N} |u|^{1^*} \, dx \leq C_1 ||grad(u)||_1^{N/(N-1)}, \qquad (31.5)$$

but does not insist on identifying what the best constant C_1 is. The relation between the two inequalities is that one can apply the last inequality to $u = \chi_A$, the characteristic function of A, assuming that A has a finite perimeter; of course, χ_A does not belong to $W^{1,1}(R^N)$, but as its partial derivatives $\frac{\partial \chi_A}{\partial x_j}$ are Radon measures, one may apply the inequality to $\chi_A \star \varrho_\varepsilon$ and then let ε tend to 0; in this way one learns that $C_0 \leq C_1$. Conversely, knowing the isoperimetric inequality, one can approach a function u by a sum of characteristic functions, using $A_n = \{x \mid n\varepsilon \leq u(x) < (n + 1)\varepsilon\}$ and deduce Sobolev's embedding theorem, so that $C_1 \leq C_0$ and the two inequalities are then essentially the same. However, the proof of the last part involves the technical study of functions of bounded variation (denoted by BV), which is classical in one dimension, but is indebted to the work of Ennio DE GIORGI[1], FEDERER[2] and Wendell FLEMING[3] for the development of the N-dimensional case.

As I observed, starting from Sobolev's embedding theorem $W^{1,1}(R^N) \subset L^{1^*}(R^N)$ (proven by Louis NIRENBERG), one can easily derive all the results already obtained, except for the question of identifying the best constants. For that, one uses the functions φ_n adapted to u, writes

$$||\varphi_n(u)||_{1^*} \leq C_0 ||\varphi_n'(u)grad(u)||_1 \leq ||\varphi_n'(u)grad(u)||_p e^{n/p'}, \qquad (31.6)$$

by Hölder's inequality, and deduces the same inequality as before, $|a_{n-1} - a_n|e^{n/p^*} \in l^p(Z)$.

---

[1] Ennio DE GIORGI, Italian mathematician, 1928–1996. He received the Wolf Prize in 1990. He worked at Scuola Normale Superiore, Pisa, Italy.

[2] Herbert FEDERER, Austrian-born mathematician, born in 1920. He worked at Brown University, Providence, RI.

[3] Wendell Helms FLEMING, American mathematician, born in 1928. He works at Brown University, Providence, RI.

However, for the case of derivatives in (different) Lorentz spaces, I could only prove it by using a multiplicative variant of the isoperimetric inequality/Sobolev's embedding theorem.

Lemma 31.1. *The Sobolev's embedding theorem $W^{1,1}(R^N) \subset L^{1^*}(R^N)$ in its additive version*

$$||u||_{1^*} \leq A \sum_{j=1}^{N} \left\|\frac{\partial u}{\partial x_j}\right\|_1 \quad \text{for all } u \in W^{1,1}(R^N), \tag{31.7}$$

is equivalent to the multiplicative version

$$||u||_{1^*} \leq N A \left(\prod_{j=1}^{N} \left\|\frac{\partial u}{\partial x_j}\right\|_1\right)^{1/N} \quad \text{for all } u \in W^{1,1}(R^N). \tag{31.8}$$

Proof: One rescales with a different scaling in different directions, i.e., one applies the additive version to v defined by

$$v(x_1,\ldots,x_N) = u\left(\frac{x_1}{\lambda_1},\ldots,\frac{x_N}{\lambda_N}\right), \tag{31.9}$$

and one obtains

$$(\lambda_1\ldots\lambda_N)^{1/1^*}||u||_{1^*} \leq A\,\lambda_1\ldots\lambda_N \sum_j \frac{1}{\lambda_j}\left\|\frac{\partial u}{\partial x_j}\right\|_1. \tag{31.10}$$

Then one notices that

if $\lambda_1\ldots\lambda_N = \mu > 0$, the minimum of $\sum_j \frac{\alpha_j}{\lambda_j}$ is attained when $\lambda_j = \beta\,\alpha_j$ for all j and the Lagrange multiplier β satisfies $\beta^N\alpha_1\ldots\alpha_N = \mu$, so the minimum is $\frac{N}{\mu}(\alpha_1\ldots\alpha_N)^{1/N}$. $\qquad$ (31.11)

One applies (31.11) to the case $\alpha_j = \left\|\frac{\partial u}{\partial x_j}\right\|_1$ and one finds the multiplicative version, as the powers of μ are identical on both sides of the inequality (because the inequality is already invariant when one rescales all the coordinates in the same way). The multiplicative version implies the additive version by the *geometric-arithmetic inequality*

$$(a_1\ldots a_N)^{1/N} \leq \frac{a_1 + \ldots + a_N}{N} \quad \text{for all } a_1,\ldots,a_N > 0, \tag{31.12}$$

which, putting $a_j = e^{z_j}$, is but the convexity property of the exponential function. $\qquad\qquad\square$

Lemma 31.2. *Let u satisfy*

$$\frac{\partial u}{\partial x_j} \in L^{p_j,q_j}(R^N), \text{ with } 1 < p_j < \infty \text{ and } 1 \leq q_j \leq \infty, \text{ for } j = 1,\ldots,N.$$
$$\tag{31.13}$$

Let p_{eff}, p_{eff}^* and q_{eff} be defined by

$$\frac{1}{p_{eff}} = \frac{1}{N}\sum_j \frac{1}{p_j}$$
$$\frac{1}{p_{eff}^*} = \frac{1}{p_{eff}} - \frac{1}{N} \tag{31.14}$$
$$\frac{1}{q_{eff}} = \frac{1}{N}\sum_j \frac{1}{q_j}.$$

Then one has

$$|a_{n-1} - a_n|e^{n/p_{eff}^*} \in l^{q_{eff}}(Z). \tag{31.15}$$

One may allow some p_j to be 1 or ∞, but using only $q_j = +\infty$ in that case.

Proof: Let $f_j = \frac{\partial u}{\partial x_j}$ for $j = 1, \dots, N$. One applies the multiplicative version to $\varphi_n(u)$, and one has to estimate $\|\varphi_n'(u)f_j\|_1$. A classical result of HARDY and LITTLEWOOD states that for all $f \in L^1(\Omega) + L^\infty(\Omega)$ and all measurable subsets $\omega \subset \Omega$ one has

$$\int_\omega |f(x)|\,dx \le \int_0^{meas(\omega)} f^*(s)\,ds, \tag{31.16}$$

and as the measure of the points where $\varphi_n'(u) \ne 0$ is at most e^n, one deduces that

$$\|\varphi_n'(u)f_j\|_1 \le K(e^n; f_j) \text{ for } j = 1, \dots, N, \tag{31.17}$$

and then, using Hölder's inequality,

$$e^{-n\theta_j}K(e^n; f_j) \in l^{q_j}(Z) \text{ with } \theta_j = \frac{1}{p_j'} \text{ for } j = 1, \dots, N, \text{ imply}$$
$$e^{-n/p_{eff}'}\|\varphi_n(u)\|_{1^*} \le N\,A\left(\prod_j e^{-n\theta_j}K(e^n; f_j)\right)^{1/N} \in l^{q_{eff}}(Z), \tag{31.18}$$

which gives (31.14). In the case where $p_j = 1$, one has (31.18) with $\theta_j = 0$ and $q_j = +\infty$, and in the case where $p_j = \infty$, one has (31.18) with $\theta_j = 1$ and $q_j = +\infty$. □

For interpreting Lemma 31.2, one assumes that

$$\text{for all } \lambda > 0, \text{ one has } meas\{x \mid |u(x)| > \lambda\} < +\infty, \tag{31.19}$$

which is a way to impose that u tends to 0 at ∞.

If $p_{eff} < N$ then it means $u \in L^{p_{eff}^*, q_{eff}}(R^N)$.

If $p_{eff} = N$ and $q_{eff} = 1$, which means that $q_j = 1$ for $j = 1, \dots, N$, then one has $|a_{n-1} - a_n| \in l^1(Z)$, so that one deduces a bound for a_n, i.e., $u \in L^\infty(R^N)$; using the density of $C_c^\infty(R^N)$ in $L^{p_j, 1}(R^N)$, one deduces that $u \in C_0(R^N)$.

If $p_{eff} = N$ and $1 < q_{eff} < \infty$, then for every $\kappa > 0$ one has $e^{\kappa|u|^{q_{eff}'}} \in L^1_{loc}(R^N)$.

If $p_{eff} = N$ and $q_{eff} = \infty$, which means that $q_j = \infty$ for $j = 1, \dots, N$, one deduces that $|a_n| \le \alpha|n| + \beta$, so that there exists $\varepsilon_0 > 0$ such that

$e^{\varepsilon_0|u|} \in L^1_{loc}(R^N)$. This is the case when all the derivatives belong to $L^{N,\infty}(R^N)$, and because $\log(|x|)$ is such a function, it is not always true that $e^{\kappa|u|} \in L^1_{loc}(R^N)$ for all $\kappa > 0$. For that particular space of functions, u belongs to $BMO(R^N)$, which Fritz JOHN and Louis NIRENBERG had introduced for studying the case of $W^{1,N}(R^N)$, and they proved that for every function in $BMO(R^N)$ there exists $\varepsilon_0 > 0$, depending upon the semi-norm of u in $BMO(R^N)$, such that $e^{\varepsilon_0|u|} \in L^1_{loc}(R^N)$.

If $p_{eff} > N$ then one has $u \in L^\infty(R^N)$. By considering $(u - \alpha)_+$ or $(u + \alpha)_-$ for $\alpha > 0$ (and letting then α tend to 0), one may assume that $u \in L^1(R^N)$, and in applying the usual rescaling argument one starts from a bound $||u||_\infty \le C(||u||_{r,s} + \sum_j ||\partial_j u||_{p_j,q_j})$, where $|| \cdot ||_p$ denotes the norm in L^p and $|| \cdot ||_{p,q}$ denotes the norm in $L^{p,q}$. Applying this inequality to $u(\frac{x_1}{\lambda_1}, \ldots, \frac{x_N}{\lambda_N})$, and writing $\mu = \lambda_1 \cdots \lambda_N$, one obtains $||u||_\infty \le C(\mu^{1/r} ||u||_{r,s} + \sum_j \mu^{1/p_j}\lambda_j^{-1}||\partial_j u||_{p_j,q_j})$; the inequality between the arithmetic mean and the geometric mean implies $\sum_j \mu^{1/p_j}\lambda_j^{-1}||\partial_j u||_{p_j,q_j} \ge N\mu^{(1/p_{eff})-(1/N)}(\prod_j ||\partial_j u||_{p_j,q_j})^{1/N}$, with equality if all $\mu^{1/p_j}\lambda_j^{-1}||\partial_j u||_{p_j,q_j}$ are equal, so that $||u||_\infty \le C(\mu^{1/r} ||u||_{r,s}+N\mu^{1/p_{eff}}\mu^{-1/N}(\prod_j ||\partial_j u||_{p_j,q_j})^{1/N})$; because minimizing $\mu^a A + \mu^{-b}B$ for $\mu > 0$ gives $a\mu^{a-1}A - b\mu^{-b-1}B = 0$ and $\mu = (b B/a A)^{1/(a+b)}$, so that the minimum is $C'A^\eta B^{1-\eta}$ with $\eta = \frac{b}{a+b}$, one deduces that $||u||_\infty \le C''||u||_{r,s}^\theta(\prod_j ||\partial_j u||_{p_j,q_j})^{(1-\theta)/N}$, with $\theta = \frac{(1/N)-(1/p_{eff})}{(1/r)+(1/N)-(1/p_{eff})}$. Choosing $r = p_i, s = q_i$, and applying the preceding inequality to the case where u is replaced by $\tau_{t e_i} u - u$, one deduces that u is Hölder-continuous of order γ_i in its ith variable, with $\gamma_i = \frac{(1/N)-(1/p_{eff})}{(1/p_i)+(1/N)-(1/p_{eff})}$.

Having different information on derivatives in different directions is usual for parabolic equations like the heat equation. For example, letting Ω be an open set of R^N, given $u_0 \in L^2(\Omega)$, one can show that there exists a unique solution u of $\frac{\partial u}{\partial t} - \Delta u = 0$ in $\Omega \times (0,T)$ satisfying the initial condition $u(x,0) = u_0(x)$ in Ω and the homogeneous Dirichlet boundary condition $\gamma_0 u = 0$ on $\partial\Omega \times (0,T)$, in the sense that $u \in C([0,T]; L^2(\Omega))$, $u \in L^2((0,T); H^1_0(\Omega))$ and $\frac{\partial u}{\partial t} \in L^2((0,T); H^{-1}(\Omega))$.

If $u_0 \in H^1_0(\Omega)$ then the solution also satisfies $u \in C^0([0,T]; H^1_0(\Omega))$, $\Delta u, \frac{\partial u}{\partial t} \in L^2((0,T); L^2(\Omega)) = L^2(\Omega \times (0,T))$; if the boundary is of class C^1 or if the open set Ω is convex (or if an inequality holds for the total curvature of the boundary), then one has $u \in L^2((0,T); H^2(\Omega))$.

If u_0 belongs to an interpolation space between $H^1_0(\Omega)$ and $L^2(\Omega)$ then one has intermediate results, but this requires enough smoothness for the boundary.

As an example, consider a function

$$u(x,t) \text{ defined on } R^N \times R \text{ and satisfying } u, \frac{\partial u}{\partial t}, \Delta u \in L^2(R^{N+1}), \quad (31.20)$$

and this implies that $\frac{\partial u}{\partial x_j} \in L^2(R^{N+1})$ for $j = 1, \ldots, N$ (by using Fourier transform, for example). Denoting the dual variables by (ξ, τ), the information is equivalent to $\mathcal{F}u, \tau \mathcal{F}u, |\xi|^2 \mathcal{F}u \in L^2(R^{N+1})$ (and therefore $\xi_j \mathcal{F}u \in L^2(R^{N+1})$ for $j = 1, \ldots, N$). One has

> $(1 + |\tau| + |\xi|^2)\mathcal{F}u \in L^2(R^{N+1})$, and if one shows $\frac{1}{1+|\tau|+|\xi|^2} \in L^{p,\infty}(R^{N+1})$
> for some $p \in (2, \infty)$, then $\mathcal{F}u \in L^{q,2}(R^{N+1})$ with $\frac{1}{q} = \frac{1}{2} + \frac{1}{p}$ and
> $u \in L^{q',2}(R^{N+1})$, $\qquad\qquad(31.21)$

the last property being due to the fact that $1 < q < 2$ and $\overline{\mathcal{F}}$ maps $L^1(R^{N+1})$ into $L^\infty(R^{N+1})$ and $L^2(R^{N+1})$ into itself, and by interpolation it maps $L^{q,2}(R^{N+1})$ into $L^{q',2}(R^{N+1})$. One has $\frac{1}{1+|\tau|+|\xi|^2} \in L^\infty(R^{N+1})$, and it is the behavior at ∞ that is interesting for obtaining the smallest value of p, so that one checks for what value of p one has $\frac{1}{|\tau|+|\xi|^2} \in L^{p,\infty}(R^{N+1})$ and one obtains the same information for the smaller function $\frac{1}{1+|\tau|+|\xi|^2}$. One uses the homogeneity properties of the function, and for $\lambda > 0$ one computes

$$meas\left\{(\xi, \tau) \mid \frac{1}{|\tau|+|\xi|^2} \geq \lambda\right\} = C\,\lambda^{-1-(N/2)}$$
$$\text{with } C = meas\left\{(\xi, \tau) \mid \frac{1}{|\tau|+|\xi|^2} \geq 1\right\}, \qquad(31.22)$$

by using the change of coordinates $\tau = \lambda^{-1}\tau'$ and $\xi = \lambda^{-1/2}\xi'$. This corresponds to $p = 1 + \frac{N}{2} = \frac{N+2}{2}$, which gives $q = \frac{2(N+2)}{N+6}$ and $q' = \frac{2(N+2)}{N-2}$ if $N \geq 3$, so one has

> for $N \geq 3$, one has $u \in L^{2(N+2)/(N-2),2}(R^{N+1}) \cap L^2(R^{N+1})$,
> for $N = 2$, one has $u \in L^r(R^3)$ for all $r \in [2, \infty)$, $\qquad(31.23)$
> for $N = 1$, one has $u \in L^\infty(R^2) \cap L^2(R^2)$.

[Taught on Friday April 7, 2000.]

Sobolev's Embedding Theorem for Besov Spaces

Using the Fourier transform one can obtain an embedding result for spaces $H^s(R^N)$ into Lorentz spaces.

Lemma 32.1. *(i) For* $0 < s < \frac{N}{2}$ *one has* $H^s(R^N) \subset L^{p(s),2}(R^N)$ *with* $\frac{1}{p(s)} = \frac{1}{2} - \frac{s}{N}$.
(ii) For $s > \frac{N}{2}$ *one has* $H^s(R^N) \subset \mathcal{F}L^1(R^N) \subset C_0(R^N)$.

Proof: (i) One has $\mathcal{F}u(\xi) = (1 + |\xi|^s)\mathcal{F}u(\xi)\frac{1}{1+|\xi|^s}$, and as $u \in H^s(R^N)$ means $(1 + |\xi|^s)\mathcal{F}u(\xi) \in L^2(R^N)$, one must check in which Lorentz space the function $\frac{1}{1+|\xi|^s}$ is. For $0 < s < \frac{N}{2}$ one has $\frac{1}{1+|\xi|^s} \leq \frac{1}{|\xi|^s} \in L^{N/s,\infty}(R^N)$, so that $\mathcal{F}u \in L^{a(s),2}(R^N)$ with $\frac{1}{a(s)} = \frac{1}{2} + \frac{s}{N}$. Because $\mathcal{F}^{-1} = \overline{\mathcal{F}}$ maps $L^1(R^N)$ into $L^\infty(R^N)$ and $L^2(R^N)$ into itself, it maps $\big(L^1(R^N), L^2(R^N)\big)_{\theta,2}$ into $\big(L^\infty(R^N), L^2(R^N)\big)_{\theta,2}$; the first space is $L^{p(\theta),2}(R^N)$ if $\frac{1}{p(\theta)} = \frac{1-\theta}{1} + \frac{\theta}{2}$, and the last space is $L^{q(\theta),2}(R^N)$ with $\frac{1}{q(\theta)} = \frac{1-\theta}{2} + \frac{\theta}{\infty} = 1 - \frac{1}{p(\theta)}$, so that $q(\theta) = p(\theta)'$. Therefore $\mathcal{F}u \in L^{a(s),2}(R^N)$ implies $u \in L^{b(s),2}(R^N)$ with $b(s) = a(s)'$, i.e., $\frac{1}{b(s)} = \frac{1}{2} - \frac{s}{N}$.
(ii) For $s > \frac{N}{2}$ one has $\frac{1}{1+|\xi|^s} \in L^2(R^N)$, so that $\mathcal{F}u \in L^1(R^N)$, and one uses the fact that $\mathcal{F}$ or $\overline{\mathcal{F}}$ map $L^1(R^N)$ into $C_0(R^N)$. □
Of course, one can improve the result for $s > \frac{N}{2}$, and show that $H^s(R^N)$ is embedded into $C^{k,\alpha}(R^N)$ when $\frac{N}{2} + k < s < \frac{N}{2} + k + 1$ with $\alpha = s - \frac{N}{2} - k$, but for Sobolev spaces corresponding to $p \neq 2$, the Fourier transform is not a good tool, and proofs must be obtained in a different way.

Definition 32.2. *For* $1 \leq p \leq \infty$ *and* $0 < s < 1$, *the Sobolev space* $W^{s,p}(R^N)$ *is defined as*

$$W^{s,p}(R^N) = \big(W^{1,p}(R^N), L^p(R^N)\big)_{1-s,p}. \tag{32.1}$$

For $1 \leq p, q \leq \infty$, *and* $0 < s < 1$ *one defines the Besov space* $B_q^{s,p}(R^N)$ *as*

$$B_q^{s,p}(R^N) = \big(W^{1,p}(R^N), L^p(R^N)\big)_{1-s,q}. \; □ \tag{32.2}$$

If k is a positive integer and $k < s < k + 1$, one may define the Sobolev space $W^{s,p}(R^N)$ or the Besov space $B_q^{s,p}(R^N)$ in at least two ways; one way is that $u \in W^{k,p}(R^N)$ and for all multi-indices α of length $|\alpha| = k$ one has $D^\alpha u \in W^{s-k,p}(R^N)$ or $B_q^{s-k,p}(R^N)$; another way is to define it as $\left(W^{m,p}(R^N), L^p(R^N)\right)_{\theta,q}$, with $q = p$ or not, with an integer $m \geq k + 1$, and with $(1 - \theta)m = s$. Of course, there are a few technical questions to check in order to show that these two definitions coincide.

Before using scaling arguments for $W^{m,p}(R^N)$, it is useful to remark that an equivalent norm is $||u||_p + \sum_{|\alpha|=m} ||D^\alpha u||_p$; it means that for $0 < |\beta| < m$ one can bound $||D^\beta u||_p$ in terms of $||u||_p$ and all the norms $||D^\alpha u||_p$ for the multi-indices of length exactly equal to m.

One starts in one dimension, noticing that if $\varphi \in C_c^\infty(R)$ is equal to 1 near 0, then $(\varphi H)' = \delta_0 + \psi$, where H is the Heaviside function, and $\psi \in C_c^\infty(R)$; one deduces that $(\varphi H) \star u'' = u' + \psi' \star u$, from which one deduces that $||u'||_p \leq ||\varphi H||_1 ||u''||_p + ||\psi'||_1 ||u||_p$. Similarly, if $0 < k < m$, one replaces the Heaviside function by the function K defined by $K(t) = \frac{t^{m-k-1}}{(m-k)!} H(t)$ for all t, so that $(\varphi K)^{(m-k)} = \delta_0 + \chi$ with $\chi \in C_c^\infty(R)$, and deriving k times and taking the convolution with u one finds that $||u^{(k)}||_p \leq ||\varphi K||_1 ||u^{(m)}||_p + ||\chi^{(k)}||_1 ||u||_p$. In order to bound $||D^\beta u||_p$, one writes $\gamma = (0, \beta_2, \ldots, \beta_N)$, $\alpha = (m + \beta_1 - |\beta|, \beta_2, \ldots, \beta_N)$ and $v = D^\gamma u$, so that for almost all $x_2, \ldots, x_N$ one can use the one-dimensional result in order to derive a bound for the norm of $D^\beta u = D_1^{\beta_1} v$ in terms of v and $D^\alpha u = D_1^{m-|\beta|} v$; then one takes the power p and integrates in $x_2, \ldots, x_N$; one finishes by using an induction argument on N.

By the usual scaling argument,

$$\text{for } |\beta| < m \text{ one has } ||D^\beta u||_p \leq C \, ||u||_p^{1-\theta} \Big(\sum_{|\alpha|=m} ||D^\alpha u||_p \Big)^\theta \text{ with } \theta = \frac{|\beta|}{m},$$
(32.3)

so that

$$\left(W^{m,p}(R^N), L^p(R^N)\right)_{(m-k)/m,1} \subset W^{k,p}(R^N) \text{ if } 0 < k < m. \quad (32.4)$$

To use the reiteration theorem 26.3, one also needs to check that one has the inclusion

$$W^{k,p}(R^N) \subset \left(W^{m,p}(R^N), L^p(R^N)\right)_{(m-k)/m,\infty}. \quad (32.5)$$

To do this, one tries the usual decomposition $u = \varrho_\varepsilon \star u + (u - \varrho_\varepsilon \star u)$, where ϱ_ε is a special smoothing sequence; for $|\alpha| = m$ one writes $D^\alpha(\varrho_\varepsilon \star u) = D^\beta \varrho_\varepsilon \star D^\gamma u$ with $\alpha = \beta + \gamma$ and $|\beta| = m - k$ and $|\gamma| = k$, so that $||\varrho_\varepsilon \star u||_{W^{m,p}(R^N)} \leq C \varepsilon^{k-m}$; one has $u(x) - (\varrho_\varepsilon \star u)(x) = \int_{R^N} \varrho_\varepsilon(y)\big(u(x) - u(x - y)\big)\, dy$, and if $k = 1$ one just uses the fact that $||u - \tau_y u||_p \leq C\, |y|\, ||grad(u)||_p$, but if $k > 1$ one must be more careful, and besides the condition $\int_{R^N} \varrho_1(y)\, dy = 1$ one also imposes the conditions $\int_{R^N} y^\gamma \varrho_1(y)\,dy = 0$ for all multi-indices γ with $1 \le |\gamma| \le k-1$. One uses Taylor's expansion formula with integral remainder $f(1) = f(0) + f'(0) + \ldots + \frac{f^{(k-1)}(0)}{(k-1)!} + \int_0^1 \frac{(1-t)^{k-1}f^{(k)}(t)}{(k-1)!}\,dt$ for the function f defined by $f(t) = u(x - ty)$, using the fact that for $1 \le |\gamma| \le k-1$ one has $\int_{R^N} \varrho_\varepsilon(y)D^\gamma u(x)y^\gamma\,dy = 0$, and one deduces that $||u - \varrho_\varepsilon \star u||_p \le C\,\varepsilon^k$, and this decomposition is valid for all $\varepsilon \in (0,1)$, which proves the assertion (because $W^{m,p}(R^N) \subset L^p(R^N)$).

Repeated applications of Sobolev's embedding theorem for $W^{1,q}(R^N)$ show that if $p > \frac{N}{m}$ one has $W^{m,p}(R^N) \subset L^\infty(R^N)$, and a scaling argument then gives $||u||_\infty \le C||u||_p^{1-\theta}\left(\sum_{|\alpha|=m}||D^\alpha u||_p\right)^\theta$ with $\theta = \frac{N}{mp}$, and this means that

$$\text{if } p > \frac{N}{m} \text{ one has } \left(W^{m,p}(R^N), L^p(R^N)\right)_{\theta,1} \subset L^\infty(R^N) \text{ with } 1 - \theta = \frac{N}{mp}.$$
(32.6)

From this, using the reiteration theorem 26.3 one deduces that

$$\text{for } 0 < s < \frac{N}{p} \text{ one has } W^{s,p}(R^N) \subset L^{p(s),p}(R^N) \text{ and } B_q^{s,p}(R^N) \subset L^{p(s),q}(R^N)$$
$$\text{with } \frac{1}{p(s)} = \frac{1}{p} - \frac{s}{N}.$$
(32.7)

Another problem where an interpolation space with second parameter 1 is useful is the question of traces of $H^s(R^N)$ spaces. For $s > \frac{1}{2}$, functions in $H^s(R^N)$ have a trace on R^{N-1}, which belongs to $H^{s-(1/2)}(R^{N-1})$, and one can reiterate this argument, so that functions in $H^s(R^N)$ have a trace on R^{N-k} if $s > \frac{k}{2}$, and the trace belongs to $H^{s-(k/2)}(R^{N-k})$; the continuity of functions in $H^s(R^N)$ for $s > \frac{N}{2}$ then appears as a natural question related to taking traces on subspaces.

Functions in $H^{1/2}(R^N)$ do not have traces on R^{N-1}, because the space of functions in $C_c^\infty(R^N)$ which vanish near R^{N-1} is dense in $H^{1/2}(R^N)$ (so that the only continuous way to define traces is to have the traces of all functions equal to 0), but the slightly smaller space $\left(H^1(R^N), L^2(R^N)\right)_{1/2,1}$ does have traces on R^{N-1}, and the space of traces is exactly $L^2(R^{N-1})$, i.e.,

$$\gamma_0\left(\left(H^1(R^N), L^2(R^N)\right)_{1/2,1}\right) = L^2(R^{N-1}).$$
(32.8)

That traces exist and belong to $L^2(R^{N-1})$ follows immediately from the standard estimate $||\gamma_0 u||_2 \le C||u||_2^{1/2}||\partial_N u||_2^{1/2}$, but I had not heard about this remark before a talk by Shmuel AGMON in 1975, where he discussed some joint work[1] with Lars HÖRMANDER where they had proven surjectivity by an argument of functional analysis, working explicitly on the transposed operator. I constructed an explicit lifting by adapting an argument which I had

---

[1] They were working on questions of scattering and they needed a space whose Fourier transform has traces on spheres, with traces belonging to L^2; they then introduced the Fourier transform of the space described here.

used for proving[2,3] the theorem of Emilio GAGLIARDO that

$$\gamma_0\big(W^{1,1}(R^N)\big) = L^1(R^{N-1}),\qquad\qquad (32.9)$$

and it is useful to describe that result first.

The idea is shown in the case of R^2, and given a function $f \in L^1(R)$ one wants to construct $u \in W^{1,1}(R^2)$ whose trace on the x axis is f. One uses a standard approximation argument used in numerical analysis, based on continuous piecewise affine functions.

For a mesh size $h > 0$, one considers the space V_h of functions in $L^1(R)$ which are affine on each interval $(k\,h, (k+1)h)$ for $k \in Z$, and continuous at the nodes $k\,h, k \in Z$; because there exists $0 < \alpha < \beta$ such that $\alpha(|g(0)| + |g(1)|) \le \int_0^1 |g(x)|\, dx \le \beta(|g(0)| + |g(1)|)$ for all affine functions on $(0,1)$, one deduces that for $g \in V_h$ one has $2\alpha\, h \sum_{k\in Z} |g(k\,h)| \le ||g||_1 = \int_R |g(x)|\, dx \le 2\beta\, h \sum_{k\in Z} |g(k\,h)|$; the important observation is that $V_h \subset W^{1,1}(R)$ and for $g \in V_h$ one has $||g'||_1 = \sum_{k\in Z} |g((k+1)h) - g(k\,h)| \le 2\sum_{k\in Z} |g(k\,h)| \le \frac{1}{\alpha h}||g||_1$.

A function $g \in V_h$ is lifted to a function in $W^{1,1}(R^2)$ by the explicit formula $G(x,y) = g(x)e^{-|y|/h}$, and one checks immediately that the trace of G is g and that $||G||_1 = 2h\,||g||_1$, $\left|\left|\frac{\partial G}{\partial y}\right|\right|_1 = 2||g||_1$ and $\left|\left|\frac{\partial G}{\partial x}\right|\right|_1 = 2h\,||g'||_1 \le \frac{2}{\alpha}||g||_1$. Once one knows that the union of all V_h for $h > 0$ is dense one has a way to lift $f \in L^1(R)$ by writing it as a series $\sum_{n=1}^\infty g_n$, choosing $0 < \varepsilon < 1$ and choosing g_1 such that $||f - g_1||_1 \le \varepsilon ||f||_1$, then g_2 such that $||(f - g_1) - g_2||_1 \le \varepsilon ||f - g_1||_1 \le \varepsilon^2 ||f||_1$, and so on, so that $\sum_{n=1}^\infty ||g_n||_1 \le \frac{1}{1-\varepsilon}||f||_1$.

The density is proven by approximating functions in $C_c(R)$, which is a dense subspace of $L^1(R)$. For $\varphi \in C_c(R)$ one constructs the interpolated[4] function $\Pi_h\varphi$ which is the function of V_h such that $\Pi_h\varphi(k\,h) = \varphi(k\,h)$ for all $k \in Z$; one checks easily that $|\Pi_h\varphi(x) - \varphi(x)| \le \omega(h)$ for all x, where ω is the modulus of uniform continuity of φ, so that when $h_n \to 0$, the sequence $\Pi_{h_n}\varphi$ converges uniformly to φ and as $support(\Pi_h\varphi) \subset support(\varphi) + [-h, +h]$ one also has $||\Pi_{h_n}\varphi - \varphi||_1 \to 0$.

Jaak PEETRE has shown that there does not exist a linear continuous lifting from $L^1(R^{N-1})$ into $W^{1,1}(R^N)$.

[Taught on Monday April 10, 2000.]

---

[2] When I was a student, I had tried to read Emilio GAGLIARDO's article, but not knowing much Italian at the time I had trouble understanding what he was doing with all these cubes which appeared in his proof. Many years after, I constructed my own proof, but I did not check if my understanding of Italian is good enough now for reading his proof; I thought that my argument must have been identical to his original idea, but Françoise DEMENGEL told me that it is different.

[3] Françoise DEMENGEL, French mathematician. She works in Cergy-Pontoise, France.

[4] Here we encounter the other meaning of the word interpolation, going back to LAGRANGE, a classical tool in numerical analysis.

The Lions–Magenes Space $H_{00}^{1/2}(\Omega)$

Again the idea is shown in the case of R^2, and given a function $f \in L^2(R)$ one wants to construct $u \in \left(H^1(R^2), L^2(R^2)\right)_{1/2,1}$ whose trace on the x axis is f. For a (positive) mesh size h one considers the space V_h of functions in $L^2(R)$ which are affine on each interval $(kh, (k+1)h)$ for $k \in Z$, and continuous at the nodes $kh, k \in Z$; because there exists $0 < \alpha < \beta$ such that $\alpha(|g(0)|^2 + |g(1)|^2)^{1/2} \le \left(\int_0^1 |g(x)|^2 \, dx\right)^{1/2} \le \beta(|g(0)|^2 + |g(1)|^2)^{1/2}$ for all affine functions on $(0,1)$, one deduces that for $g \in V_h$ one has $\sqrt{2}\alpha \sqrt{h}\left(\sum_{k \in Z} |g(kh)|^2\right)^{1/2} \le ||g||_2 = \left(\int_R |g(x)|^2 \, dx\right)^{1/2} \le \sqrt{2}\beta \sqrt{h}\left(\sum_{k \in Z} |g(kh)|^2\right)^{1/2}$; the important observation is that $V_h \subset H^1(R)$ and for $g \in V_h$ one has $||g'||_2 = \frac{1}{\sqrt{h}}\left(\sum_{k \in Z} |g((k+1)h) - g(kh)|^2\right)^{1/2} \le \frac{2}{\sqrt{h}}\left(\sum_{k \in Z} |g(kh)|^2\right)^{1/2} \le \frac{C}{h}||g||_2$.

A function $g \in V_h$ is lifted to a function in $H^1(R^2)$ by the explicit formula $G(x,y) = g(x)e^{-|y|/h}$, and one checks immediately that the trace of G is g and that $||G||_2 = \sqrt{h} ||g||_2$, $\left\|\frac{\partial G}{\partial y}\right\|_2 = \frac{1}{\sqrt{h}}||g||_2$ and $\left\|\frac{\partial G}{\partial x}\right\|_2 = \sqrt{h} ||g'||_2 \le \frac{C}{\sqrt{h}}||g||_2$. One then has $||G||_{L^2(R^2)} \le \sqrt{h} ||g||_2$ and $||G||_{H^1(R^2)} \le \frac{C}{\sqrt{h}}||g||_2$ for $0 < h < 1$, so that $||G||_{(H^1(R^2), L^2(R^2))_{1/2,1}} \le C||G||_{H^1(R^2)}^{1/2}||G||_{L^2(R^2)}^{1/2} \le C'||g||_2$. Then one uses the fact that the union of all V_h for $h > 0$ is dense, so that any $f \in L^2(R)$ can be written as a series $\sum_{n=1}^{\infty} g_n$, with $\sum_{n=1}^{\infty} ||g_n||_2 \le C ||f||_2$.

Although functions in $H^{1/2}(R)$ are not continuous, as they are not even bounded, piecewise smooth functions which are discontinuous at a point do not belong to $H^{1/2}(R)$. For example, let $\varphi \in C_c^{\infty}(R)$ with $\varphi = 1$ near 0 and let $u = \varphi H$ where H is the Heaviside function; then $u' = \delta_0 + \psi$ with $\psi \in C_c^{\infty}(R)$, so that $2i\pi\xi\mathcal{F}u(\xi) = 1 + \mathcal{F}\psi(\xi)$ so that $|\mathcal{F}u|$ behaves like $\frac{1}{2\pi|\xi|}$ near ∞, so that $(1 + |\xi|^{1/2})|\mathcal{F}u| \notin L^2(R)$, i.e., $u \notin H^{1/2}(R)$.

As it seems that functions in $H^{1/2}(R)$ cannot have discontinuities at a point, one expects some kind of continuity, but of a different nature as the value at a point does not make sense. The following ideas have been introduced

by Jacques-Louis LIONS and Enrico MAGENES, and some related work has
been done by Pierre GRISVARD.[1]

Lemma 33.1. *If* $u \in H^{1/2}(R)$ *then one has* $\frac{|u(x)-u(-x)|}{\sqrt{|x|}} \in L^2(R)$.

Proof: By Hardy's inequality (13.3) one has $\left\| \frac{|u(x)-u(0)|}{|x|} \right\|_2 \leq 2\|u'\|_2$ for $u \in$
$H^1(R)$, and similarly $\left\| \frac{|u(-x)-u(0)|}{|x|} \right\|_2 \leq 2\|u'\|_2$, so that $\left\| \frac{|u(x)-u(-x)|}{|x|} \right\|_2 \leq$
$4\|u'\|_2$ for all $u \in H^1(R)$. One considers the mapping $u \mapsto u - \check{u}$ which
maps $H^1(R)$ into L^2 for the measure $\frac{dx}{|x|^2}$ and $L^2(R)$ into L^2 for the measure
dx, so that it maps $H^{1/2}(R) = \big(H^1(R), L^2(R)\big)_{1/2,2}$ into the corresponding
interpolation space, which is L^2 for the measure $\frac{dx}{|x|}$. $\quad\square$

Similarly, Jacques-Louis LIONS and Enrico MAGENES noticed that when
considering the interpolation spaces $\big(H_0^1(\Omega), L^2(\Omega)\big)_{\theta,2}$ for a bounded open
set with a Lipschitz boundary, it does give $H_0^{1-\theta}(\Omega)$ for $\theta \neq \frac{1}{2}$ (and one has
$H_0^s(\Omega) = H^s(\Omega)$ for $0 \leq s \leq \frac{1}{2}$), but for $\theta = \frac{1}{2}$ it gives a new space, which
they denoted by $H_{00}^{1/2}(\Omega)$.

Lemma 33.2. *If* $u \in H_{00}^{1/2}(R_+) = \big(H_0^1(R_+), L^2(R_+)\big)_{1/2,2}$, *then* $\frac{u}{\sqrt{x}} \in$
$L^2(R_+)$.

Proof: As $u \in H_0^1(R_+)$ implies $\frac{u}{x} \in L^2(R_+)$ by Hardy's inequality (13.3), one
has

$$\big(H_0^1(R_+), L^2(R_+)\big)_{1/2,2} \subset \big(L^2\big(R_+; \tfrac{dx}{x^2}\big), L^2\big(R_+; dx\big)\big)_{1/2,2} = \atop L^2\big(R_+; \tfrac{dx}{x}\big). \;\square \tag{33.1}$$

A related result is that if $u \in H_0^1(\Omega)$ the extension $\tilde{u}$ of u by 0 outside Ω
belongs to $H^1(R^N)$ and similarly if $u \in H_0^s(\Omega)$ (closure of $C_c^\infty(\Omega)$ in $H^s(\Omega)$)
then $\tilde{u} \in H^s(R^N)$ for $0 \leq s \leq 1$ if $s \neq \frac{1}{2}$, but not for $s = \frac{1}{2}$, because
$1 \in H_0^{1/2}(\Omega) = H^{1/2}(\Omega)$ and the extension by 0 is piecewise smooth and
discontinuous, so that it is not in $H^{1/2}(R^N)$. Actually, $H_{00}^{1/2}(\Omega)$ is characterized either as the space of functions in $H^{1/2}(\Omega)$ such that $\frac{u}{\sqrt{d(x)}} \in L^2(\Omega)$,
where $d(x)$ is the distance of x to the boundary $\partial\Omega$, or as the space of functions
$u \in H^{1/2}(\Omega)$ such that $\tilde{u} \in H^{1/2}(R^N)$.

A related difficulty is that any partial derivative $\frac{\partial}{\partial x_j}$ maps $H^1(\Omega)$ into
$L^2(\Omega)$ and $L^2(\Omega)$ into $H^{-1}(\Omega)$, and it does map $H^{1-s}(\Omega)$ into $H^{-s}(\Omega) =$
$\big(H_0^s(\Omega)\big)'$ for $0 \leq s \leq 1$ and $s \neq \frac{1}{2}$, but not for $s = \frac{1}{2}$, because in this case it
maps $H^{1/2}(\Omega)$ into the dual of $H_{00}^{1/2}(\Omega)$.

This technical difficulty appears when one solves boundary value problems
like $-\Delta u = f$ in Ω with the boundary $\partial\Omega$ made of two disjoint pieces, Γ_D
where a Dirichlet condition is imposed, and Γ_N where a Neumann condition

[1] Pierre GRISVARD, French mathematician, 1940–1994. He worked in Nice, France.

is imposed; in the case where the parts Γ_D and Γ_N have a common boundary, the precise pairs of data allowed are a little technical to characterize. If all the boundary is Γ_N then the precise space is $H^{-1/2}(\partial\Omega)$, the dual of $H^{1/2}(\partial\Omega)$ (traces of functions of $H^1(\Omega)$), but the important point is that one cannot restrict an element of $H^{-1/2}(\partial\Omega)$ to a part Γ_N whose boundary is smooth enough, because restriction is the transpose of the operator of extension by 0, and this extension operation $\tilde{\ }$ does not act on $H^{1/2}(\Gamma_N)$ if the $(N-1)$-dimensional Hausdorff measure of Γ_D is positive.

These technical difficulties may seem quite academic, but some models in continuum mechanics lead to using operations which are not defined in an obvious way, and it is important to understand if one should reject some laws as being nonphysical or if one should try to overcome the mathematical difficulty that they represent. One such example is the static law of friction due to COULOMB,[2] which involves a sign of a normal force at the boundary and an inequality on the strength of a tangential force at the boundary; if one uses linearized elasticity, the natural information coming from the finiteness of the stored elastic energy gives the various forces as normal traces of functions in $H(div;\Omega)$, i.e., elements in $H^{-1/2}(\partial\Omega)$; unfortunately, one cannot define the absolute value of an arbitrary element in $H^{-1/2}(\partial\Omega)$. However, one can define what a nonnegative element is by stating that it is a nonnegative measure, so the question is to find a way to express *Coulomb's law* which makes sense from a mathematical point of view, although there are indications that Coulomb's law is not exactly what real materials follow, because dynamics does play a role.

[Taught on Wednesday April 12, 2000.]

---

[2] Charles Augustin DE COULOMB, French engineer, 1736–1806.

34

Defining Sobolev Spaces and Besov Spaces for Ω

One has proven that for all integers $0 < k < m$ and all $1 \leq p \leq \infty$ one has $\left(W^{m,p}(R^N), L^p(R^N)\right)_{\theta,1} \subset W^{k,p}(R^N) \subset \left(W^{m,p}(R^N), L^p(R^N)\right)_{\theta,\infty}$ with $m(1-\theta) = k$. For $s > 0$ which is not an integer one defines the Sobolev space $W^{s,p}(R^N)$ as $\left(W^{m,p}(R^N), L^p(R^N)\right)_{\eta,p}$ with $m > s$ and $(1-\eta)m = s$, and one deduces from the reiteration theorem 26.3 that one also has $W^{s,p}(R^N) = \left(W^{m_1,p}(R^N), W^{m_2,p}(R^N)\right)_{\zeta,p}$ with equivalent norms if m_1, m_2 are nonnegative integers such that $s = (1-\zeta)m_1 + \zeta m_2$ for some $\zeta \in (0,1)$ (i.e., either $m_1 > s > m_2$ or $m_1 < s < m_2$), and this is still true even if m_1, m_2 are nonnegative real numbers which are not necessarily integers, always under the condition $\zeta \in (0,1)$, of course.

Similarly, if for $s > 0$ and $1 \leq q \leq \infty$ one defines the Besov space $B_q^{s,p}(R^N)$ as $\left(W^{m,p}(R^N), L^p(R^N)\right)_{\eta,q}$ with $m > s$ and $(1-\eta)m = s$, one deduces that for nonnegative reals s_1, s_2 such that $s = (1-\zeta)s_1 + \zeta s_2$ with $\zeta \in (0,1)$, one has $B_q^{s,p}(R^N) = \left(W^{s_1,p}(R^N), W^{s_2,p}(R^N)\right)_{\zeta,q}$ and $B_q^{s,p}(R^N) = \left(B_{q_1}^{s_1,p}(R^N), B_{q_2}^{s_2,p}(R^N)\right)_{\zeta,q}$ with equivalent norms, for all $1 \leq q_1, q_2 \leq \infty$.

If s is not an integer, one has $W^{s,p}(R^N) = B_p^{s,p}(R^N)$.

Although one has $H^1(R^N) = \left(H^2(R^N), L^2(R^N)\right)_{1/2,2}$, the case for $p = \infty$ is different and the space $W^{1,\infty}(R^N)$ which is the space $Lip(R^N)$ of Lipschitz continuous functions is a proper subspace of $\left(W^{2,\infty}(R^N), L^\infty(R^N)\right)_{1/2,\infty}$, which coincides with a space introduced by Antoni ZYGMUND and denoted by $\Lambda_1(R^N)$,

$$\Lambda_1(R^N) = \{u \in L^\infty(R^N) \mid \text{ there exists } C \text{ such that } \\ |u(x-h) + u(x+h) - 2u(x)| \leq C\,|h| \text{ for all } x, h \in R^N, \tag{34.1}$$

or equivalently $u \in L^\infty(R^N)$ and $||\tau_h u + \tau_{-h} u - 2u||_\infty \leq C|h|$ for all $h \in R^N$.

One deduces also that if $s > k$ where k is a positive integer, then for every multi-index α with $|\alpha| = k$ the derivation D^α is linear continuous from $W^{s,p}(R^N)$ into $W^{s-|\alpha|,p}(R^N)$ and also from $B_q^{s,p}(R^N)$ into $B_q^{s-|\alpha|,p}(R^N)$, if $1 \leq p, q \leq \infty$. This follows immediately from the fact that for any integer $m \geq k$ the derivation D^α is linear continuous from $W^{m,p}(R^N)$ into $W^{m-|\alpha|,p}(R^N)$, and after choosing m_1, m_2 such that $k \leq m_1 < s < m_2$ and computing $\zeta \in (0,1)$ such that $s = (1-\zeta)m_1 + \zeta\, m_2$, one applies the interpolation property, and D^α maps continuously $\left(W^{m_1,p}(R^N), W^{m_2,p}(R^N)\right)_{\zeta,q} = B_q^{s,p}(R^N)$ into $\left(W^{m_1-|\alpha|,p}(R^N), W^{m_2-|\alpha|,p}(R^N)\right)_{\zeta,q} = B_q^{s-|\alpha|,p}(R^N)$.

For Ω an open subset of R^N, one may define the Sobolev $W^{s,p}(\Omega)$ for all positive real s which are not integers in at least two different ways; the first one will be denoted by $X^{s,p}$ for the discussion,

$$W^{s,p}(\Omega) = \text{ space of restrictions of functions from } W^{s,p}(R^N), \qquad (34.2)$$

with the quotient norm

$$||u||_{W^{s,p}(\Omega)} = \inf_{U|_\Omega = u} ||U||_{W^{s,p}(R^N)}; \qquad (34.3)$$

the second one will be denoted by $Y^{s,p}$ for the discussion,

$$W^{s,p}(\Omega) = \left(W^{m_1,p}(\Omega), W^{m_2,p}(\Omega)\right)_{\zeta,p} \text{ with } 0 < \zeta < 1$$
$$\text{and } m_1, m_2 \text{ nonnegative integers such that } s = (1-\zeta)m_1 + \zeta\, m_2. \qquad (34.4)$$

Of course, one can also give two definitions for defining the Besov spaces $B_q^{s,p}(\Omega)$.

Because the restriction to Ω is linear continuous from $W^{m_1,p}(R^N)$ into $W^{m_1,p}(\Omega)$ and also linear continuous from $W^{m_2,p}(R^N)$ into $W^{m_2,p}(\Omega)$, it is continuous from $W^{s,p}(R^N) = \left(W^{m_1,p}(R^N), W^{m_2,p}(R^N)\right)_{\zeta,p}$ into $Y^{s,p} = \left(W^{m_1,p}(\Omega), W^{m_2,p}(\Omega)\right)_{\zeta,p}$, and as every element of $X^{s,p}$ is the restriction of an element from $W^{s,p}(R^N)$, one deduces that $X^{s,p} \subset Y^{s,p}$.

If the boundary of Ω is smooth enough so that there exists a continuous extension P which maps $W^{m_1,p}(\Omega)$ into $W^{m_1,p}(R^N)$ and $W^{m_2,p}(\Omega)$ into $W^{m_2,p}(R^N)$, then it maps $Y^{s,p} = \left(W^{m_1,p}(\Omega), W^{m_2,p}(\Omega)\right)_{\zeta,p}$ into $W^{s,p}(R^N) = \left(W^{m_1,p}(R^N), W^{m_2,p}(R^N)\right)_{\zeta,p}$, so that every element of $Y^{s,p}$ is the restriction to Ω of an element of $W^{s,p}(R^N)$, i.e., one has $Y^{s,p} \subset X^{s,p}$.

The extension property has been shown for $W^{1,p}(\Omega)$ if Ω is bounded with a Lipschitz boundary (Lemma 12.4), and an analogous situation has been described for $W^{k,p}(R_+^N)$ and $0 \leq k \leq m$ (Lemma 12.5), and it extends to the case of bounded open sets with smooth boundary. As was mentioned before, STEIN has constructed an extension valid for all $W^{m,p}(\Omega)$ if Ω is bounded with a Lipschitz boundary.

[Taught on Monday April 17, 2000.]

Characterization of $W^{s,p}(R^N)$

Before giving a characterization of $W^{s,p}(\Omega)$ in the case of a bounded open set with a Lipschitz boundary, it is useful to begin with the case of R^N. One starts with a preliminary result.

Lemma 35.1. *If* $0 < s < 1$ *and* $1 \leq p \leq \infty$ *one has*

$$B_\infty^{s,p}(R^N) = \big(W^{1,p}(R^N), L^p(R^N)\big)_{1-s,\infty} = \{u \in L^p(R^N) \mid ||u - \tau_h u||_p \leq C\,|h|^s$$
$$\text{for all } h \in R^N\}.$$

(35.1)

Proof: If $u \in W^{1,p}(R^N)$ one has $||u - \tau_h u||_p \leq |h|\,||grad(u)||_p$ and if $u \in L^p(R^N)$ one has $||u - \tau_h u||_p \leq 2||u||_p$, and as the mapping $u \mapsto u - \tau_h u$ is linear, one finds that for $\theta \in (0,1)$ one has $||u - \tau_h u||_p \leq C\,|h|^{1-\theta}||u||_{(W^{1,p}(R^N), L^p(R^N))_{\theta,\infty}}$ for all $u \in \big(W^{1,p}(R^N), L^p(R^N)\big)_{\theta,\infty}$.

Conversely, assume that $u \in L^p(R^N)$ and $||u - \tau_h u||_p \leq C\,|h|^s$ for all $h \in R^N$. One decomposes

$$u = \varrho_\varepsilon \star u + (u - \varrho_\varepsilon \star u),$$

(35.2)

where ϱ_ε is a special smoothing sequence, i.e., $\varrho_\varepsilon(x) = \frac{1}{\varepsilon^N}\varrho_1\big(\frac{x}{\varepsilon}\big)$, with $\varrho_1 \in C_c^\infty(R^N)$ and $\int_{R^N} \varrho_1(x)\,dx = 1$, but one adds the hypothesis that ϱ_1 is an even function. One has

$u(x) - (\varrho_\varepsilon \star u)(x) = \int_{R^N} \varrho_\varepsilon(y)\big(u(x) - u(x-y)\big)\,dy$, so that $||u - \varrho_\varepsilon \star u||_p \leq$

$\int_{R^N} |\varrho_\varepsilon(y)|\,||u - \tau_y u||_p\,dy \leq \int_{R^N} C\,|\varrho_\varepsilon(y)|\,|y|^s\,dy = C'\,\varepsilon^s.$

(35.3)

One has $||\varrho_\varepsilon \star u||_p \leq ||\varrho_\varepsilon||_1 ||u||_p = C\,||u||_p$, and for any derivative $\partial_j = \frac{\partial}{\partial x_j}$ one has $\partial_j(\varrho_\varepsilon \star u) = (\partial_j \varrho_\varepsilon) \star u$, but one needs a better estimate than $||\partial_j(\varrho_\varepsilon \star u)||_p \leq ||(\partial_j \varrho_\varepsilon)||_1 ||u||_p = \frac{C}{\varepsilon}||u||_p$, which has not used all the information on u. Changing y into $-y$ in the integral, one can write $\partial_j(\varrho_\varepsilon \star u)(x)$ as either $\int_{R^N} (\partial_j \varrho_\varepsilon)(y)u(x-y)\,dy$ or $\int_{R^N} (\partial_j \varrho_\varepsilon)(-y)u(x+y)\,dy$ or as the half sum of these two terms, and this is where having chosen for ϱ_1 an even function is useful, as $\partial_j \varrho_\varepsilon$ is an odd function, so that

$$\partial_j(\varrho_\varepsilon \star u)(x) = \frac{1}{2} \int_{R^N} (\partial_j \varrho_\varepsilon)(y)\big(u(x-y) - u(x+y)\big)\, dy, \qquad (35.4)$$

from which one deduces

$$
\begin{aligned}
&||\partial_j(\varrho_\varepsilon \star u)||_p \le \tfrac{1}{2} \int_{R^N} |(\partial_j \varrho_\varepsilon)(y)|\, ||\tau_y u - \tau_{-y} u||_p \, dy \le \\
&C \int_{R^N} |(\partial_j \varrho_\varepsilon)(y)|\, |h|^s \, dy = C'' \varepsilon^{s-1}.
\end{aligned}
\qquad (35.5)
$$

For $E_0 = W^{1,p}(R^N)$ and $E_1 = L^p(R^N)$, this shows that $K(t;u) \le C_1(1 + \varepsilon^{s-1}) + C_2 t\,\varepsilon^s$, and choosing $\varepsilon = \frac{1}{t}$ for $t \ge 1$, one obtains $K(t;u) \le C_3 t^{1-s}$, and for $0 < t < 1$ one has $K(t;u) \le t\,||u||_p$ (because in the case $E_0 \subset E_1$ one always has the decomposition $u = 0 + u$), so that $t^{-\theta} K(t;u) \in L^\infty(0,\infty)$ with $\theta = 1 - s$. $\qquad\square$

Lemma 35.2. *For $0 < s < 1$ and $1 \le p < \infty$,*

$$W^{s,p}(R^N) = \left\{ u \in L^p(R^N) \mid \iint_{R^N \times R^N} \frac{|u(x) - u(y)|^p}{|x-y|^{N+sp}}\, dx\, dy < \infty \right\}, \quad (35.6)$$

or equivalently $u \in L^p(R^N)$ and $\int_{R^N} \big(\frac{||u - \tau_y u||_p}{|y|^s}\big)^p \frac{dy}{|y|^N} < \infty$.

Proof: Let $E_0 = W^{1,p}(R^N)$ and $E_1 = L^p(R^N)$. If $u \in W^{s,p}(R^N) = (E_0, E_1)_{\theta,p}$ with $\theta = 1-s$, one has a decomposition $u = u_0 + u_1$ with $u_0 \in E_0, u_1 \in E_1$ and $||grad(u_0)||_p + t\,||u_1||_p \le 2K(t;u)$ and $t^{-\theta} K(t;u) \in L^p\big(R_+; \frac{dt}{t}\big)$. For $y \in R^N$ one has

$$
\begin{aligned}
&||u - \tau_y u||_p \le ||u_0 - \tau_y u_0||_p + ||u_1 - \tau_y u_1||_p \le |y|\, ||grad(u_0)||_p + 2||u_1||_p \le \\
&C\big(|y| + \tfrac{1}{t}\big) K(t;u),
\end{aligned}
\qquad (35.7)
$$

and the choice $t = \frac{1}{|y|}$ gives $||u - \tau_y u||_p \le C\,|y|\, K\big(\frac{1}{|y|}; u\big)$. Denoting by σ_{N-1} the $(N-1)$-Hausdorff measure of the unit sphere, this gives

$$
\begin{aligned}
&\int_{R^N} \big(\tfrac{||u - \tau_y u||_p}{|y|^s}\big)^p \tfrac{dy}{|y|^N} \le C \int_{R^N} \big(\tfrac{|y|\, K(\frac{1}{|y|};u)}{|y|^s}\big)^p \tfrac{dy}{|y|^N} = \\
&C\,\sigma_{N-1} \int_0^\infty \big[|y|^{1-s} K\big(\tfrac{1}{|y|}; u\big)\big]^p \tfrac{dy}{|y|} = C\,\sigma_{N-1} \int_0^\infty \big(t^{-\theta} K(t;u)\big)^p \tfrac{dt}{t} < \infty.
\end{aligned}
\qquad (35.8)
$$

Conversely, assume that $u \in L^2(R^N)$ and $\iint_{R^N \times R^N} \frac{|u(x)-u(y)|^p}{|x-y|^{N+sp}}\, dx\, dy < \infty$, so that if one writes $F(z) = ||u - \tau_z u||_p$ for $z \in R^N$, one has $\int_{R^N} \frac{F(z)}{|z|^{N+sp}}\, dz < \infty$. Let

$$\overline{F}(r) = \text{ the average of } F \text{ on the sphere } |y| = r, \text{ so that}$$

$$\int_{|y|=r} |\overline{F}(y)|^p \, dH^{N-1} \le \int_{|y|=r} |F(y)|^p \, dH^{N-1} \qquad (35.9)$$

by Hölder's inequality, so that

$$\int_0^\infty \frac{|\overline{F}(r)|^p}{r^{sp}} \frac{dr}{r} = \frac{1}{\sigma_{N-1}} \int_{R^N} \frac{\overline{F}(z)}{|z|^{N+sp}} \, dz \le \int_{R^N} \frac{F(z)}{|z|^{N+sp}} \, dz < \infty. \quad (35.10)$$

Then as for the preceding lemma one decomposes $u = \varrho_\varepsilon \star u + (u - \varrho_\varepsilon \star u)$ for an even special smoothing sequence with $support(\varrho_1) \subset \overline{B(0,1)}$, and one obtains

$$||u - \varrho_\varepsilon \star u||_p \le \int_{R^N} |\varrho_\varepsilon(y)| \, ||u - \tau_y u||_p \, dy = \int_{R^N} |\varrho_\varepsilon(y)| \, F(y) \, dy \le$$
$$\frac{C}{\varepsilon^N} \int_{|y| \le \varepsilon} F(y) \, dy = \frac{C}{\varepsilon^N} \int_{|y| \le \varepsilon} \overline{F}(y) \, dy = \frac{C\sigma_{N-1}}{\varepsilon} \int_0^\varepsilon \overline{F}(r) \, dr, \quad (35.11)$$

and using the fact that $\partial_j \varrho_\varepsilon$ is odd, one obtains

$$||\partial_j(\varrho_\varepsilon \star u)||_p \le \tfrac{1}{2} \int_{R^N} |\partial_j \varrho_\varepsilon(y)| \, ||\tau_y u - \tau_{-y} u||_p \, dy \le$$
$$\tfrac{1}{2} \int_{R^N} |\partial_j\big(F(y) + F(-y)\big) \, dy \le \frac{C}{\varepsilon^{N-1}} \int_{|y| \le \varepsilon} \overline{F}(y) \, dy \le \frac{C\sigma_{N-1}}{\varepsilon^2} \int_0^\varepsilon \overline{F}(r) \, dr. \quad (35.12)$$

One deduces that $K(t; u) \le C\big(1 + \frac{1}{\varepsilon^2} \int_0^\varepsilon \overline{F}(r) \, dr\big) + \frac{Ct}{\varepsilon} \int_0^\varepsilon \overline{F}(r) \, dr$, so that for $t > 1$ one takes $\varepsilon = \frac{1}{t}$ and one has $K(t; u) \le C + C t^2 \int_0^{1/t} \overline{F}(r) \, dr$ (and $K(t; u) \le t \, ||u||_p$ for $t < 1$). Then with $\theta = 1 - s$, one has $t^{-\theta} K(t; u) = C t^{-\theta} + C t^s t \int_0^{1/t} \overline{F}(r) \, dr$, and as the desired condition $t^{-\theta} K(t; u) \in L^p\big(R_+; \frac{dt}{t}\big)$ is invariant by the change of t into $\frac{1}{t}$, one has to show that the function G defined on $(0,1)$ by $G(\varepsilon) = \frac{1}{\varepsilon} \int_0^\varepsilon \frac{\overline{F}(r)}{\varepsilon^s} \, dr$ belongs to $L^p\big((0,1); \frac{d\varepsilon}{\varepsilon}\big)$, as a consequence of the hypothesis that $\frac{\overline{F}(r)}{r^s} \in L^p\big(R_+; \frac{d\varepsilon}{\varepsilon}\big)$, and this is like deriving Hardy's inequality (13.3). $\square$

Of course, the proof also applies to the case $p = \infty$, but is covered by the previous lemma.

[Taught on Wednesday April 19, 2000.]

Characterization of $W^{s,p}(\Omega)$

One can now deduce a characterization of $W^{s,p}(\Omega)$ in the case of a bounded open set of R^N with a Lipschitz boundary, with $0 < s < 1$ and $1 \le p \le \infty$.

Lemma 36.1. *If Ω is a bounded open set of R^N with a Lipschitz boundary, then for $0 < s < 1$ and $1 \le p < \infty$,*

$$u \in W^{s,p}(\Omega) \text{ if and only if } u \in L^p(\Omega) \text{ and } \int\int_{\Omega \times \Omega} \frac{|u(x) - u(y)|^p}{|x-y|^{N+sp}} \, dx \, dy < \infty.$$
(36.1)

Proof: If $u \in W^{s,p}(\Omega)$, defined as the restriction to Ω of a function $U \in W^{s,p}(R^N)$, then one has $U \in L^p(R^N)$ and therefore by restriction $u \in L^p(\Omega)$, and

$$\begin{aligned}
&\int\int_{\Omega \times \Omega} \frac{|u(x)-u(y)|^p}{|x-y|^{N+sp}} \, dx \, dy = \int\int_{\Omega \times \Omega} \frac{|U(x)-U(y)|^p}{|x-y|^{N+sp}} \, dx \, dy \le \\
&\int\int_{R^N \times R^N} \frac{|U(x)-U(y)|^p}{|x-y|^{N+sp}} \, dx \, dy < \infty.
\end{aligned}$$
(36.2)

Let $u \in L^p(\Omega)$ satisfy $\int\int_{\Omega \times \Omega} \frac{|u(x)-u(y)|^p}{|x-y|^{N+sp}} \, dx \, dy < \infty$. For a partition of unity $\theta_i, i \in I$, each $\theta_i u$ satisfies a similar property, because

$$\begin{aligned}
&|(\theta_i u)(x) - (\theta_i u)(y)| \le |\theta_i(x)| \, |u(x) - u(y)| + |\theta_i(x) - \theta_i(y)| \, |u(y)| \le \\
&C \, |u(x) - u(y)| + C \, |x - y| \, |u(y)|,
\end{aligned}$$
(36.3)

using the assumption that each θ_i is a Lipschitz continuous function. One deduces that $|(\theta_i u)(x) - (\theta_i u)(y)|^p \le C' \, |u(x) - u(y)|^p + C' \, |x - y|^p \, |u(y)|^p$, and one must show that $\int\int_{\Omega \times \Omega} \frac{|x-y|^p |u(y)|^p}{|x-y|^{N+sp}} \, dx \, dy < \infty$; this is a consequence of $\int_\Omega \frac{|x-y|^p}{|x-y|^{N+sp}} \, dx \le M$ uniformly in $y \in \Omega$, which holds because if Ω has diameter d, the integral is bounded by $\int_{|z| \le d} \frac{|z|^p}{|z|^{N+sp}} \, dz = C \int_0^d \frac{dt}{t^\alpha} < \infty$, as $\alpha = 1 + sp - p < 1$. Using a local change of basis, one is led to consider the case of $\Omega_F = \{(x', x_N) \mid x_N > F(x')\}$ for F Lipschitz continuous, and one extends u by symmetry, defining $Pu(x', x_N) = u(x', x_N)$

if $x_N > F(x')$ and $P u(x', x_N) = u(x', 2F(x') - x_N)$ if $x_N < F(x')$. If for $x = (x', x_N)$ one defines $\overline{x} = (x', 2F(x') - x_N)$, so that $\overline{(\overline{x})} = x$, then the integral $\int\int_{R^N \times R^N} \frac{|P u(x) - P u(y)|^p}{|x-y|^{N+sp}} \, dx \, dy$ can be cut into four parts, one is $I = \int\int_{\Omega_F \times \Omega_F} \frac{|u(x) - u(y)|^p}{|x-y|^{N+sp}} \, dx \, dy$, which is finite by hypothesis; two parts have the form $\int\int_{\Omega_F \times (R^N \setminus \Omega_F)} \frac{|u(x) - u(\overline{y})|^p}{|x-y|^{N+sp}} \, dx \, dy$, which is $\leq K^p I$ because $|x - \overline{y}| \leq K |x - y|$ for all $x \in \Omega_F, y \in R^N \setminus \Omega_F$, and the fourth part is $\int\int_{(R^N \setminus \Omega_F) \times (R^N \setminus \Omega_F)} \frac{|u(\overline{x}) - u(\overline{y})|^p}{|x-y|^{N+sp}} \, dx \, dy$, which is $\leq K^p I$ because $|\overline{x} - \overline{y}| \leq K |x - y|$ for $x, y \in R^N \setminus \Omega_F$; indeed, the map $x = (x', x_N) \mapsto (x', x_N - F(x'))$ is Lipschitz continuous from Ω_F onto R_+^N, and its inverse is Lipschitz continuous, as it is $z = (z', z_N) \mapsto (z', z_N + F(z'))$, and one is reduced to studying the same inequalities for R_+^N, i.e., in the case $F = 0$, where one has $|x - \overline{y}| \leq |x - y|$ if $x_N > 0 > y_N$, and $|\overline{x} - \overline{y}| = |x - y|$ for all x, y.   □

Of course, the proof adapts to the case $p = \infty$, the only difference coming from the fact that the norm is not expressed by an integral. It is important to notice that without any regularity hypothesis on a set $A \subset R^N$, any Lipschitz continuous function defined on A can be extended to R^N, and the same result is true for Hölder continuous functions. To show this, one assumes that $C_1 \leq u(x) \leq C_2$ for all $x \in A$, and that there exists $\alpha \in (0, 1]$ and $K \geq 0$ such that $|u(x) - u(y)| \leq K |x - y|^\alpha$ for all $x, y \in A$, then one defines $v(x) = \sup_{y \in A} u(y) - K |x - y|^\alpha$, and this gives a Hölder continuous function which coincides with u on A, and then one truncates it by $w(x) = \min\{C_2, \max\{C_1, v(x)\}\}$.

One should notice that $W^{1,\infty}(\Omega)$ contains the space $Lip(\Omega)$ of Lipschitz continuous functions, but may be different if Ω is not a bounded open set with a Lipschitz boundary, because $u \in W^{1,\infty}(\Omega)$ only implies that there exists K such that $|u(x) - u(y)| \leq K d_\Omega(x, y)$ where $d_\Omega(x, y)$ is the geodesic distance from x to y, the infimum of the lengths of paths from x to y which stay in Ω, and the geodesic distance from x to y could be much larger than the Euclidean distance from x to y.

As a way to ascertain the importance of the regularity of the boundary in proving some properties of Sobolev spaces, I describe a counter-example which I had constructed in order to answer (partially) a question that Sergei VODOP'YANOV[1] had asked a few years ago in a talk at Carnegie Mellon University; I mentioned my result to my good friend Edward FRAENKEL,[2] who had studied domains with irregular boundaries, and he later mentioned it to O'FARRELL,[3] who gave a more general construction. After learning about

---

[1] Sergei Konstantinovitch VODOP'YANOV, Russian mathematician. He works in the Sobolev Institute of Mathematics in Novosibirsk, Russia.

[2] Ludwig Edward FRAENKEL, German-born mathematician, born in 1927. He worked in London, in Cambridge, in Brighton, and in Bath, England.

[3] Anthony G. O'FARRELL, Irish mathematician, born in 1947. He worked at UCLA (University of California at Los Angeles), Los Angeles, CA, and at National University of Ireland, Maynooth, Ireland.

O'FARRELL's construction, I contacted Sergei VODOP'YANOV, who mentioned that he had also solved the case $p > 2$.

Lemma 36.2. *For $N \geq 2$, there exists a (bounded) connected open set Ω such that $W^{1,\infty}(\Omega)$ is not dense in $W^{1,p}(\Omega)$ for $1 \leq p < \infty$.*

Proof: I only give the proof for $N = 2$ and for the case $2 < p < \infty$; for the case $1 \leq p \leq 2$, O'FARRELL has introduced a more technical construction. One defines $\Omega = (A \cup B) \bigcup_n C_n$, where A is the open set $\{(x,y) \mid x > 0, y < 0, x^2 + y^2 < 1\}$, B is the open set $\{(x,y) \mid x < 0, y < 0, x^2 + y^2 < 1\}$, and for $n \geq 1$ the passage C_n is defined in polar coordinates by $\{(x,y) \mid 0 \leq \theta \leq \pi, 2^{-n} - \varepsilon_n < r < 2^{-n}\}$, and the sequence ε_n is chosen to satisfy $0 < \varepsilon_n < 2^{-n-1}$, so that the passages do not overlap, but also $\varepsilon_n \to 0$ sufficiently rapidly so that $\sum_n \varepsilon_n 2^{np} < \infty$ for every $p \in [1, \infty)$.

One checks immediately that the function u_* defined by $u_* = 0$ in A, $u_* = \pi$ in B and $u_* = \theta$ in all C_n belongs to $W^{1,p}(\Omega)$ for $1 \leq p < \infty$. This function u_* cannot be approached in $W^{1,p}(\Omega)$ by functions in $W^{1,\infty}(\Omega)$. Indeed, because A is a bounded open set with a Lipschitz boundary, functions in $W^{1,p}(A)$ have an extension in $W^{1,p}(R^2)$ (Lemma 12.4), which is a continuous function as $p > N = 2$, so that $u \in W^{1,p}(\Omega)$ implies $u \in C^0(\overline{A})$, and the linear form $L_A u = u(0)$ is continuous, where $u(0)$ is computed from the side of A. The same argument applies for B, and the linear form $L_B u = u(0)$ is continuous, where $u(0)$ is now computed from the side of B. One has $L_B(u_*) - L_A(u_*) = \pi$, and u_* cannot be approached by functions from $W^{1,\infty}(\Omega)$ because $L_B(u) = L_A(u)$ for all $u \in W^{1,\infty}(\Omega)$, so that any function v belonging to the closure of $W^{1,\infty}(\Omega)$ must satisfy $L_B(v) = L_A(v)$. Indeed, if $u \in W^{1,\infty}(\Omega)$ one has $|u(x) - u(y)| \leq ||grad(u)||_\infty d_\Omega(x,y)$ where $d_\Omega(x,y)$ is the shortest distance from x to y when one stays in Ω, and letting x tend to 0 from the side of A and y tend to 0 from the side of B, one has $d_\Omega(x,y) \to 0$ because there are arbitrary short paths by using the passages C_n for large n.   □

For the general case $1 \leq p < \infty$, O'FARRELL starts with a *Cantor set of positive measure*[4] in a segment embedded in an open set of R^2, and he constructs passages through the complement of the Cantor set in such a way that going from one side to the other by using the passages can be done with $d_\Omega(x,y) \leq C\, d(x,y)$ for all $x, y \in \Omega$; then he defines a function taking different values on both sides of the Cantor set, which cannot be approached by functions in $W^{1,\infty}(\Omega)$ because they take the same value on both sides of the Cantor set, and for functions in $W^{1,1}(\Omega)$ the restriction to any side of the Cantor set is continuous.

The classical Cantor set has measure zero, and is used in the construction of a nondecreasing function f_∞ on $[0,1]$, with $f_\infty(0) = 0, f_\infty(1) = 1$, but is constant on disjoint intervals $I_n, n \geq 1$ such that $\sum_n length(I_n) = 1$; this

---

[4] Georg Ferdinand Ludwig Philipp CANTOR, Russian-born mathematician, 1845–1918. He worked in Halle, Germany.

construction is often called the "devil's staircase", as it has infinitely many flat levels (steps) and it goes up from 0 to 1 without having any jump.

The particular function f_∞ constructed satisfies $f_\infty(1 - x) = 1 - f_\infty(x)$ for all $x \in [0,1]$, and $f_\infty(3x) = 2f_\infty(x)$ for all $0 \le x \le \frac{1}{3}$, and it is Hölder continuous of order $\alpha = \frac{\log 3}{\log 2}$; it is the unique fixed point of the mapping T defined on functions φ which are continuous on $[0,1]$ with $\varphi(0) = 0$ and $\varphi(1) = 1$, and $T(\varphi) = \psi$ means $\psi(x) = \frac{1}{2}\varphi(3x)$ for $0 \le x \le \frac{1}{3}$, $\psi(y) = \frac{1}{2}$ for $\frac{1}{3} \le y \le \frac{2}{3}$, and $\psi(z) = 1 - \psi(1 - z)$ for $\frac{2}{3} \le z \le 1$; one usually starts from $f_0(x) = x$ for all x and one defines $f_n = T(f_{n-1})$ for $n \ge 1$ and f_n converges to f_∞ uniformly.

[Taught on Friday April 21, 2000.]

Variants with BV Spaces

The characterizations of $W^{s,p}(\Omega)$ and $W^{s,p}(R^N)$ provide a characterization of which functions $u \in W^{s,p}(\Omega)$ are such that the extension of u by 0 outside Ω, denoted by $\tilde{u}$, belong to $W^{s,p}(R^N)$.

Lemma 37.1. *Let Ω be a bounded open set of R^N with a Lipschitz boundary. Then, for $0 < s < 1$ and $1 \le p \le \infty$ one has $\tilde{u} \in W^{s,p}(R^N)$ if and only if $u \in W^{s,p}(\Omega)$ and $d^{-s}u \in L^p(\Omega)$, where $d(x)$ denotes the distance from x to the boundary $\partial\Omega$.*

Proof: One shows the case $p < \infty$, the case $p = \infty$ being easier, using the fact that the functions used are Hölder continuous of order s. $\tilde{u} \in L^p(R^N)$ is equivalent to $u \in L^p(\Omega)$ and $\int\int_{R^N \times R^N} \frac{|\tilde{u}(x)-\tilde{u}(y)|^p}{|x-y|^{N+sp}} \, dx \, dy < \infty$ is equivalent to $\int\int_{\Omega \times \Omega} \frac{|u(x)-u(y)|^p}{|x-y|^{N+sp}} \, dx \, dy < \infty$ and $\int\int_{\Omega \times (R^N \setminus \Omega)} \frac{|u(x)|^p}{|x-y|^{N+sp}} \, dx \, dy < \infty$, i.e., $u \in W^{s,p}(\Omega)$ and $\varphi u \in L^p(\Omega)$ where $|\varphi(x)|^p = \int_{R^N \setminus \Omega} \frac{1}{|x-y|^{N+sp}} \, dy$ for $x \in \Omega$. Because $B(0,d(x)) \subset \Omega$ one has $|\varphi(x)|^p \le \int_{|z| \ge d(x)} \frac{1}{|z|^{N+sp}} \, dz = \frac{c_N}{d(x)^{sp}}$, so that $|\varphi(x)| \le C \, d(x)^{-s}$. This shows that if $u \in W^{s,p}(\Omega)$ and $d^{-s}u \in L^p(\Omega)$ then $\tilde{u} \in W^{s,p}(R^N)$.

In order to prove the other implication, one uses a partition of unity, $\theta_i, i \in I$, and one notices that $v_i = \theta_i u \in W^{s,p}(\Omega)$ and if $\tilde{u} \in W^{s,p}(R^N)$ one has $\tilde{v}_i = \theta_i \tilde{u} \in W^{s,p}(R^N)$, and as I is finite it is enough to show that for each i one has $d^{-s}v_i \in L^p(\Omega)$. This corresponds to proving that $|\varphi| \ge C \, d^{-s}$ in the case of Ω_F when F is Lipschitz continuous; by using the mapping $(x', x_N) \mapsto (x', x_N - F(x'))$ whose inverse is $(x', x_N) \mapsto (x', x_N + F(x'))$ which are both Lipschitz continuous, one has to consider the case $F = 0$, and in that case one has $\int_{y_N < 0} \frac{1}{|x-y|^{N+sp}} \, dy = K \, |x_N|^{-s}$ if $x_N > 0$, by an argument of homogeneity, and K is the value of the integral for $x_N = 1$. $\square$

The devil's staircase is an example of a function which is not absolutely continuous, a term equivalent to having the derivative in L^1, and the derivative in the sense of distribution is actually a nonnegative measure whose support is the Cantor set, and it is useful to show Laurent SCHWARTZ' proof that nonnegative distributions are Radon measures.

Lemma 37.2. *If a distribution $T \in \mathcal{D}'(\Omega)$ is nonnegative in the sense that $\langle T, \varphi \rangle \geq 0$ for all $\varphi \in C_c^\infty(\Omega)$ such that $\varphi \geq 0$, then T is a nonnegative Radon measure.*

Proof: Let K be a compact in Ω and let $\theta \in C_c^\infty(\Omega)$ satisfy $\theta \geq 0$ everywhere and $\theta = 1$ on K. Then for every $\varphi \in C_c^\infty(\Omega)$ with $support(\varphi) \subset K$, one has $-||\varphi||_\infty \theta \leq \varphi \leq ||\varphi||_\infty \theta$, so that $-||\varphi||_\infty \langle T, \theta \rangle \leq \langle T, \varphi \rangle \leq ||\varphi||_\infty \langle T, \theta \rangle$, i.e., $|\langle T, \varphi \rangle| \leq C_K ||\varphi||_\infty$, with $C_K = \langle T, \theta \rangle$. Then one extends this inequality to the case where $\varphi \in C_c(\Omega)$ with $support(\varphi) \subset K$, showing that T is a Radon measure, by applying the preceding inequality to $\varrho_\varepsilon \star \varphi$ for a smoothing sequence ϱ_ε (and using $C_{K'}$ for a larger compact set). Of course one also proves in this way that $\langle T, \varphi \rangle \geq 0$ for all $\varphi \in C_c(\Omega)$ such that $\varphi \geq 0$.   □

In one dimension, one says that a function f has bounded variation if there exists a constant C such that for all N and all increasing sequences $x_1 < x_2 < \ldots < x_N$ one has $\sum_{i=1}^{N-1} |f(x_i) - f(x_{i+1})| \leq C$. One then proves that such a function f has a limit on the left and a limit on the right at every point, with at most a countable number of points of discontinuity, and that $f = g - h$ where g and h are nondecreasing, and as one checks easily by regularization that the derivative of a nondecreasing function is a nonnegative Radon measure, one finds that if f has bounded variation then f' is a Radon measure with finite total mass, and the converse is true, because any Radon measure μ can be written as $\mu_+ - \mu_-$ for nonnegative Radon measures μ_+, μ_-, and every nonnegative Radon measure is the derivative in the sense of distributions of a nondecreasing function.

In order to define functions of bounded variation in more than one space dimension, one needs to define the space $\mathcal{M}_b(R^N)$ of bounded Radon measures.

If K is a compact, then $C(K)$ the space of continuous functions on K equipped with the sup norm is a Banach space, whose dual $\mathcal{M}(K)$ is the space of Radon measures on K, equipped with the dual norm $||\mu|| = \sup_{||\varphi||_\infty \leq 1} |\langle \mu, \varphi \rangle|$. If M_n are distinct points in K then $\mu = \sum_n c_n \delta_{M_n}$ belongs to $\mathcal{M}(K)$ if and only if $\sum_n |c_n| < \infty$ and one has $||\mu|| = \sum_n |c_n|$.

If Ω is open, then $C_c(\Omega)$ and its dual $\mathcal{M}(\Omega)$ of all the Radon measures in Ω are not Banach spaces (nor Fréchet spaces either), and if a sequence M_n of points tends to the boundary $\partial\Omega$ then $\mu = \sum_n c_n \delta_{M_n}$ belongs to $\mathcal{M}(\Omega)$ for all coefficients c_n (as each compact K only contains a finite number of these points).

The space $\mathcal{M}_b(\Omega)$, the space of bounded Radon measures, also called the space of measures with finite (total) mass or the space of measures with finite total variation, is the space of $\mu \in \mathcal{M}(\Omega)$ such that there exists C with $|\langle \mu, \varphi \rangle| \leq C||\varphi||_\infty$ for all $\varphi \in C_c(\Omega)$, and such a μ can be extended to all $\varphi \in C_0(\Omega)$, the space of bounded continuous functions in Ω which tend to 0 at the boundary $\partial\Omega$ of Ω, which is a Banach space with the sup norm, so that

$\mathcal{M}_b(\Omega)$ is the dual of $C_0(\Omega)$ and is a Banach space. If M_n are distinct points in Ω then $\mu = \sum_n c_n \delta_{M_n}$ belongs to $\mathcal{M}_b(\Omega)$ if and only if $\sum_n |c_n| < \infty$ and one has $||\mu|| = \sum_n |c_n|$.

The generalization to more than one space dimension has been studied by Ennio DE GIORGI, FEDERER and Wendell FLEMING, but the earlier names TONELLI[1] and Lamberto CESARI[2,3] are often mentioned.

Definition 37.3. *For an open set $\Omega \subset R^N$, a function u belongs to $BV(\Omega)$, the space of functions of bounded variation in Ω if $u \in L^1(\Omega)$ and $\frac{\partial u}{\partial x_j} \in \mathcal{M}_b(\Omega)$ for $j = 1, \ldots, N$.* □

Lemma 37.4. *$BV(R^N)$ is continuously embedded in $L^{1^*,1}(R^N)$ for $N \geq 2$, and in $L^\infty(R)$ for $N = 1$. If ϱ_ε is a smoothing sequence and $u \in BV(R^N)$ then $\varrho_\varepsilon \star u$ is bounded in $W^{1,1}(R^N)$. Moreover*

$$BV(R^N) = \{u \in L^1(R^N) \mid ||\tau_h u - u||_1 \leq C\,|h| \text{ for all } h \in R^N\}. \quad (37.1)$$

Proof: For $j = 1, \ldots, N$, and $\varphi \in C_c^\infty(R^N)$ one has

$$\begin{aligned}
\langle \tfrac{\partial(\varrho_\varepsilon \star u)}{\partial x_j}, \varphi \rangle &= -\langle \varrho_\varepsilon \star u, \tfrac{\partial \varphi}{\partial x_j} \rangle = -\langle u, \check{\varrho}_\varepsilon \star \tfrac{\partial \varphi}{\partial x_j} \rangle = \\
&-\langle u, \tfrac{\partial(\check{\varrho}_\varepsilon \star \varphi)}{\partial x_j} \rangle = \langle \tfrac{\partial u}{\partial x_j}, \check{\varrho}_\varepsilon \star \varphi \rangle,
\end{aligned} \quad (37.2)$$

and as $\check{\varrho}_\varepsilon \star \varphi$ is continuous with compact support and has a sup norm $\leq C\,||\varphi||_\infty$ one deduces that $|\langle \tfrac{\partial(\varrho_\varepsilon \star u)}{\partial x_j}, \varphi \rangle| \leq C\,||\varphi||_\infty$; as $\tfrac{\partial(\varrho_\varepsilon \star u)}{\partial x_j} \in C^\infty(\Omega)$, it means that $||\tfrac{\partial(\varrho_\varepsilon \star u)}{\partial x_j}||_1 \leq C$. By my improvement of Sobolev's embedding theorem for $p = 1$ (proven by Louis NIRENBERG), $\varrho_\varepsilon \star u$ stays in a bounded set of the Lorentz space $L^{1^*,1}(R^N)$ if $N \geq 2$ (or a bounded set of $L^\infty(R)$ if $N = 1$); of course, $\varrho_\varepsilon \star u$ converges to u in $L^1(R^N)$ strong, and because $L^{1^*,1}(R^N)$ is a dual (as will be proven later), one has $u \in L^{1^*,1}(R^N)$. For $u \in BV(R^N)$ one has $\tau_h u - u \in L^1(R^N)$, and $\varrho_\varepsilon \star (\tau_h u - u) \to \tau_h u - u$ in $L^1(R^N)$ as $\varepsilon \to 0$, but as it is also $\tau_h(\varrho_\varepsilon \star u) - (\varrho_\varepsilon \star u)$ one has $||\tau_h(\varrho_\varepsilon \star u) - (\varrho_\varepsilon \star u)||_1 \leq |h|\,||grad(\varrho_\varepsilon \star u)||_1 \leq C\,|h|$, which gives $||\tau_h u - u||_1 \leq C\,|h|$ for all $h \in R^N$. Conversely if $u \in L^1(R^N)$ and $||\tau_h u - u||_1 \leq C\,|h|$ for all $h \in R^N$, then for $h = t\,e_j$ one has $\frac{\tau_{t\,e_j} u - u}{t} \rightharpoonup \frac{\partial u}{\partial x_j}$ in the sense of distributions as $t \to 0$, but as $\frac{\tau_{t\,e_j} u - u}{t}$ is bounded in $L^1(R^N)$ and $L^1(R^N) \subset \left(C_0(R^N)\right)' = \mathcal{M}_b(R^N)$, the limit belongs to $\mathcal{M}_b(R^N)$. □

The fact that $u \in W^{1,p}(R^N)$ is equivalent to $u \in L^p(R^N)$ and $||\tau_h u - u||_p \leq C\,|h|$ for all $h \in R^N$ is true for $1 < p \leq \infty$, but not for $p = 1$.

---

[1] Leonida TONELLI, Italian mathematician, 1885–1946. He worked in Cagliari, in Parma, in Bologna, and in Pisa, Italy. The department of mathematics of Università di Pisa, Pisa, Italy, is named after him.

[2] Lamberto CESARI, Italian-born mathematician, 1910–1990. He worked in Pisa, in Bologna, Italy, at Purdue University, West Lafayette, IN, and at University of Michigan, Ann Arbor, MI.

[3] John PURDUE, American industrialist, 1802–1876.

However, the difference between $W^{1,1}(R^N)$ and $BV(R^N)$ is not seen for some interpolation spaces defined with these two spaces.

Lemma 37.5. *For $0 < \theta < 1$ and $1 \leq q \leq \infty$ one has*

$$\left(\mathcal{M}_b(R^N), L^\infty(R^N)\right)_{\theta,q} = \left(L^1(R^N), L^\infty(R^N)\right)_{\theta,q} = L^{p,q}(R^N) \ \text{for } p = \frac{1}{1-\theta},$$
$$(37.3)$$

and

$$\left(BV(R^N), \mathcal{M}_b(R^N)\right)_{\theta,q} = \left(BV(R^N), L^1(R^N)\right)_{\theta,q} = B_q^{1-\theta,1}(R^N)$$
$$\left(W^{1,1}(R^N), \mathcal{M}_b(R^N)\right)_{\theta,q} = \left(W^{1,1}(R^N), L^1(R^N)\right)_{\theta,q} = B_q^{1-\theta,1}(R^N).$$
$$(37.4)$$

Proof: One uses the fact that $L^1(R^N) \subset \mathcal{M}_b(R^N)$ and for $u \in L^1(R^N)$ the norm of u in $L^1(R^N)$ and the norm of u in $\mathcal{M}_b(R^N)$ coincide. Let $E_0 = \mathcal{M}_b(R^N)$, $E_1 = L^\infty(R^N)$, and $F_0 = L^1(R^N)$, then $F_0 \subset E_0$, so that $(F_0, E_1)_{\theta,q} \subset (E_0, E_1)_{\theta,q}$; conversely $a \in (E_0, E_1)_{\theta,q}$ means $a = \int_0^\infty u(t) \frac{dt}{t}$ with $u(t) \in E_0 \cap E_1$ a.e. $t \in (0,\infty)$ and $t^{-\theta} \max\{||u(t)||_{E_0}, t\,||u(t)||_{E_1}\} \in L^q\left(R_+; \frac{dt}{t}\right)$, but as $E_0 \cap E_1 = F_0 \cap E_1$ and $||u(t)||_{E_0} = ||u(t)||_{F_0}$, one has $u(t) \in F_0 \cap E_1$ a.e. $t \in (0,\infty)$ and $t^{-\theta} \max\{||u(t)||_{F_0}, t\,||u(t)||_{E_1}\} \in L^q\left(R_+; \frac{dt}{t}\right)$, so that $a \in (F_0, E_1)_{\theta,q}$. The same argument gives

$$\left(BV(R^N), \mathcal{M}_b(R^N)\right)_{\theta,q} = \left(BV(R^N), L^1(R^N)\right)_{\theta,q}$$
$$\left(W^{1,1}(R^N), \mathcal{M}_b(R^N)\right)_{\theta,q} = \left(W^{1,1}(R^N), L^1(R^N)\right)_{\theta,q}.$$
$$(37.5)$$

One observes that $\left(BV(R^N), L^1(R^N)\right)_{\theta,\infty} \subset \left(W^{1,1}(R^N), L^1(R^N)\right)_{\theta,\infty}$, so that these two spaces are equal, and one then uses the reiteration theorem 26.3. Indeed, the linear map $u \mapsto \tau_h u - u \in L^1(R^N)$ has a norm $\leq C\,|h|$ in $BV(R^N)$ and a norm ≤ 2 on $L^1(R^N)$ and therefore a norm $\leq C\,|h|^{1-\theta}$ on $\left(BV(R^N), L^1(R^N)\right)_{\theta,\infty}$, i.e., one has $||\tau_h u - u||_1 \leq C\,|h|^{1-\theta}$ on this space, which is the characterization of elements of $\left(W^{1,1}(R^N), L^1(R^N)\right)_{\theta,\infty}$. $\square$

[Taught on Monday April 24, 2000.]

38

Replacing *BV* by Interpolation Spaces

Solving nonlinear partial differential equations sometimes requires a careful use of adapted functional spaces, and knowing the theory of interpolation spaces is helpful for creating a large family of such spaces, some of them quite useful.

Many of the nonlinear partial differential equations which are studied have their origin in continuum mechanics or physics, but very few mathematicians take time to try to understand what the right equations and the right questions should be, and many work for years on distorted equations without knowing it; there are unfortunately many who know the defects of the models that they use but prefer to hide them in order to pretend that they are working on some useful realistic problem. It is wiser to be aware of the defects of the models, but it happens that very honest mathematicians are unaware of some practical limitations of the equations that they study, and when Jean LERAY told me that he did not want the Germans to know that he had worked on questions of fluid dynamics,[1] and that he had said that he was a topologist, I first thought that it was for fear that his results on the Navier–Stokes equation could be used by the enemy, which would have been very naive, as they were much too theoretical to be of any practical use, but more likely he had meant that he did not want to be forced to work on practical problems in fluid dynamics. Many do not seem to realize that the equations that mathematicians work with under the name of the Navier–Stokes equation are oversimplified and therefore not so realistic, but the motivation of a mathematician for working

---

[1] In 1984, I had mentioned the political difficulties that I was encountering in the French university system, and Jean LERAY had explained to me the origin of the political difficulties that he had encountered himself almost forty years earlier. As an officer in the French army, he had been taken prisoner and he had spent most of World War II in a German camp, while a famous member of the Bourbaki group had dodged the draft; inside the camp, he had continued to do research, and he had even organized a university, of which he was the chancellor. He worked on topology, and soon after introduced the basic ideas for sheaf theory, which another member of the Bourbaki group plagiarized afterward.

on these simplified equations is that Jean LERAY had not been able to solve some particular questions about them; however, I have no good explanation for the fact that a good mathematician who had to describe what the equations are about could show such an appalling level of understanding of continuum mechanics as to forget mentioning Galilean[2] invariance, or the isotropy of the fluid involved, leading to an invariance of the equations under rotations. It would be wiser to know about all these properties, and to mention them, of course, but also to explain what one tries to achieve by working on simplified or deficient models; indeed, if a technical difficulty has been identified on a realistic model, it is easier to try to overcome it by working first on a simplified model, even if it has lost some of the realistic features of the problem that one would like to solve; having explained what one is trying to achieve should help when coming back to more realistic questions afterward.

The space *BV* has been widely used in situations where solutions are discontinuous, but there are reasons to think that this functional space is not adapted to most nonlinear partial differential equations where discontinuous solutions are found; indeed, outside cases where the maximum principle can be used, or in one space dimension, spaces modeled on $W^{1,1}$ or *BV* are not adapted to linear partial differential equations with constant coefficients, so that it is unlikely that they could serve in a linearized version for studying stability of solutions under perturbations.

Of course, there are problems where a space like *BV* is asked for, and geometric measure theory is often the framework that one imposes to look for a domain with finite perimeter (i.e., whose characteristic function belongs to *BV*), or to look for a set with finite $(N-1)$-dimensional Hausdorff measure. In "applications" to minimal surfaces, one often hears about soap bubbles, but hardly anyone working on problems of that type seems to know anything about the chemistry involved in a soapy layer. In "applications" to image processing, hardly anyone working on problems of that type seems to know anything about the physical processes leading to the creation of the images, either optical devices, radar applications, or NMR (nuclear magnetic resonance), whose name has been changed to MRI (medical resonance imaging) because of the fear that the term "nuclear" generates in the grossly uninformed[3] public. However, it is worth repeating that for numerical applications, any nonphysical approach may be used if it leads to an efficient algorithm, but why should one be so interested in some part of mathematics which only leads to inefficient algorithms?

The main class of partial differential equations where discontinuous solutions appear for intrinsic reasons is that of hyperbolic conservation

---

[2] Galileo GALILEI, Italian mathematician, 1564–1642. He worked in Siena, in Pisa, in Padova (Padua), and again in Pisa, Italy.

[3] Obviously, politicians like to keep everyone uninformed, to maintain control of the population, but scientists could certainly do better when it comes to explaining what the resonance of nuclei is, and why it has nothing to do with fission, or with fusion.

laws; this class covers important situations in continuum mechanics, but too little is understood about it from a mathematical point of view. Because of their practical importance, numerical simulations of these problems are performed, for example in order to compute the flow of (compressible) air around an airplane. The (Franco-British) Concorde was the only commercial plane which flew over Mach[4-6] 1, but most commercial planes fly fast enough to require computations of transonic flows; indeed, the speed of sound depends upon the temperature and the pressure, and the shape given to the wings of the plane makes more air go below the wing and creates a slight surpression below the wing and a high depression above the wing; at the cruise velocity of large commercial jets, the speed of sound in that depression (which sucks the plane upward) is then less than the velocity of the plane. Numerical simulations have become much less expensive than using small-scale planes in wind tunnels, and the shape of the plane can be improved (mostly for diminishing the fuel consumption of the plane, rarely for diminishing the noise that it makes), but the mathematical knowledge of these questions, mostly due to the work of GARABEDIAN[7] and of Cathleen MORAWETZ,[8-10] is still insufficient for corroborating the intuition of the engineers.

---

[4] Ernst MACH, Czech-born physicist, 1838–1916. He worked in Graz, Austria, at Charles University in Prague (then in Austria, now capital of the Czech republic), and in Vienna, Austria.

[5] CHARLES IV of Luxembourg, 1316–1378. German king and King of Bohemia (in 1346) and Holy Roman Emperor (in 1355) as Karl IV; he founded the University of Prague in 1348.

[6] The Mach number is the ratio of the velocity of the plane to the speed of sound.

[7] Paul Roesel GARABEDIAN, American mathematician. He works at NYU (New York University), New York, NY.

[8] Cathleen SYNGE MORAWETZ, Canadian-born mathematician, 1923. She works at NYU (New York University), New York, NY. Her father was John Lighton SYNGE, Irish mathematician, 1897–1995, who worked in Dublin, Ireland, but he had also worked in Toronto, Ontario (Canada), at OSU (Ohio State University), Columbus, Ohio, and at Carnegie Tech (Carnegie Institute ot Technology), now CMU (Carnegie Mellon University), Pittsburgh, PA, where he was the head of the mathematics department from 1946 to 1948.

[9] Andrew CARNEGIE, Scottish-born businessman and philanthropist, 1835–1919. Besides endowing the school that became Carnegie Institute ot Technology and later Carnegie Mellon University when it merged with the Mellon Institute of Industrial Research, he funded about three thousand public libraries, named Carnegie libraries in United States.

[10] Andrew William MELLON, American financier and philanthropist, 1855–1937. He funded the Mellon Institute of Industrial Research in Pittsburgh, PA, which merged in 1967 with the Carnegie Institute ot Technology to form Carnegie Mellon University.

For $1 < p < \infty$, the Besov space

$$B_\infty^{1/p,p}(R) = \left(W^{1,p}(R), L^p(R)\right)_{1/p',\infty} = \{u \in L^p(R) \mid ||\tau_h u - u||_p \leq C |h|^{1/p}$$
$$\text{for all } h \in R\}$$

(38.1)

contains discontinuous (piecewise smooth) functions (while for $1 < q < \infty$, the space $B_q^{1/p,p}(R)$ does not contain piecewise smooth discontinuous functions). The excluded case $p = 1$ corresponds to $BV(R)$, and $p = \infty$ corresponds to $L^\infty(R)$, so why think that the case $1 < p < \infty$ is better?

Linear partial differential equations with constant coefficients can be solved by using elementary solutions, and for elliptic equations this leads to singular integrals which can be studied using the Calderón–Zygmund theorem, which requires $1 < p < \infty$. For example, solving $\Delta u = f$ for $f \in L^p(R^N)$ gives $\frac{\partial^2 u}{\partial x_i \partial x_j} \in L^p(R^N)$ for $i, j = 1, \ldots, N$, if $1 < p < \infty$, but the result is false for $p = 1$ or $p = \infty$, and $f \in L^1(R^N)$ does not imply that the derivatives $\frac{\partial u}{\partial x_i}$ belong to $BV(R^N)$.

There are, however, other spaces which can be used for replacement, using the Hardy space $\mathcal{H}^1(R^N)$ instead of $L^1(R^N)$, and the space $BMO(R^N)$ instead of $L^\infty(R^N)$, and indeed, operators defined by singular integrals act from $\mathcal{H}^1(R^N)$ into itself and from $BMO(R^N)$ into itself; actually, the interpolation spaces between $\mathcal{H}^1(R^N)$ and $BMO(R^N)$ are the same as the already studied interpolation spaces between $L^1(R^N)$ and $L^\infty(R^N)$, but these results cannot be derived so easily. However, there is another obstacle, which suggests that the choice $p = 2$ is the only right one.

The (scalar) wave equation in a general medium has the form

$$\varrho(x)\frac{\partial^2 u}{\partial t^2} - \sum_{i,j=1}^{N} \frac{\partial}{\partial x_i}\left(a_{ij}(x)\frac{\partial u}{\partial x_j}\right) = 0, \tag{38.2}$$

where the coefficients ϱ and a_{ij}, $i, j = 1, \ldots, N$ belong to $L^\infty(R^N)$, satisfy the symmetry condition $a_{ji}(x) = a_{ij}(x)$ a.e. $x \in R^N$ for all $i, j = 1, \ldots, N$, satisfy the positivity property $\varrho(x) \geq \varrho_- > 0$ a.e. $x \in R^N$, and also satisfy the ellipticity property that for some $\alpha > 0$ one has $\sum_{i,j=1}^{N} a_{ij}(x)\xi_i\xi_j \geq \alpha |\xi|^2$ for all $\xi \in R^N$, a.e. $x \in R^N$. Under these conditions, the Cauchy problem is well posed if one imposes $u|_{t=0} = u_0 \in H^1(R^N)$ and $\frac{\partial u}{\partial t}\big|_{t=0} = u_1 \in L^2(R^N)$, and one has conservation of total energy, sum of the kinetic energy $\frac{1}{2}\int_{R^N} \varrho(x)\big|\frac{\partial u}{\partial t}\big|^2 dx$, and of the potential energy $\frac{1}{2}\int_{R^N}\left(\sum_{i,j=1}^{N} a_{ij}(x)\frac{\partial u}{\partial x_i}\frac{\partial u}{\partial x_j}\right) dx$. It would seem natural to expect that with smooth coefficients, like for the simplified wave equation in a homogeneous isotropic material $\frac{\partial^2 u}{\partial t^2} - c^2\Delta u = 0$, one could take $u_0 \in W^{1,p}(R^N)$ and $u_1 = 0$ for example and find the solution $u(\cdot, t) \in W^{1,p}(R^N)$ for $t > 0$ (or for $t < 0$ as the wave equation is invariant through time reversal), but Walter LITTMAN[11] has shown that this only happens for $p = 2$.

Could the space $B_\infty^{1/2,2}(R^N)$ then be used for quasi-linear hyperbolic equations, where one expects some discontinuities to occur?

It is useful to compare $BV(R^N)$ with the Besov space $B_\infty^{1/p,p}(R^N)$, which is the space of $u \in L^p(R^N)$ such that $||\tau_h u - u||_p \le C\,|h|^{1/p}$ for all $h \in R^N$.

Lemma 38.1. *One has* $BV(R^N) \cap L^\infty(R^N) \subset B_\infty^{1/p,p}(R^N)$, *and more precisely*

$$\left(BV(R^N), L^\infty(R^N)\right)_{1/p',p} \subset B_\infty^{1/p,p}(R^N). \tag{38.3}$$

If χ *is a characteristic function, then* $\chi \in B_\infty^{1/p,p}(R^N)$ *implies* $\chi \in BV(R^N)$ *(and conversely).*

Proof: The linear mapping $u \mapsto \tau_h u - u$ is continuous from $BV(R^N)$ into $L^1(R^N)$ with norm $\le C\,|h|$, and from $L^\infty(R^N)$ into itself with norm ≤ 2, so that

$$u \mapsto \tau_h u - u \text{ is continuous from } \left(BV(R^N), L^\infty(R^N)\right)_{1/p',p} \text{ into}$$
$$\left(L^1(R^N), L^\infty(R^N)\right)_{1/p',p} = L^p(R^N) \text{ with norm } \le C\,|h|^{1/p}. \tag{38.4}$$

If $\chi \in B_\infty^{1/p,p}(R^N)$ one has $\int_{R^N} |\chi(x-h)-\chi(x)|^p\,dx \le C\,|h|$ for all $h \in R^N$; because χ is a characteristic function, then $\chi(x-h) - \chi(x)$ can only take the values $-1,0,1$, so that $|\chi(x-h)-\chi(x)|^p = |\chi(x-h)-\chi(x)|$, from which one deduces that $||\tau_h \chi - \chi||_1 \le C\,|h|$ for every $h \in R^N$, and as one also has $||\chi||_1 = ||\chi||_p^p$, one deduces that $\chi \in BV(R^N)$ (the first part implying the converse). $\square$

[Taught on Wednesday April 26, 2000.]

---

[11] Walter LITTMAN, American mathematician. He works at University of Minnesota Twin Cities, Minneapolis, MN.

Shocks for Quasi-Linear Hyperbolic Systems

Quasi-linear hyperbolic equations have properties quite different from semi-linear hyperbolic equations, mostly because discontinuities may appear even when initial data are very smooth. For second-order equations modeled on the *linear* wave equation

$$\frac{\partial^2 u}{\partial t^2} - c^2 \Delta u = 0, \tag{39.1}$$

semi-linear equations are of the form

$$\frac{\partial^2 u}{\partial t^2} - c^2 \Delta u = F\left(u, \frac{\partial u}{\partial t}, \frac{\partial u}{\partial x_1}, \dots, \frac{\partial u}{\partial x_N}\right), \tag{39.2}$$

where the higher-order part is linear with constant coefficients, while for *quasi-linear* equations the higher-order part is still linear but with coefficients which depend upon lower-order derivatives; in the case $N = 1$ one has for example

$$\frac{\partial^2 u}{\partial t^2} - f\left(\frac{\partial u}{\partial x}\right) \frac{\partial^2 u}{\partial x^2} = 0. \tag{39.3}$$

This equation was first studied by POISSON in 1807, as a model for a compressible gas; the classical relation $p = c\varrho$ leads to an incorrect value of the velocity of sound (which had been estimated by NEWTON), and LAPLACE seems to have proposed the use of the relation $p = c\varrho^\gamma$; POISSON had found some special solutions (which are called simple waves now), but he left them in an implicit form, so that it took forty years before someone pointed out that there was a problem, which STOKES explained in 1848 by the formation of discontinuities. STOKES computed the correct jump conditions to impose for discontinuous solutions, by using the conservation of mass and the conservation of momentum, and these conditions were rediscovered by RIEMANN in 1860, but instead of being called the Stokes conditions or the Stokes–Riemann conditions, they are now known as the Rankine–Hugoniot conditions.[1,2] The defects of these (isentropic) models were not obvious then, as thermodynamics was barely in its infancy at the time, and even STOKES was wrongly convinced later by THOMSON (who later became Lord Kelvin) and by Rayleigh,[3] who told him that his discontinuous solutions were not physical because they did not conserve energy.[4]

The appearance of discontinuities is more easily seen in first-order equations, for which a classical model[5] is the Burgers equation $\frac{\partial u}{\partial t} + u \frac{\partial u}{\partial x} = 0$. Using the method of characteristic curves, one finds easily the critical time of existence of a smooth solution.

Lemma 39.1. *Let u_0 be a smooth bounded function on R. If u_0 is nondecreasing, there exists a unique smooth solution of $\frac{\partial u}{\partial t} + u \frac{\partial u}{\partial x} = 0$ for $t > 0$ satisfying $u(x,0) = u_0(x)$ for $x \in R$. If $\inf_{x \in R} \frac{du_0}{dx}(x) = -\alpha < 0$, there is a unique smooth solution for $0 < t < T_c = \frac{1}{\alpha}$, and there is no smooth solution over an interval $(0, T)$ with $T > T_c$.*

Proof: Assume that there exists a smooth solution for $0 < t < T$. For $y \in R$ one defines the characteristic curve with initial point y by $\frac{dz(t)}{dt} = u(z(t), t)$ and $z(0) = y$, and one deduces that $\frac{d[u(z(t),t)]}{dt} = \frac{\partial u}{\partial x}(z(t),t) \frac{dz(t)}{dt} + \frac{\partial u}{\partial t}(z(t),t) = 0$, so that $u(z(t), t) = u(z(0), 0) = u_0(y)$ for $0 < t < T$; this gives $\frac{dz(t)}{dt} = u_0(y)$, i.e., $z(t) = y + t\,u_0(y)$. This shows that on the line $x = y + t\,u_0(y)$ for $0 < t < T$ the smooth solution is given by $u(x,t) = u_0(y)$. If u_0 is nondecreasing,

---

[1] William John Macquorn RANKINE, Scottish engineer, 1820–1872. He worked in Glasgow, Scotland.

[2] Pierre Henri HUGONIOT, French engineer, 1851–1887.

[3] John William STRUTT, third baron Rayleigh, English physicist, 1842–1919. He received the Nobel Prize in Physics in 1904. He held the Cavendish professorship at Cambridge, England, 1879–1884.

[4] Around 1880, STOKES was editing his works, but he did not reproduce his 1848 derivation of the jump condition, and he apologized for his "mistake", because he had been (wrongly) convinced by Rayleigh and THOMSON (who became Lord Kelvin in 1892) that his solutions were not physical because they did not conserve energy, so that one deduces that none of them understood well that mechanical energy can be transformed into heat. If one has learnt thermodynamics, one should not disparage these great scientists of the 19th century for their curious mistake, and one should recognize that there are things which take time to understand. One learns now that thermodynamics is not about dynamics, and it is still not so well understood a subject, and mathematicians should pay more attention to it; ignoring thermodynamics, and publishing too much on isentropic equations, for example, tends to make engineers and physicists believe that mathematicians do not know what they are talking about, but believing all the rules of thermodynamics shows that one is lacking critical judgment, as some of the rules are obviously wrong and should be changed into a better theory, that one cannot yet describe, as some parts still have to be discovered.

[5] The function u has the dimension $L\,T^{-1}$ of a velocity; some physicists prefer to write $\frac{\partial u}{\partial t} + c\,u \frac{\partial u}{\partial x} = 0$, where c is a characteristic velocity and u has no dimension.

the mapping $y \mapsto y + t\,u_0(y)$ is a global diffeomorphism from R onto R for any $t > 0$ (it is increasing), so that there exists a unique global smooth solution for all $t > 0$. If there exist $y_1 < y_2$ with $u_0(y_1) > u_0(y_2)$, then the characteristic line with initial point y_1 catches upon the characteristic line with initial point y_2 and a smooth solution cannot exist up to the time of encounter of the two characteristic lines as it would have to take two different values at their intersection; one can easily check that for any $T > T_c$ one can find two characteristic lines which intersect before T.

One can check directly that the solution is well defined for $0 < t < T$ because the mapping $y \mapsto y + t\,u_0(y)$ is a global diffeomorphism from R onto R for $0 < t < T_c$, and in order to show that there is no solution on $(0,T)$ with $T > T_c$, one defines $v = \frac{\partial u}{\partial x}$ and one checks that $\frac{d[v(z(t),t)]}{dt} = [v(z(t),t)]^2$, so that $v(z(t),t) = \frac{u_0'(y)}{1 + t\,u_0'(y)}$, so that the smooth solution satisfies $\lim_{t \to T_c} \sup_{x \in R} \frac{\partial u}{\partial x} = +\infty$. $\square$

The analog of the implicit formula used by POISSON would be to say that the solution must satisfy $u(x,t) = u_0\big(x - t\,u(x,t)\big)$, and seeing the limitation in time on this formula is less obvious.

After the critical time T_c one cannot have smooth solutions, and the correctly defined solution will be discontinuous at some points, so that the product of u by $\frac{\partial u}{\partial x}$ is not defined, and it is important to write the equation in conservation form, $\frac{\partial u}{\partial t} + \frac{\partial (u^2/2)}{\partial x} = 0$, and to consider the weak solutions, i.e., the solutions in the sense of distributions. The jump conditions which STOKES had computed correspond to saying that if $\frac{\partial u}{\partial t} + \frac{\partial v}{\partial x} = 0$ and u,v are continuous on each side of a curve with equation $x = \varphi(t)$ and $s = \varphi'(t)$ denotes the velocity of the (possible) discontinuity at time t, then one has $jump(v) = s\,jump(u)$, where $jump(w) = w\big(\varphi(t)+,t\big) - w\big(\varphi(t)-,t\big) = (w+) - (w-)$ (noticing that changing x into $-x$ changes v into $-v$, changes the sign of $jump(u)$ and of $jump(v)$, but changes also the sign of s).

Unfortunately, there are too many weak solutions; for example, for $u_0 = 0$ there is a global smooth solution which is $u = 0$, but there are infinitely many weak solutions by taking x_0 arbitrary, $a > 0$ arbitrary, and defining u by

$$u(x,t) = \begin{cases} 0 \text{ for } x < x_0 - t\,a \\ u(x,t) = -2a \text{ for } x - t\,a < x_0 \\ u(x,t) = 2a \text{ for } 0 < x < t\,a \\ u(x,t) = 0 \text{ for } t\,a < x. \end{cases} \qquad (39.4)$$

In order to reject all nonphysical weak solutions, one then decides to keep only the discontinuities for which $u_- > u_+$; the method of characteristic curves gives a local Lipschitz continuous solution when one starts from a nondecreasing function and by continuity it gives such a Lipschitz continuous solution in the other case $u_- < u_+$, i.e., when one starts from the discontinuous function jumping from u_- up to u_+ (called a rarefaction wave). This selection rule, called an "entropy" condition,[6] and related to a more general Lax condition, extends to the more general equation

$$\frac{\partial u}{\partial t} + \frac{\partial f(u)}{\partial x} = 0 \tag{39.5}$$

only if the nonlinearity f is a convex function; the case of a general (smooth) f has been solved by Olga OLEINIK,[7] and the Oleinik condition is that one accepts or rejects a discontinuity according to the position of the chord joining $\big(u_-, f(u_-)\big)$ and $\big(u_+, f(u_+)\big)$ with respect to the graph of the function f:

one accepts the discontinuity $u_- > u_+$ if and only if
the chord is above the graph,
one accepts the discontinuity $u_- < u_+$ if and only if   (39.6)
the chord is below the graph.

Under the Oleinik condition there is a unique piecewise smooth solution, but in order to study weak solutions without having to assume that they are piecewise smooth, Eberhard HOPF[8] derived the equivalent Hopf condition

$$\frac{\partial \varphi(u)}{\partial t} + \frac{\partial \psi(u)}{\partial x} \leq 0 \text{ in the sense of distributions for all convex } \varphi, \tag{39.7}$$

and these φ are called "entropy" functions[9] and ψ the corresponding "entropy flux", defined by $\psi' = \varphi' f'$. Peter LAX extended the idea to systems of conservation laws,

$$\frac{\partial U}{\partial t} + \frac{\partial F(U)}{\partial x} = 0, \text{ with } U(x,t) \in R^p, \tag{39.8}$$

but not all functions φ on R^p are entropies, because $\nabla \varphi \nabla F = \nabla \psi$ implies compatibility conditions that φ must satisfy, i.e., $curl(\nabla \varphi \nabla F) = 0$, and the trivial entropies $\varphi(U) = \pm U_i$ for $i = 1, \ldots, p$ just correspond to the given system of equations.[10]

---

[6] By analogy with conditions imposed by thermodynamics for the system of compressible gas dynamics.

[7] Olga Arsen'evna OLEINIK, Ukrainian-born mathematician, 1925–2001. She worked in Moscow, Russia.

[8] Eberhard Frederich Ferdinand HOPF, Austrian-born mathematician, 1902–1983. He worked at MIT (Massachusetts Institute of Technology), Cambridge, MA, in Leipzig, in München (Munich), Germany, and at Indiana University, Bloomington, IN, where I met him in 1980.

[9] Again, this is only by analogy with thermodynamics, and these "entropies" are rarely related with the thermodynamic entropy.

[10] It is not clear if the right notion of solution has been found, but all the physical examples seem to be endowed with a strictly convex "entropy", which sometimes is the total energy!

One way to construct such admissible weak solutions is to consider a regularization by artificial viscosity

$$\frac{\partial U_\varepsilon}{\partial t} + \frac{\partial F(U_\varepsilon)}{\partial x} - \varepsilon \frac{\partial^2 U_\varepsilon}{\partial x^2} = 0 \text{ for } \varepsilon > 0. \tag{39.9}$$

The scalar case with $f(u) = \frac{u^2}{2}$ (also called the Burgers–Hopf equation) was first solved by Eberhard HOPF using a nonlinear change of function which transforms the equation into the linear heat equation, and this transformation was also found by Julian COLE,[11,12] and is now known as the Hopf–Cole transform. The scalar case with a general f was solved by Olga OLEINIK, and the scalar case with more than one space variable,[13] $\frac{\partial u}{\partial t} + \sum_{j=1}^N \frac{\partial f_j(u)}{\partial x_j} = 0$ was obtained by KRUZHKOV.[14]

The important difference between the scalar case and the vectorial case (regularized by artificial viscosity), is that there are simple BV estimates for the scalar case, which are unknown for the vector case; the BV estimates are used to prove convergence by a compactness argument, but any uniform estimate in a Besov space $(B_q^{s,p})_{loc}(R^N)$ with $s > 0$ would be sufficient. Unfortunately, the estimates for the scalar case are based on the maximum principle, and the same argument cannot be extended to systems.

There is another method due to James GLIMM,[15] which proves existence for some systems if the total variation is small enough.

I have introduced another approach, based on the compensated compactness method which I had partly developed with François MURAT, which does not need estimates in Besov spaces, but which requires a special understanding of how to use entropies to generate a kind of compactness; Ron DIPERNA[16,17] was the first to find a way to apply my method to systems.

---

[11] Julian D. COLE, American mathematician, 1925–1999. He worked at Caltech (California Institute of Technology), Pasadena, CA, at UCLA (University of California at Los Angeles), Los Angeles, CA, and at RPI (Rensselaer Polytechnic Institute), Troy, NY.

[12] Kilean VAN RENSSELAER, Dutch diamond merchant, c. 1580–1644.

[13] A scalar equation in N variables is not a good physical model for $N > 1$, because it implies a very strong anisotropy of the space (due to a particular direction of propagation).

[14] Stanislav Nikolaevich KRUZHKOV, Russian mathematician, 1936–1997. He worked in Moscow, Russia.

[15] James G. GLIMM, American mathematician, born in 1934. He worked at MIT (Massachusetts Institute of Technology), Cambridge, MA, at NYU (New York University), New York, NY, and at SUNY (State University of New York), Stony Brook, NY.

[16] Ronald J. DIPERNA, American mathematician, 1947–1989. He worked at Brown University, Providence, RI, at University of Michigan, Ann Arbor, MI, at University of Wisconsin, Madison, WI, at Duke University, Durham, NC, and at UCB (University of California at Berkeley), Berkeley, CA.

[17] Washington DUKE, American industrialist, 1820–1905.

Of course, the preceding list of methods is not exhaustive, and there has been other partially successful approaches.

The BV estimate in the scalar case can be linked to a $L^1(R)$ contraction property, which was noticed by Barbara KEYFITZ[18] in one dimension and by KRUZHKOV in dimension N. This property is strongly related to the maximum principle, and I have noticed with Michael CRANDALL[19] that if a map S from $L^1(\Omega)$ into itself satisfies $\int_\Omega S(f)\,dx = \int_\Omega f\,dx$ for all f, then S is an L^1 contraction if and only if S is order preserving; as order-preserving properties do not occur for realistic systems, one cannot expect L^1 contraction properties for systems; Michael CRANDALL and Andrew MAJDA[20] later applied the same idea for some discrete approximations, the simplest of which for $\frac{\partial u}{\partial t} + \frac{\partial f(u)}{\partial x} = 0$ being the *Lax–Friedrichs scheme*,

$$\frac{1}{\Delta t}\left(U_i^{n+1} - \frac{1}{2}(U_{i-1}^n + U_{i+1}^n)\right) + \frac{1}{2\Delta x}\left(f(U_{i+1}^n) - f(U_{i-1}^n)\right) = 0, \quad (39.10)$$

where U_i^n is expected to approximate $U(i\,\Delta x, n\,\Delta t)$; starting with a bounded initial data u_0, satisfying $\alpha \leq u_0(x) \leq \beta$ a.e. $x \in R$, one chooses for example $U_i^0 = \frac{1}{\Delta x}\int_{i\,\Delta x}^{(i+1)\Delta x} u_0(y)\,dy$ for all i, and the explicit scheme generates the numbers U_i^n for $n > 0$, but one must impose the Courant–Friedrichs–Lewy[21,22] condition (known as the CFL condition)

$$\frac{\Delta t}{\Delta x} \sup_{v \in [\alpha,\beta]} |f'(v)| \leq 1, \quad (39.11)$$

which imposes that the numerical velocity of propagation $\frac{\Delta x}{\Delta t}$ is at least equal to the real velocity of propagation; under this condition one has $\alpha \leq U_i^n \leq \beta$ for all i and all $n > 0$, and it is exactly the condition which imposes that

---

[18] Barbara Lee KEYFITZ, Canadian-born mathematician, born in 1944. She worked at Columbia University, New York, NY, in Princeton, NJ, at Arizona State University, Tempe, AZ, in Houston, TX, and at the FIELDS Institute for Research in Mathematical Sciences, Toronto, Ontario (Canada).

[19] Michael Grain CRANDALL, American mathematician, born in 1940. He worked at Stanford University, Stanford, CA, at UCLA (University of California at Los Angeles), Los Angeles, CA, at University of Wisconsin, Madison, WI, and he works now at UCSB (University of California at Santa Barbara), Santa Barbara, CA.

[20] Andrew Joseph MAJDA, American mathematician, born in 1949. He worked at UCB (University of California at Berkeley), Berkeley, CA, in Princeton, NJ, and at NYU (New York University), New York, NY.

[21] Richard COURANT, German-born mathematician, 1888–1972. He worked in Göttingen, Germany, and at NYU (New York University), New York, NY. The department of mathematics of NYU (New York University) is now named after him, the Courant Institute of Mathematical Sciences.

[22] Hans LEWY, German-born mathematician, 1904–1988. He received the Wolf Prize in 1984. He worked in Göttingen, Germany, at Brown University, Providence, RI, and at UCB (University of California at Berkeley), Berkeley, CA.

U_i^{n+1} is a nondecreasing function of U_{i-1}^n and of U_{i+1}^n, and the $l^1(Z)$ contraction property follows. Of course this approach creates solutions such that $||\tau_h u(\cdot, t) - u(\cdot, t)||_1 \leq ||\tau_h u_0 - u_0||_1$, which gives a BV estimate if $u_0 \in BV(R)$. This scheme is only of order 1 and tends to smooth out the discontinuities too much, but higher-order schemes are not order preserving; there is, however, a class of higher-order schemes, called TVD schemes (total variation diminishing), for which the total variation is not increasing.

Obtaining BV estimates for general systems of conservation laws, or more generally obtaining some estimates on fractional derivatives using interpolation spaces is certainly a difficult open question, and some new ideas or some new functional spaces may be needed for that important question.

[Taught on Friday April 28, 2000.]

Interpolation Spaces as Trace Spaces

The possibility of defining the spaces $(E_0, E_1)_{\theta,p}$ for $0 < \theta < 1$ and $1 \le p \le \infty$ as spaces of traces has been mentioned, and it is time to explain the proof. One notices that if $t^{-\theta}K(t;a)$ is bounded then one can decompose $a = a_0(t) + a_1(t)$ with $||a_0(t)||_0 + t\,||a_1(t)||_1 \le 2K(t;a) \le C\,t^\theta$, so that $a_1(t) \to 0$ in E_1 as $t \to \infty$, and because the traces are taken as $t \to 0$, there will be a change of t into $\frac{1}{t}$.

Lemma 40.1. *Let $0 < \theta < 1$ and $1 \le p \le \infty$. If $v(t) \in E_0$ and $v'(t) \in E_1$ with $t^\theta||v(t)||_0 \in L^p(R_+; \frac{dt}{t})$ and $t^\theta||v'(t)||_1 \in L^p(R_+; \frac{dt}{t})$, then $v(0) \in (E_0, E_1)_{\theta,p}$. Conversely, every element of $(E_0, E_1)_{\theta,p}$ can be written as $v(0)$ with v satisfying the preceding properties.*

Proof: Let $a \in (E_0, E_1)_{\theta,p}$, i.e., $e^{-n\theta}K(e^n; a) \in l^p(Z)$. For $n \in Z$ one chooses a decomposition $a = a_{0,n} + a_{1,n}$ with $a_{0,n} \in E_0, a_{1,n} \in E_1$ such that

$$||a_{0,n}||_0 + e^n||a_{1,n}||_1 \le 2K(e^n; a), \qquad (40.1)$$

and one notices that

$$||a_{0,n+1} - a_{0,n}||_1 = ||a_{1,n+1} - a_{1,n}||_1 \le$$
$$2e^{-(n+1)}K(e^{n+1}; a) + 2e^{-n}K(e^n; a) \le 4e^{-n}K(e^n; a). \qquad (40.2)$$

One defines the function u with values in E_0 by $u(e^n) = a_{0,n}$, and extends u to be affine in each interval (e^n, e^{n+1}); this gives

$$||u(t)||_0 \le \max\{||a_{0,n}||_0, ||a_{0,n+1}||_0\} \le 2K(e^{n+1}; a) \le 2e\,K(e^n; a) \qquad (40.3)$$

on the interval (e^n, e^{n+1}); in the case $p = \infty$, one deduces that $t^{-\theta}||u(t)||_0 \le e^{-n\theta}2e\,K(e^n; a)$ on the interval (e^n, e^{n+1}) and $t^{-\theta}||u(t)||_0 \in L^\infty(R_+; \frac{dt}{t})$, while if $1 \le p < \infty$ one has $\int_{e^n}^{e^{n+1}} t^{-\theta p}||u(t)||_0^p \frac{dt}{t} \le e^{-n\theta p}(2e)^p K(e^n; a)^p$ and

$t^{-\theta}||u(t)||_0 \in L^p(R_+; \frac{dt}{t})$. On the interval (e^n, e^{n+1}) one has $u' = \frac{a_{0,n+1}-a_{0,n}}{e^{n+1}-e^n}$, and

$$||u'(t)||_1 \leq \frac{4e^{-n}K(e^n;a)}{e^{n+1}-e^n} = \frac{4}{e-1}e^{-2n}K(e^n;a). \qquad (40.4)$$

In the case $p = \infty$, one deduces that

$$t^{2-\theta}||u'(t)||_1 \leq e^{(2-\theta)(n+1)}\frac{4}{e-1}e^{-2n}K(e^n;a) = \frac{4e^{2-\theta}}{e-1}e^{-\theta n}K(e^n;a) \quad (40.5)$$

on the interval (e^n, e^{n+1}) and $t^{2-\theta}||u'(t)||_1 \in L^\infty(R_+; \frac{dt}{t})$, while if $1 \leq p < \infty$ one has

$$\int_{e^n}^{e^{n+1}} t^{(2-\theta)p}||u'(t)||_1^p \frac{dt}{t} \leq e^{(2-\theta)(n+1)p}\frac{4^p}{(e-1)^p}e^{-2np}K(e^n;a)^p = \\ C e^{-\theta n p}K(e^n;a)^p, \qquad (40.6)$$

so that $t^{2-\theta}||u'(t)||_1 \in L^p(R_+; \frac{dt}{t})$. Defining $v(s) = u(\frac{1}{s})$ gives $v(s) \to a$ as $s \to 0$, $t^\theta||v(t)||_0 \in L^p(R_+; \frac{dt}{t})$ and $t^\theta||v'(t)||_1 \in L^p(R_+; \frac{dt}{t})$.

Conversely, assume that $t^\theta||v(t)||_0$ and $t^\theta||v'(t)||_1$ belong to $L^p(R_+; \frac{dt}{t})$. Then for $t > \varepsilon > 0$ one has $||v(t) - v(\varepsilon)||_1 \leq \int_\varepsilon^t ||v'(s)||_1 \, ds \leq C t^{1-\theta}$, so that $v(t)$ tends to a limit in $E_0 + E_1$ as s tends to 0. Then using the decomposition $v(0) = v(t) + (v(0) - v(t))$, one has $t^\theta||v(t)||_0 \in L^p(R_+; \frac{dt}{t})$, and from $\frac{||v(0)-v(t)||_1}{t} \leq \frac{1}{t}\int_0^t ||v'(s)||_1 \, ds$ one deduces by Hardy's inequality (13.3) that $t^{\theta-1}||v(0) - v(t)||_1 \in L^p(R_+; \frac{dt}{t})$; then one changes t into $\frac{1}{t}$, or one notices that it says that $v(0) \in (E_1, E_0)_{1-\theta,p}$, which is $(E_0, E_1)_{\theta,p}$. $\qquad \square$

The initial definition of trace spaces by Jacques-Louis LIONS and Jaak PEETRE used four parameters and considered functions satisfying $t^{\alpha_0}u \in L^{p_0}(0, \infty; E_0)$ with $t^{\alpha_1}u' \in L^{p_1}(0, \infty; E_1)$, for suitable parameters $\alpha_0, \alpha_1, p_0, p_1$; they had noticed that the family depended on at most three parameters by changing t into t^λ, but they had also introduced the important parameter θ; it was Jaak PEETRE who later[1] found that the family depended upon only two parameters, and developed the simpler K-method and J-method that have been followed in this course.

---

[1] Jacques-Louis LIONS's interests had switched to other questions concerning the use of functional analysis in linear and then nonlinear partial differential equations, in optimization and control problems and in their numerical approximations. After writing his books with Enrico MAGENES, his interest in interpolation spaces became marginal, but he used the ideas when necessary; after finding a nonlinear framework for interpolating regularity for variational inequalities, he thought of a generalization and he probably found that it was a good problem for a student instead of investigating the question himself. Developing this idea made the first part of my thesis, and the second part answered another (slightly academic) question that he had thought of, and I characterized the traces of functions satisfying $u^3 \in L^2((0,T); H^1(\Omega))$ and $\frac{\partial u}{\partial t} \in L^2((0,T); L^2(\Omega))$.

With the same arguments used in the preceding proposition, the characterization of trace spaces is similar to studying the following variant, where one defines $L_{p_0,p_1}(t;a)$ as

$$L_{p_0,p_1}(t;a) = \inf_{a=a_0+a_1} \left(||a_0||_0^{p_0} + t\,||a_1||_1^{p_1} \right), \qquad (40.7)$$

and one defines the space $(E_0,E_1)_{\theta,p;L}$ as

$$(E_0,E_1)_{\theta,p;L} = \left\{ a \in E_0 + E_1 \mid t^{-\theta} L_{p_0,p_1}(t;a) \in L^p\left(R_+; \frac{dt}{t} \right) \right\}. \qquad (40.8)$$

The lack of homogeneity may look strange, and if trace spaces had not been defined before it would not even be obvious that $(E_0,E_1)_{\theta,p;L}$ is a vector subspace, and it is actually a space already defined.

Lemma 40.2. *For $0 < \theta < 1$ and $1 \le p_0,p_1,p \le \infty$, one has $(E_0,E_1)_{\theta,p;L} = (E_0,E_1)_{\overline{\theta},\overline{p}}$, with $\overline{\theta}$ defined by $\frac{1-\overline{\theta}}{\overline{\theta}} = \frac{1-\theta}{\theta}\frac{p_0}{p_1}$, and $\overline{p} = ((1-\theta)p_0 + \theta\,p_1)p$.*

Proof: If one defines $K_q(t;a) = \left[\inf_{a=a_0+a_1} \left(||a_0||_0^q + t^q||a_1||_1^q \right) \right]^{1/q}$ and one lets q tend to ∞, one obtains $K_\infty(t;a) = \inf_{a=a_0+a_1} (\max\{||a_0||_0, t\,||a_1||_1\})$, and one has $K_\infty(t;a) \le K(t;a) \le 2K_\infty(t;a)$ for all $a \in E_0 + E_1$. Geometrically, for the Gagliardo set associated to a, i.e., $\{(x_0,x_1) \mid$ there exists a decomposition $a = a_0 + a_1$ with $||a_0||_0 \le x_0, ||a_1||_1 \le x_1\}$, the line $x_0 = t\,x_1$ intersects it at $x_0 = K_\infty(t;a), x_1 = \frac{K_\infty(t;a)}{t}$. Similarly one defines $L_\infty(s;a) = \inf_{a=a_0+a_1}(\max\{||a_0||_0^{p_0}, s\,||a_1||_1^{p_1}\})$, and one has $L_\infty(s;a) \le L_{p_0,p_1}(s;a) \le 2L_\infty(s;a)$ for all $a \in E_0 + E_1$, and in order to find the same point of the Gagliardo set one chooses $s = t^{p_1} K_\infty(t;a)^{p_0-p_1}$, so that $K_\infty(t;a)^{p_0} = s\frac{K_\infty(t;a)^{p_1}}{t^{p_1}}$, and one deduces $L_\infty(s;a) = K_\infty(t;a)^{p_0}$; then one notices that $t \mapsto s = t^{p_1} K_\infty(t;a)^{p_0-p_1}$ is a good change of variable, because $t \mapsto K_\infty(t;a)$ is nondecreasing and $t \mapsto \frac{K_\infty(t;a)}{t}$ is nonincreasing, and one deduces that $\frac{ds}{s} = C(t)\frac{dt}{t}$ with $\min\{p_0,p_1\} \le C(t) \le \max\{p_0,p_1\}$ for all $t > 0$, so that $\int_0^\infty s^{-\theta\,p}L_\infty(s;a)^p\,\frac{ds}{s} < \infty$ is equivalent to $\int_0^\infty t^{-\theta\,p\,p_1} K_\infty(s;a)^{p\,p_0-\theta\,p(p_0-p_1)}\,\frac{dt}{t} < \infty$. This gives the condition $\overline{p} = p\,p_0 - \theta\,p(p_0-p_1) = (1-\theta)p_0\,p + \theta\,p_1\,p$, and $\overline{\theta}\overline{p} = \theta\,p_1\,p = \overline{p} - (1-\theta)p_0p$, so that $(1-\overline{\theta})\overline{p} = (1-\theta)p_0p$ and eliminating $\overline{p}$ between these last two formulas gives the desired formula for $\overline{\theta}$. $\qquad\square$

Lemma 40.3. *If $u \in W^{1,p}(R^N)$, then its trace $\gamma_0 u$ on $x_N = 0$ satisfies*

$$\gamma_0 u \in W^{1/p',p}(R^{N-1}) = B_p^{1/p',p}(R^{N-1}) = \left(W^{1,p}(R^{N-1}), L^p(R^{N-1}) \right)_{1/p,p}. \qquad (40.9)$$

Proof: Using x_N as the variable t (and using only the fact that $u \in W^{1,p}(R_+^N)$), one finds that $u \in W^{1,p}(R^N)$ implies $u \in L^p\left(0,\infty; W^{1,p}(R^{N-1})\right)$ and

$u' \in L^p(0, \infty; L^p(R^{N-1}))$, and this corresponds to $\theta = \frac{1}{p}$ for $E_0 = W^{1,p}(R^{N-1})$ and $E_1 = L^p(R^{N-1})$.   □

In defining interpolation spaces, one has not really used the fact that E_0 and E_1 are normed vector spaces, and one can extend the theory to the case of commutative (Abelian) groups, and moreover the norm can be replaced by a quasi-norm, satisfying $[a] \geq 0, [-a] = [a]$ for all a, $[a] = 0$ if and only if $a = 0$, and the c-triangle inequality $[a + b] \leq c([a] + [b])$ for all a, b (one calls then $[\cdot]$ a c-norm). One notices that if one defines ϱ by $(2c)^{\varrho} = 2$ then there is a 1-norm $|| \cdot ||$ such that $||a|| \leq [a]^{\varrho} \leq 2||a||$ for all a, and such a norm is defined by $||a|| = \inf_{a = \sum_{i=1}^{n} a_i} [a_i]^{\varrho}$, where the infimum is taken over all n and all decompositions of a.

One can define the space $(E_0, E_1)_{\theta, p}$ if E_0, E_1 are quasi-normed Abelian groups, and with $0 < \theta < 1$ as usual, but for a larger family of p, as one may take $0 < p \leq \infty$, obtaining a quasi-normed space in the case $0 < p < 1$, even if E_0 and E_1 are normed spaces.

If $|| \cdot ||$ is a 1-norm and for $\alpha > 0$ one defines $[u] = ||u||^{\alpha}$, then $[\cdot]$ is a c-norm if one has $(a + b)^{\alpha} \leq c(a^{\alpha} + b^{\alpha})$ for all $a, b \geq 0$, and one checks easily that one may take $c = 1$ if $\alpha \leq 1$ and $c = 2^{\alpha-1}$ if $\alpha \geq 1$. One may then consider the variant $L_{p_0,p_1}(t; a)$ for $0 < p_0, p_1 \leq \infty$, as a particular case of using quasi-norms.

[Taught on Monday May 1, 2000.]

Duality and Compactness for Interpolation Spaces

To characterize the dual of $(E_0, E_1)_{\theta,p}$ for $0 < \theta < 1$ and $1 \leq p < \infty$, one needs a few technical results.

First of all there is a new hypothesis, that $E_0 \cap E_1$ is dense in E_0 and dense in E_1; this implies that E_0' and E_1' are subspaces of $(E_0 \cap E_1)'$, because if j_0 is the injection of $E_0 \cap E_1$ into E_0, then j_0^T is a continuous mapping from E_0' into $(E_0 \cap E_1)'$ whose range is dense, while the hypothesis that the range of j_0 is dense implies that j_0^T is injective. Therefore $E_0' \cap E_1'$ and $E_0' + E_1'$ are well defined. One denotes by $|| \cdot ||_k$ the norm on E_k and by $|| \cdot ||_{*k}$ the norm on E_k' for $k = 1, 2$.

Lemma 41.1. *Assume that $E_0 \cap E_1$ is dense[1] in E_0 and dense in E_1. For $t > 0$, the dual space of $E_0 \cap E_1$ equipped with the norm $J(t; a; E_0, E_1) = \max\{||a||_0, t\,||a||_1\}$ is $E_0' + E_1'$ equipped with the norm $K\left(\frac{1}{t}; b; E_0', E_1'\right) = \inf_{b_0 + b_1 = b}\left(||b_0||_{*0} + \frac{1}{t}\,||b_1||_{*1}\right)$.*

Proof: If $a \in E_0 \cap E_1$ and $b \in E_0' + E_1'$ with a decomposition $b = b_0 + b_1, b_0 \in E_0', b_1 \in E_1'$, then one has $|\langle b, a \rangle| \leq |\langle b_0, a \rangle| + |\langle b_1, a \rangle| \leq ||b_0||_{*0}||a||_0 + \frac{1}{t}||b_1||_{*1}t\,||a||_1 \leq \left(||b_0||_{*0} + \frac{1}{t}||b_1||_{*1}\right)\max\{||a||_0, t\,||a||_1\}$, so that by minimizing among all the decompositions of b one deduces that

$$|\langle b, a \rangle| \leq K\left(\frac{1}{t}, b; E_0', E_1'\right) J(t; a; E_0, E_1) \text{ for all } a \in E_0 \cap E_1, b \in E_0' + E_1', t > 0.$$

(41.1)

Conversely, let L be a linear continuous form on $E_0 \cap E_1$ equipped with the norm $J(t; a; E_0, E_1)$, and let $M = ||L||$, so that $|L(a)| \leq M \max\{||a||_0, t\,||a||_1\}$ for all $a \in E_0 \cap E_1$; one uses the Hahn–Banach theorem in order to find a linear continuous form L_* on $E_0 \times E_1$, equipped with the norm $||(a_0, a_1)|| = \max\{||a_0||_0, t\,||a_1||_1\}$, which extends the linear form $\widetilde{L}$ defined on the diagonal

---

[1] Of course, this precludes using $E_0 = L^1(\Omega)$ and $E_1 = L^\infty(\Omega)$ directly, but the same result is valid, and proven using the reiteration theorem 26.3 together with Lemma 41.1 applied to $E_0 = L^1(\Omega)$ and $E_1 = L^p(\Omega)$ for some $p < \infty$.

of $(E_0 \cap E_1) \times (E_0 \cap E_1)$ by $\widetilde{L}(a,a) = L(a)$, for which one has $|\widetilde{L}(a,a)| = |L(a)| \le M \max\{||a||_0, t\,||a||_1\} = M\,||(a,a)||$, so that $||\widetilde{L}|| \le M$ and therefore there exists an extension L_* on $E_0 \times E_1$ satisfying $||L_*|| \le M$. Any linear continuous form on $E_0 \times E_1$ can be written as $(a_0, a_1) \mapsto \langle b_0, a_0 \rangle + \langle b_1, a_1 \rangle$ with $b_0 \in E_0'$ and $b_1 \in E_1'$, and the norm of that linear continuous form is obviously $\le ||b_0||_{*0} + \frac{1}{t}||b_1||_{*1}$, but it is actually equal to that quantity because one can choose $a_0 \in E_0$ with $||a_0||_0 = 1$ and $\langle b_0, a_0 \rangle = ||b_0||_{*0}$ and $a_1 \in E_1$ with $||a_1||_1 = \frac{1}{t}$ and $\langle b_1, a_1 \rangle = \frac{1}{t}||b_1||_{*1}$ (again by an application of the Hahn–Banach theorem); one then deduces that $L(a) = \widetilde{L}(a,a) = \langle b_0, a \rangle + \langle b_1, a \rangle = \langle b_0 + b_1, a \rangle$ for all $a \in E_0 \cap E_1$, and $||b_0||_{*0} + \frac{1}{t}||b_1||_{*1} \le M$, so that L is given by the element $b = b_0 + b_1 \in E_0' + E_1'$, which satisfies $K\left(\frac{1}{t}; b; E_0', E_1'\right) \le M$.

It remains to show that b is defined in a unique way, i.e., that $\langle b, a \rangle = 0$ for all $a \in E_0 \cap E_1$ implies $b = 0$; indeed, because $|\langle b_0, a \rangle| = |\langle b_1, a \rangle| \le C\,||a||_1$ for all $a \in E_0 \cap E_1$ and $E_0 \cap E_1$ is dense in E_1, b_0 extends in a unique way to an element of E_1', which then coincides with $-b_1$ on the dense subspace $E_0 \cap E_1$ of E_1, so that one has $b_0, b_1 \in E_0' \cap E_1'$ and $b_0 + b_1 = 0$.   □

Lemma 41.2. *Assume that $E_0 \cap E_1$ is dense in E_0 and dense in E_1. For $s > 0$, the dual space of $E_0 + E_1$ equipped with the norm $K(s; a; E_0, E_1) = \inf_{a_0 + a_1 = a}\left(||a_0||_0 + s\,||a_1||_1\right)$ is $E_0' \cap E_1'$ equipped with the norm $J\left(\frac{1}{s}; b; E_0', E_1'\right) = \max\{||b||_{*0}, \frac{1}{s}||b||_{*1}\}$.*

Proof: If $a \in E_0 + E_1$ and $b \in E_0' \cap E_1'$ with a decomposition $a = a_0 + a_1$, then one has $|\langle b, a \rangle| \le |\langle b, a_0 \rangle| + |\langle b, a_1 \rangle| \le ||b||_{*0}||a_0||_0 + \frac{1}{s}||b||_{*1}s\,||a_1||_1 \le \max\{||b||_{*0}, \frac{1}{s}||b||_{*1}\}(||a||_0 + s\,||a||_1)$, so that by minimizing among all the decompositions of a one deduces that

$$|\langle b, a \rangle| \le J\left(\frac{1}{s}, b; E_0', E_1'\right) K(s; a; E_0, E_1) \text{ for all } a \in E_0 + E_1, b \in E_0' \cap E_1', s > 0.$$

(41.2)

Conversely, if L is a linear continuous form on $E_0 + E_1$ equipped with the norm $K(s; a; E_0, E_1)$, then it is a linear continuous form on E_0 and also a linear continuous form on E_1, so that L is given by an element $b \in E_0' \cap E_1'$.

For computing the norm of L, for $0 < \varepsilon < 1$, $x, y > 0$ one chooses $a_0 \in E_0$ with $||a_0||_0 = x$ and $\langle b, a_0 \rangle \ge (1 - \varepsilon)x\,||b||_{*0}$ and $a_1 \in E_1$ with $||a_1||_1 = y$ and $\langle b, a_1 \rangle \ge (1 - \varepsilon)y\,||b||_{*1}$, so that $a = a_0 + a_1$ satisfies $K(s; a; E_0, E_1) \le x + sy$ and $|\langle b, a \rangle| \ge (1 - \varepsilon)(x\,||b||_{*0} + y\,||b||_{*1})$, so that $||L|| \ge (1 - \varepsilon)\frac{x\,||b||_{*0} + y\,||b||_{*1}}{x + sy}$, and letting ε tend to 0 and either x or y tend to 0, one finds $||L|| \ge \max\{||b||_{*0}, \frac{1}{s}||b||_{*1}\}$.   □

The main result about duality for interpolation spaces, which has already been mentioned for Lorentz spaces, is the following.

Lemma 41.3. *Assume that $E_0 \cap E_1$ is dense in E_0 and dense in E_1. Then, for $0 < \theta < 1$ and $1 \le q < \infty$, one has $\left((E_0, E_1)_{\theta,q}\right)' = (E_0', E_1')_{\theta,q'}$ with equivalent norms, where $\frac{1}{q} + \frac{1}{q'} = 1$.*

Proof: We shall prove that

$$\left((E_0, E_1)_{\theta,q;J}\right)' \subset (E_0', E_1')_{\theta,q';K} \quad \text{and} \quad (E_0', E_1')_{\theta,q';J} \subset \left((E_0, E_1)_{\theta,q;K}\right)',$$
(41.3)

and the result will follow from the identity (with equivalent norms) between the interpolation spaces defined by the J-method and the K-method.

In order to prove the first inclusion, one takes $a' \in \left((E_0, E_1)_{\theta,q;J}\right)'$, and as a' defines a linear continuous form on $E_0 \cap E_1$, a preceding lemma asserts that for every $\varepsilon > 0$ there exists a sequence $b_n \in E_0 \cap E_1$ such that $J(2^n; b_n; E_0, E_1) = 1$ and $K(2^{-n}; a'; E_0', E_1') - \varepsilon \min\{1, 2^{-n}\} \le \langle a', b_n \rangle$. For a sequence α_n such that $2^{-\theta n}\alpha_n \in l^q(Z)$, one defines $a(\alpha) = \sum_{n \in Z} \alpha_n b_n$, and one has $a(\alpha) \in (E_0, E_1)_{\theta,q;J}$ with $||a(\alpha)||_{\theta,q;J} \le ||2^{-\theta n}\alpha_n||_q$ because of the normalization chosen for the sequence b_n; the particular choice for the sequence b_n implies $\sum_{n \in Z} \alpha_n\big(K(2^{-n}; a'; E_0', E_1') - \varepsilon \min\{1, 2^{-n}\}\big) \le \langle a', a(\alpha)\rangle \le ||a'|| \, ||a(\alpha)||_{\theta,q;J} \le ||a'|| \, ||2^{-\theta n}\alpha_n||_q$ (notice that $\alpha_n \min\{1, 2^{-n}\} \in l^1$); by letting ε tend to 0 one deduces that $\sum_{n \in Z} \alpha_n K(2^{-n}; a'; E_0', E_1') \le ||a'|| \, ||2^{-\theta n}\alpha_n||_q$ for all such sequences α_n. As $\sum_{n \in Z} \alpha_n \beta_n \le M \, ||2^{-\theta n}\alpha_n||_q$ for all sequences α_n is equivalent to $||2^{\theta n}\beta_n||_{q'} \le M$, one deduces that $||2^{\theta n}K(2^{-n}; a'; E_0', E_1')||_{q'} \le ||a'||$, i.e., $a' \in (E_0', E_1')_{\theta,q';K}$.

In order to prove the second inclusion, one takes $a' \in (E_0', E_1')_{\theta,q';J}$, and one writes $a' = \sum_n a_n'$ with $a_n' \in E_0' \cap E_1'$ and $2^{-\theta n}J(2^n; a_n'; E_0', E_1') \in l^{q'}(Z)$; then for $a \in (E_0, E_1)_{\theta,q;K}$ one has

$$|\langle a', a\rangle| \le \sum_n |\langle a_n', a\rangle| \le \sum_n J(2^n; a_n'; E_0', E_1')K(2^{-n}; a; E_0, E_1) \le$$
$$M \, ||2^{\theta n}K(2^{-n}; a; E_0, E_1)||_{\ell^q(Z)} \le M \, ||a||_{\theta,q;K}. \square$$
(41.4)

When I was a student I had noticed that for $0 < \theta < 1$ the space $(E_0', E_1')_{\theta,1}$ is actually a dual, although not the dual of $(E_0, E_1)_{\theta,\infty}$; I had mentioned that to my advisor, Jacques-Louis LIONS, and he had told me that Jaak PEETRE had already made that observation.[2] The idea is to observe that l^1 is the dual of c_0, and that one can define a new[3] interpolation space modeled on c_0, considering that the usual interpolation space indexed by θ, p is actually modeled on $l^p(Z)$. For $0 < \theta < 1$ and two Banach spaces E_0, E_1 continuously embedded into a common topological vector space, one defines the space $(E_0, E_1)_{\theta;c_0}$

---

[2] It is quite natural that in the process of doing research one finds results which have already been found before, and Ennio DE GIORGI had once said "chi cerca trova, chi ricerca ritrova" (the first part reminds one of the "seek and you will find" from the gospels, but the play on the prefix does not work well in English, although one could replace seek and find by search and discover in order to use research and rediscover). Sometimes, one may find the result by a different method and it may be worth publishing if the new proof is simpler than the previous one, or if it contains ideas which could be useful for other problems; of course, one should mention the author of the first proof, even if he/she has not published it.

[3] Obviously, one can describe more general classes of interpolation spaces, and Jaak PEETRE has actually developed a quite general framework for doing that.

as the space of $a \in E_0 + E_1$ such that $2^{-\theta n} K(2^n; a) \in c_0(Z)$, equipped with same norm as $(E_0, E_1)_{\theta,\infty}$ (this space is the closure of $E_0 \cap E_1$ in $(E_0, E_1)_{\theta,\infty}$). The proof of the previous proposition easily extends to show that the dual of $(E_0, E_1)_{\theta;c_0}$ is $(E_0', E_1')_{\theta,1}$ with an equivalent norm.

Another useful result concerning interpolation spaces is the question of compactness.

Lemma 41.4. *If A is a linear mapping from a normed space F into $E_0 \cap E_1$, such that A is linear continuous from F into E_0 and compact from F into E_1, then for $0 < \theta < 1$ the mapping A is compact from F into $(E_0, E_1)_{\theta,1}$ (and therefore compact from F into $(E_0, E_1)_{\theta,p}$ for $1 \le p \le \infty$).*

If B is a linear mapping from $E_0 + E_1$ into a normed space G, such that B is linear continuous from E_0 into G and compact from E_1 into G, then for $0 < \theta < 1$ the mapping B is compact from $(E_0, E_1)_{\theta,\infty}$ into G (and therefore compact from $(E_0, E_1)_{\theta,p}$ into G for $1 \le p \le \infty$).

Proof: If $\|f_n\|_F \le 1$, then $A f_n$ is bounded in E_0 and belongs to a compact subset of E_1, so that a subsequence f_m converges in E_1, so that it is a Cauchy sequence; one has $\|x\|_{\theta,1} \le C \|x\|_0^{1-\theta} \|x\|_1^{\theta}$ for all $x \in (E_0, E_1)_{\theta,1}$, and applying this inequality to $x = f_m - f_{m'}$ one deduces that $A f_m$ is a Cauchy sequence in $(E_0, E_1)_{\theta,1}$, so that A is compact from F into $(E_0, E_1)_{\theta,1}$.

Let $\|e_n\|_{\theta,\infty} \le 1$ so that for each $\varepsilon > 0$ there exists a decomposition $e_n = e_n^0 + e_n^1$ with $\|e_n^0\|_0 + \varepsilon \|e_n^1\|_1 \le 2K(\varepsilon; e_n) \le 2\varepsilon^{\theta}$; then $\|B e_n^0\|_G \le 2\|B\|_{\mathcal{L}(E_0,G)}\varepsilon^{\theta}$, and $B e_n^1$ belongs to a compact subset of G so that a subsequence $B e_m^1$ converges in G, and therefore for this subsequence one has $\limsup_{m,m'\to\infty} \|B e_m^1 - B e_{m'}^1\|_G \le 4\|B\|_{\mathcal{L}(E_0,G)}\varepsilon^{\theta}$; using Cantor diagonal subsequence argument one finds that $B e_n$ contains a converging subsequence in G, so that B is compact from $(E_0, E_1)_{\theta,\infty}$ into G.   □

[Taught on Wednesday May 3, 2000.]

Miscellaneous Questions

During 1974–1975, I spent a year at University of Wisconsin, Madison, WI, and I often had discussions with Michael CRANDALL; among our joint results that we did not publish there was a question related to interpolation. The motivation for looking at the problem was some kind of generalization which had been published, for which it was not clear if there was any example showing that it was indeed a genuine[1] generalization, and as our result did not cover the same situations as the published theorem, it might well have been more general than the previous ones in some cases. Although we started by proving some observations for linear mappings, and then extended the method to a nonlinear setting, I present the results in reverse order.

Lemma 42.1. *Let Ω be a bounded (Lebesgue) measurable subset of R^N and let F be a nonlinear mapping from $L^\infty(\Omega)$ into itself satisfying the following properties:*

(a) F is Lipschitz continuous from $L^\infty(\Omega)$ into itself,

(b) F is monotone (in the L^2 sense), i.e., $\int_\Omega \big(F(u_2) - F(u_1)\big)(u_2 - u_1)\,dx \geq 0$ for all $u_1, u_2 \in L^\infty(\Omega)$.

Then for every $p > 2$, F is Lipschitz continuous from $L^p(\Omega)$ into itself (i.e., F is Lipschitz continuous with respect to the distance L^p, and hence it extends in a unique way as a Lipschitz continuous mapping from $L^p(\Omega)$ into itself).

---

[1] My advisor had once mentioned that when reading a highly abstract article one should first look at the examples. On a previous occasion, I had applied his advice and shown that a generalization of the Lax–Milgram lemma was not a genuine one, and that all the examples of the proposed new theory could be dealt with in a classical way, once a particular observation had been made; a friend insisted that I publish the observation, and it became my shortest article. Of course, in such situations, it is better to avoid mentioning names, and one should remember that even the best mathematicians have made mistakes (I was told that a great mathematician had his ego a little bruised after publishing a perfectly good proof, when he realized that no object satisfied all the hypotheses of his theorem, which therefore was a quite useless one).

Proof: Let M_∞ be the Lipschitz constant for F with respect to the distance L^∞. For $0 < \varepsilon < \frac{1}{M_\infty}$ and $v \in L^\infty(\Omega)$ there is a unique $u \in L^\infty(\Omega)$ solution of the equation $u + \varepsilon F(u) = v$, as it is the unique fixed point of the mapping $u \mapsto v - \varepsilon F(u)$, which is a strict contraction; moreover the mapping $v \mapsto u$ is Lipschitz continuous with a constant $\leq \frac{1}{1-\varepsilon M_\infty}$, and if one defines the mapping G_ε by $G_\varepsilon(v) = v - u = \varepsilon F(u)$ then G_ε is Lipschitz continuous with a constant $\leq \frac{\varepsilon M_\infty}{1-\varepsilon M_\infty}$.

For $v_1, v_2 \in L^\infty(\Omega)$, one subtracts $u_1 + \varepsilon F(u_1) = v_1$ from $u_2 + \varepsilon F(u_2) = v_2$ and one multiplies by $F(u_2) - F(u_1)$, giving $\varepsilon \int_\Omega |F(u_2) - F(u_1)|^2 \, dx \leq \varepsilon \int_\Omega |F(u_2) - F(u_1)|^2 \, dx + \int_\Omega (F(u_2) - F(u_1))(u_2 - u_1) \, dx = \int_\Omega (F(u_2) - F(u_1))(v_2 - v_1) \, dx$, so that $||G_\varepsilon(v_2) - G_\varepsilon(v_1)||_2 \leq ||v_2 - v_1||_2$. In particular G_ε has a unique extension to $L^2(\Omega)$ (which is a contraction).

Having shown that G_ε is Lipschitz continuous on $L^\infty(\Omega)$ with a constant $\leq \frac{\varepsilon M_\infty}{1-\varepsilon M_\infty}$ and Lipschitz continuous on $L^2(\Omega)$ with a constant ≤ 1, one deduces by nonlinear interpolation that G_ε is Lipschitz continuous on $L^p(\Omega)$ with a constant $\leq \left(\frac{\varepsilon M_\infty}{1-\varepsilon M_\infty}\right)^\theta$ if θ is defined by $\frac{1}{p} = \frac{\theta}{\infty} + \frac{1-\theta}{2}$, and as one has $\theta = \frac{p-2}{p} > 0$ one deduces that the constant may be made small by taking ε small. Assuming then that ε has been chosen small enough so that the Lipschitz constant of G_ε in $L^p(\Omega)$ is $\leq K < 1$, then for every $u_1, u_2 \in L^\infty(\Omega)$, one defines $v_j = u_j + \varepsilon F(u_j)$ for $j = 1, 2$, and one deduces that $\varepsilon ||F(u_2) - F(u_1)||_p = ||G(v_2) - G(v_1)||_p \leq K ||v_2 - v_1||_p \leq K (||u_2 - u_1||_p + \varepsilon ||F(u_2) - F(u_1)||_p)$, so that $||F(u_2) - F(u_1)||_p \leq M_p ||u_2 - u_1||_p$ with $M_p = \frac{K}{\varepsilon(1-K)}$. □

In the case of linear mappings, the idea is that one may use unbounded (densely defined) closed operators by considering their resolvents, i.e., the bounded operators $(A - \lambda I)^{-1}$ for some $\lambda \in C$. In particular, if for a real Hilbert space H a closed unbounded operator A has a dense domain and satisfies $\langle A u, u \rangle \geq 0$ for all $u \in D(A)$, then for $\lambda < 0$ the resolvent exists and is a contraction. Then for $\varepsilon > 0$ the bounded operator $I - (I + \varepsilon A)^{-1}$ is also a contraction, while if A is bounded its norm is $O(\varepsilon)$, and by interpolation it will have a small norm in an interpolation space and it will show that the operator is bounded in that space. The preceding proposition has followed the same scenario in a nonlinear setting, but one can deduce more in a linear setting by using the spectral radius of an operator.[2]

Lemma 42.2. *Let A be a linear mapping from $E_0 + E_1$ into itself satisfying $A \in \mathcal{L}(E_0; E_0)$ with spectral radius ϱ_0 and $A \in \mathcal{L}(E_1; E_1)$ with spectral radius ϱ_1 (where E_0 and E_1 are Banach spaces). Then for $0 < \theta < 1$ and $1 \leq p \leq \infty$, the spectral radius $\varrho(\theta, p)$ of A on $(E_0, E_1)_{\theta,p}$ satisfies the inequality $\varrho(\theta, p) \leq \varrho_0^{1-\theta} \varrho_1^\theta$.*

Proof: One uses the fact that $\varrho(A) = \lim_{n \to \infty} ||A^n||^{1/n}$, and the fact that $||A^n||_{\mathcal{L}((E_0;E_1)_{\theta,p},(E_0,E_1)_{\theta,p})} \leq C ||A^n||_{\mathcal{L}(E_0;E_0)}^{1-\theta} ||A^n||_{\mathcal{L}(E_1;E_1)}^\theta$ with C depending

---

[2] If $A \in \mathcal{L}(E; E)$ for a Banach space E, the spectrum of A is the nonempty closed set of $\lambda \in C$ for which $A - \lambda I$ is not invertible, and the spectral radius $\varrho(A)$ is the maximum of $|\lambda|$ for λ in the spectrum of A.

upon which equivalent norm is used for $(E_0, E_1)_{\theta,p}$ (but not on n); taking the power $\frac{1}{n}$ and letting n tend to ∞ gives the result.   □

If f is holomorphic, then the spectrum of $f(A)$ is the image by f of the spectrum of A, and using the preceding lemma one deduces that if K_0 is the spectrum of A in E_0, K_1 is the spectrum of A in E_1, and $K_{\theta,p}$ is the spectrum of A in $(E_0, E_1)_{\theta,p}$, then for every holomorphic function f one has $\max_{z \in K_{\theta,p}} |f(z)| \le (\max_{z \in K_0} |f(z)|)^{1-\theta} (\max_{z \in K_1} |f(z)|)^{\theta}$. I once asked Ciprian FOIAS[3] if he knew some situation where the spectrum strongly depends upon the space used, but I did not understand his answer.

I do not know any good example of applications of these results obtained with Michael CRANDALL, which is one reason why I never wrote that proof before, but I find interesting the fact that with respect to interpolation a monotonicity property is almost like a Lipschitz condition. Actually, the last inequality giving a localization of the spectrum suggests that one could develop notions of interpolation of sets.

Finally, I want to mention a result which I found a few years ago, while I was teaching a graduate course on mathematical methods in control, because I wanted to explain the following result of Yves MEYER,[4] which he had used in connection with a control problem.[5]

Lemma 42.3. *Let $d\mu$ be a Radon measure on R and $T > 0$. A necessary and sufficient condition that there exists a constant $C(T)$ such that*

$$\int_R |\mathcal{F}f(\xi)|^2 \, d\mu \le C(T) \int_0^T |f(x)|^2 \, dx \tag{42.1}$$

for all functions $f \in L^2(R)$ which vanish outside $(0, T)$

is that

$$\sup_{k \in Z} \mu([k, k+1]) \le C' < \infty. \quad □ \tag{42.2}$$

---

[3] Ciprian Ilie FOIAS, Romanian-born mathematician, born in 1933. He worked in Bucharest, Romania, at Université Paris Sud XI, Orsay, France (where he was my colleague in 1978–1979), at Indiana University, Bloomington, IN, and at Texas A&M, College Station, TX.

[4] Yves François MEYER, French mathematician, born in 1939. He worked at Université Paris Sud XI, Orsay (where he was my colleague from 1975 to 1979), at Université Paris IX-Dauphine, Paris, and at ENS-Cachan (Ecole Normale Supérieure de Cachan), Cachan, France.

[5] The title of his article mentioned the control of deformable structures in space, but only contained a result of control for the scalar wave equation, although a little idealistic, as the control was applied at a point inside the domain. I guess that Jacques-Louis LIONS had understood that the control of flexible structures in space is an important question, but because elasticity with large displacement is too difficult a subject, and even the linearized version of elasticity is a complicated hyperbolic system, he had started by considering questions related to a scalar wave equation, but he had probably forgotten to point out how far these questions really are from controlling large deformable structures.

After looking at his proof, I found that with very little change one could prove the following variant.

Lemma 42.4. *Let $d\mu$ be a Radon measure on R. The condition (42.2) is equivalent to the existence of a constant C such that*

$$\int_R |\mathcal{F}f(\xi)|^2 \, d\mu \le C \left(\int_R |f(x)|^2 \, dx \right)^{1/2} \left(\int_R (1+x^2)|f(x)|^2 \, dx \right)^{1/2}, \quad (42.3)$$

for all functions f such that $\int_R (1+x^2)|f(x)|^2 \, dx < \infty$. □

Using the characterization of the space $(E_0, E_1)_{1/2,1}$ of Jacques-Louis LIONS and Jaak PEETRE, it means that one can replace the right side of the inequality by the norm of f in a corresponding interpolation space; here E_1 is $L^2(R)$ for the Lebesgue measure dx while E_0 is L^2 for the measure $(1+x^2) \, dx$, and it is not difficult to characterize $(E_0, E_1)_{1/2,1}$ as

$$(E_0, E_1)_{1/2,1} = \left\{ f \in L^2(R) \mid \sum_{k \ge 1} \left(\int_{2^k \le |x| \le 2^{k+1}} 2^k |f(x)|^2 \, dx \right)^{1/2} < \infty \right\}.$$
$$(42.4)$$

To prove that the condition is necessary, Yves MEYER considers a function $\varphi \in L^2(R)$ whose Fourier transform does not vanish on $(0,1)$ (if φ_n converges to δ_0 then $\mathcal{F}\varphi_n$ converges to 1), and applies the inequality to f defined by $f(x) = e^{2i\pi k x}\varphi(x)$ so that for $\xi \in [k, k+1]$ one has $|\mathcal{F}f(\xi)| \ge \gamma = \min_{\eta \in (0,1)} |\mathcal{F}\varphi(\eta)|$. That the condition is sufficient is a consequence of the following lemma.

Lemma 42.5. *There exists a constant C'' such that*

$$\sum_{k \in Z} \sup_{\xi \in [k,k+1]} |\mathcal{F}f(\xi)|^2 \le C'' \left(\int_R |f(x)|^2 \, dx \right)^{1/2} \left(\int_R (1+x^2)|f(x)|^2 \, dx \right)^{1/2},$$
$$(42.5)$$

for all functions $f \in L^2(R)$ such that $\int_R (1+x^2)|f(x)|^2 \, dx < \infty$.

Proof: Let $a_k = \left(\int_{[k,k+1]} |\mathcal{F}f(\xi)|^2 \, d\xi \right)^{1/2}$ and $b_k = \left(\int_{[k,k+1]} \left| \frac{d\mathcal{F}f(\xi)}{d\xi} \right|^2 \, d\xi \right)^{1/2}$; the usual proof of continuity of functions of the Sobolev space $H^1((0,1))$ gives $\sup_{\xi \in [k,k+1]} |\mathcal{F}f(\xi)|^2 \le K^2 a_k (a_k^2 + b_k^2)^{1/2}$, and summing in k and using Cauchy–Schwarz inequality gives

$$\sum_{k \in Z} \sup_{\xi \in [k,k+1]} |\mathcal{F}f(\xi)|^2 \le K^2 \left(\int_R |\mathcal{F}f(\xi)|^2 \, d\xi \right)^{1/2}$$
$$\times \left(\int_R \left(|\mathcal{F}f(\xi)|^2 + \left| \frac{d\mathcal{F}f(\xi)}{d\xi} \right|^2 \right) d\xi \right)^{1/2}, \quad (42.6)$$

which is essentially the desired result. □

With the remark concerning interpolation, and exchanging the roles of f and $\mathcal{F}f$, this lemma says a little more about the fact that the functions from $\left(H^1(R), L^2(R)\right)_{1/2,1}$ are continuous and tend to 0 at ∞, as it gives some precise way how the functions tend to 0 at ∞, as it implies that

$$\sum_{n \in Z} \sup_{x \in [n,n+1]} |u(x)|^2 < \infty. \tag{42.7}$$

In the case of functions with support in $(0, T)$, my variant gives the same growth in $1+T$ found by Yves MEYER, who notices that the growth is optimal for large values of T, because if f is the characteristic function of $(0, T)$, then $\int_R |\mathcal{F}f(\xi)|^2 \, d\xi = T$. In the case of R^N one can easily generalize the proof and consider Radon measures $d\mu \geq 0$ for which there exists a constant C such that $\mu(Q) \leq C$ for every cube of size 1, and consider functions f such that $\sum_{k \geq 1} \left(\int_{2^k \leq |x| \leq 2^{k+1}} 2^{kN} |f(x)|^2\right)^{1/2} dx < \infty$; for functions with support in a bounded set K, one obtains a growth like $\left(1 + diameter(K)\right)^N$, but it is not clear to me if the diameter of K is the correct geometric quantity to use in such an inequality.

[Taught on Friday May 5, 2000.]

Biographical Information

[In a reference a–b, a is the lecture number, 0 referring to the Preface, and b the footnote number in that lecture.]

44

Abbreviations and Mathematical Notation

Abbreviations for states: For those not familiar with geography, I have mentioned a few states in the United States of America: AZ = Arizona, CA = California, CT = Connecticut, IL = Illinois, IN = Indiana, KS = Kansas, KY = Kentucky, MA = Massachusetts, MD = Maryland, MI = Michigan, MN = Minnesota, NC = North Carolina, NJ = New Jersey, NY = New York, OH = Ohio, PA = Pennsylvania, RI = Rhode Island, TX = Texas, WI = Wisconsin.

- a.e.: almost everywhere.
- $B(x,r)$: open ball centered at x and radius $r > 0$, i.e., $\{y \in E \mid ||x-y||_E < r\}$ (in a normed space E).
- BMO(R^N): space of functions of bounded mean oscillation on R^N, i.e., semi-norm $||u||_{BMO} < \infty$, with $||u||_{BMO} = \sup_{cubes\ Q} \frac{\int_Q |u-u_Q|\,dx}{|Q|} < \infty$ ($u_Q = \frac{\int_Q u\,dx}{|Q|}$, $|Q| = meas(Q)$).
- $BV(\Omega)$: space of functions of bounded variation in Ω, whose partial derivatives (in the sense of distributions) belong to $\mathcal{M}_b(\Omega)$, i.e., have finite total mass.
- $C(\Omega)$: space of scalar continuous functions in an open set $\Omega \subset R^N$ ($\mathcal{E}_0(\Omega)$ in the notation of L. SCHWARTZ).
- $C(\Omega; R^m)$: space of continuous functions from an open set $\Omega \subset R^N$ into R^m.
- $C(\overline{\Omega})$: space of scalar continuous and bounded functions on $\overline{\Omega}$, for an open set $\Omega \subset R^N$.
- $C_0(\Omega)$: space of scalar continuous bounded functions tending to 0 at the boundary of an open set $\Omega \subset R^N$, equipped with the sup norm.
- $C_c(\Omega)$: space of scalar continuous functions with compact support in an open set $\Omega \subset R^N$.
- $C_c^k(\Omega)$: space of scalar functions of class C^k with compact support in an open set $\Omega \subset R^N$.

- $C^k(\Omega)$: space of scalar continuous functions with continuous derivatives up to order k in an open set $\Omega \subset R^N$.
- $C^k(\overline{\Omega})$: restrictions to $\overline{\Omega}$ of functions in $C^k(R^N)$, for an open set $\Omega \subset R^N$.
- $C^{0,\alpha}(\Omega)$: space of scalar Hölder continuous functions of order $\alpha \in (0,1)$ (Lipschitz continuous functions if $\alpha = 1$), i.e., bounded functions for which there exist M such that $|u(\mathbf{x}) - u(\mathbf{y})| \leq M\,|\mathbf{x} - \mathbf{y}|^\alpha$ for all $\mathbf{x}, \mathbf{y} \in \Omega \subset R^N$; it is included in $C(\overline{\Omega})$.
- $C^{k,\alpha}(\Omega)$: space of functions of $C^k(\Omega)$ whose derivatives of order k belong to $C^{0,\alpha}(\Omega) \subset C(\overline{\Omega})$, for an open set $\Omega \subset R^N$.
- curl: rotational operator $(curl(u))_i = \sum_{jk} \varepsilon_{ijk} \frac{\partial u_j}{\partial x_k}$, used for open sets $\Omega \subset R^3$.
- D^α: $\frac{\partial^{\alpha_1}}{\partial x_1^{\alpha_1}} \ldots \frac{\partial^{\alpha_N}}{\partial x_N^{\alpha_N}}$ (for a multi-index $\boldsymbol{\alpha}$ with α_j nonnegative integers, $j = 1, \ldots, N$).
- $\mathcal{D}'(\Omega)$: space of distributions T in Ω, dual of $C_c^\infty(\Omega)$ ($\mathcal{D}(\Omega)$ in the notation of L. SCHWARTZ, equipped with its natural topology), i.e., for every compact $K \subset \Omega$ there exists $C(K)$ and an integer $m(K) \geq 0$ with $|\langle T, \varphi \rangle| \leq C(K) \sup_{|\alpha| \leq m(K)} ||D^\alpha \varphi||_\infty$ for all $\varphi \in C_c^\infty(\Omega)$ with support in K.
- div: divergence operator $div(u) = \sum_i \frac{\partial u_i}{\partial x_i}$.
- $\mathcal{F}$: Fourier transform, $\mathcal{F}f(\boldsymbol{\xi}) = \int_{R^N} f(\mathbf{x}) e^{-2i\pi(\mathbf{x}.\boldsymbol{\xi})} \, d\mathbf{x}$.
- $\overline{\mathcal{F}}$: inverse Fourier transform, $\overline{\mathcal{F}}f(\boldsymbol{\xi}) = \int_{R^N} f(\mathbf{x}) e^{+2i\pi(\mathbf{x}.\boldsymbol{\xi})} \, d\mathbf{x}$.
- $grad(u)$: gradient operator, $grad(u) = \left(\frac{\partial u}{\partial x_1}, \ldots \frac{\partial u}{\partial x_N} \right)$.
- $H^s(R^N)$: Sobolev space of tempered distributions ($\in \mathcal{S}'(R^N)$), or functions in $L^2(R^N)$ if $s \geq 0$, such that $(1 + |\boldsymbol{\xi}|^2)^{s/2} \mathcal{F}u \in L^2(R^N)$ ($L^2(R^N)$ for $s = 0$, $W^{s,2}(R^N)$ for s a positive integer).
- $H^s(\Omega)$: space of restrictions to Ω of functions from $H^s(R^N)$ (for $s \geq 0$), for an open set $\Omega \subset R^N$.
- $H_0^s(\Omega)$: for $s \geq 0$, closure of $C_c^\infty(\Omega)$ in $H^s(\Omega)$, for an open set $\Omega \subset R^N$.
- $H^{-s}(\Omega)$: for $s \geq 0$, dual of $H_0^s(\Omega)$, for an open set $\Omega \subset R^N$.
- $H(div; \Omega)$: space of functions $u \in L^2(\Omega; R^N)$ with $div(u) \in L^2(\Omega)$, for an open set $\Omega \subset R^N$.
- $H(curl; \Omega)$: space of functions $u \in L^2(\Omega; R^3)$ with $curl(u) \in L^2(\Omega; R^3)$, for an open set $\Omega \subset R^3$.
- $\mathcal{H}^1(R^N)$: Hardy space of functions $f \in L^1(R^N)$ with $R_j f \in L^1(R^N)$, $j = 1, \ldots, N$, where R_j, $j = 1, \ldots, N$ are the (M.) Riesz operators.
- $\mathcal{H}(\theta)$: class of Banach spaces satisfying $(E_0, E_1)_{\theta,1;J} \subset E \subset (E_0, E_1)_{\theta,\infty;K}$.
- $J(t; a)$: for $a \in E_0 \cap E_1$, $J(t, a) = \max\{||a||_{E_0}, t\,||a||_{E_1}\}$.
- $\mathcal{J}(\theta)$: class of Banach spaces satisfying $(E_0, E_1)_{\theta,1;J} \subset E \subset E_0 + E_1$.
- $ker(A)$: kernel of a linear operator $A \in \mathcal{L}(E, F)$, i.e., $\{e \in E \mid Ae = 0\}$.
- $K(t; a)$: for $a \in E_0 + E_1$, $K(t, a) = \inf_{a=a_0+a_1}(||a_0||_{E_0} + t\,||a_1||_{E_1})$.
- $\mathcal{K}(\theta)$: class of Banach spaces satisfying $E_0 \cap E_1 \subset E \subset (E_0, E_1)_{\theta,\infty;K}$.
- $\mathcal{L}(E; F)$: space of linear continuous operators M from the normed space E into the normed space F, i.e., with $||M||_{\mathcal{L}(E;F)} = \sup_{e \neq 0} \frac{||M\,e||_F}{||e||_E} < \infty$.

- $L^p(A)$, $L^\infty(A)$: Lebesgue space of (equivalence classes of a.e. equal) measurable functions u with $||u||_p = \left(\int_A |u(\mathbf{x})|^p \, d\mathbf{x}\right)^{1/p} < \infty$ if $1 \le p < \infty$, with $||u||_\infty = \inf\{M \mid |u(\mathbf{x})| \le M \text{ a.e. in } A\} < \infty$, for a Lebesgue measurable set $A \subset R^N$ (spaces also considered for the induced $(N-1)$-dimensional Hausdorff measure if $A = \partial\Omega$ for an open set $\Omega \subset R^N$ with a smooth boundary).
- $L^p_{loc}(A)$: (equivalence classes of) measurable functions whose restriction to every compact $K \subset A$ belongs to $L^p(K)$ (for $1 \le p \le \infty$), for a Lebesgue measurable set $A \subset R^N$.
- $L^p\big((0,T); E\big)$: (weakly or strongly) measurable functions u from $(0,T)$ into a separable Banach space E, such that $t \mapsto ||u(t)||_E$ belongs to $L^p(0,T)$ (for $1 \le p \le \infty$).
- $|\boldsymbol{\alpha}|$: length of a multi-index $\boldsymbol{\alpha} = (\alpha_1, \dots, \alpha_N)$, $|\boldsymbol{\alpha}| = |\alpha_1| + \dots + |\alpha_N|$.
- $Lip(\Omega)$: space of scalar Lipschitz continuous functions, also denoted by $C^{0,1}(\Omega)$, i.e., bounded functions for which there exists M such that $|u(\mathbf{x}) - u(\mathbf{y})| \le M|\mathbf{x} - \mathbf{y}|$ for all $\mathbf{x}, \mathbf{y} \in \Omega \subset R^N$; it is included in $C(\overline{\Omega})$.
- loc: for any space Z of functions in an open set $\Omega \subset R^N$, Z_{loc} is the space of functions u such that $\varphi u \in Z$ for all $\varphi \in C_c^\infty(\Omega)$.
- Mf: maximal function of f, i.e., $Mf(\mathbf{x}) = \sup_{r>0} \frac{\int_{B(\mathbf{x},r)} |f(\mathbf{y})| \, d\mathbf{y}}{|B(\mathbf{x},r)|}$.
- $\mathcal{M}(\Omega)$: space of Radon measures μ in an open set $\Omega \subset R^N$, dual of $C_c(\Omega)$ (equipped with its natural topology), i.e., for every compact $K \subset \Omega$ there exists $C(K)$ with $|\langle\mu,\varphi\rangle| \le C(K)||\varphi||_\infty$ for all $\varphi \in C_c(\Omega)$ with support in K.
- $\mathcal{M}_b(\Omega)$: space of Radon measures μ with finite total mass in an open set $\Omega \subset R^N$, dual of $C_0(\Omega)$, the space of continuous bounded functions tending to 0 at the boundary of Ω (equipped with the sup norm), i.e., there exists C with $|\langle\mu,\varphi\rangle| \le C||\varphi||_\infty$ for all $\varphi \in C_c(\Omega)$.
- meas(A): Lebesgue measure of A, sometimes written $|A|$.
- $|\cdot|$: norm in H, or sometimes the Lebesgue measure of a set.
- $||\cdot||$: norm in V.
- $||\cdot||_*$: dual norm in V'.
- p': conjugate exponent of $p \in [1,\infty]$, i.e., $\frac{1}{p} + \frac{1}{p'} = 1$.
- p^*: Sobolev exponent of $p \in [1,N)$, i.e., $\frac{1}{p^*} = \frac{1}{p} - \frac{1}{N}$ for $\Omega \subset R^N$ and $N \ge 2$.
- R_+: $(0,\infty)$.
- R_+^N: $\{x \in R^N \mid x_N > 0\}$.
- $R(A)$: range of a linear operator $A \in \mathcal{L}(E;F)$, i.e., $\{f \in F \mid f = Ae \text{ for some } e \in E\}$.
- R_j: Riesz operators, $j = 1, \dots, N$, defined by $\mathcal{F}(R_j u)(\boldsymbol{\xi}) = \frac{i\xi_j \mathcal{F}u(\boldsymbol{\xi})}{|\boldsymbol{\xi}|}$ on $L^2(R^N)$; natural extensions to R^N of the Hilbert transform, they map $L^p(R^N)$ into itself for $1 < p < \infty$, and $L^\infty(R^N)$ into $BMO(R^N)$.
- $\mathcal{S}(R^N)$: Schwartz space of functions $u \in C^\infty(R^N)$ with $x^\alpha D^\beta u$ bounded for all multi-indices $\boldsymbol{\alpha}, \boldsymbol{\beta}$ with α_j, β_j nonnegative integers for $j = 1, \dots, N$.
- $\mathcal{S}'(R^N)$: tempered distributions, dual of $\mathcal{S}(R^N)$, i.e., $T \in \mathcal{D}'(R^N)$ and there exists C and an integer $m \ge 0$ with $|\langle T, \psi\rangle| \le C \sup_{|\alpha|,|\beta| \le m} ||x^\alpha D^\beta \psi||_\infty$ for all $\psi \in \mathcal{S}(R^N)$.

- $\star$: convolution product $(f \star g)(\mathbf{x}) = \int_{\mathbf{y} \in R^N} f(\mathbf{x} - \mathbf{y}) g(\mathbf{y}) \, d\mathbf{y}$.
- $supp(\cdot)$: support; for a continuous function u from a topological space into a vector space, it is the closure of $\{x \mid u(x) \neq 0\}$, but for a locally integrable function f, a Radon measure μ, or a distribution T defined on an open set $\Omega \subset R^N$, it is the complement of the largest open set ω where f, μ, or T is 0, i.e., where $\int_\omega \varphi f \, dx = 0$, or $\langle \mu, \varphi \rangle = 0$ for all $\varphi \in C_c(\Omega)$, or $\langle T, \varphi \rangle = 0$ for all $\varphi \in C_c^\infty(\Omega)$.
- $W^{m,p}(\Omega)$: Sobolev space of functions in $L^p(\Omega)$ whose derivatives (in the sense of distributions) of length $\leq m$ belong to $L^p(\Omega)$, for an open set $\Omega \subset R^N$.
- $W^{m,p}(\Omega; R^m)$: Sobolev space of functions from Ω into R^m whose components belong to $W^{m,p}(\Omega)$, for an open set $\Omega \subset R^N$.
- x': in R^N, $x = (x', x_N)$, i.e., $x' = (x_1, \ldots, x_{N-1})$.
- x^α: $x_1^{\alpha_1} \ldots x_N^{\alpha_N}$ for a multi-index α with α_j nonnegative integers for $j = 1, \ldots, N$, for $\mathbf{x} \in R^N$.
- Δ: Laplacian $\sum_{j=1}^N \frac{\partial^2}{\partial x_j^2}$, defined on any open set $\Omega \subset R^N$.
- δ_{ij}: Kronecker symbol, equal to 1 if $i = j$ and equal to 0 if $i \neq j$ (for $i, j = 1, \ldots, N$).
- ε_{ijk}: for $i, j, k \in \{1, 2, 3\}$, completely antisymmetric tensor, equal to 0 if two indices are equal, and equal to the signature of the permutation $123 \mapsto ijk$ if indices are distinct (i.e., $\varepsilon_{123} = \varepsilon_{231} = \varepsilon_{312} = +1$ and $\varepsilon_{132} = \varepsilon_{321} = \varepsilon_{213} = -1$).
- γ_0: trace operator, defined for smooth functions by restriction to the boundary $\partial\Omega$, for an open set $\Omega \subset R^N$ with a smooth boundary, and extended by density to functional spaces in which smooth functions are dense.
- Λ_1: Zygmund space, $|u(\mathbf{x}+\mathbf{h}) + u(\mathbf{x}-\mathbf{h}) - 2u(\mathbf{x})| \leq M|\mathbf{h}|$ for all $\mathbf{x}, \mathbf{h} \in R^N$.
- ν: exterior normal to $\Omega \subset R^N$, open set with Lipschitz boundary.
- ϱ_ε: smoothing sequence, with $\varrho_\varepsilon(\mathbf{x}) = \frac{1}{\varepsilon^N} \varrho_1\left(\frac{\mathbf{x}}{\varepsilon}\right)$ with $\varepsilon > 0$ and $\varrho_1 \in C_c^\infty(R^N)$ with $\int_{\mathbf{x} \in R^N} \varrho_1(\mathbf{x}) \, d\mathbf{x} = 1$, and usually $\varrho_1 \geq 0$.
- $\tau_\mathbf{h}$: translation operator of $\mathbf{h} \in R^N$, acting on a function $f \in L_{loc}^1(R^N)$ by $\tau_\mathbf{h} f(\mathbf{x}) = f(\mathbf{x} - \mathbf{h})$ a.e. $\mathbf{x} \in R^N$.
- Ω_F: $\{x \in R^N \mid x_N \geq F(x')\}$, for a continuous function F, where $x' = (x_1, \ldots, x_{N-1})$.

References

[1] AGMON Shmuel, *Lectures on elliptic boundary value problems*, William Marsh Rice University, Houston, TX. Summer Institute for Advanced Graduate Students, 1963.

[2] BERGH J. & LÖFSTRÖM J., *Interpolation spaces. An introduction*, x+207 pp., Grundlehren der Mathematischen Wissenschaften, No. 223. Springer-Verlag, Berlin–New York, 1976.

[3] BIRKHOFF Garrett, *A source book in classical analysis*, xii+470 pp., Harvard University Press, Cambridge, MA, 1973.

[4] DAUTRAY Robert & LIONS Jacques-Louis, *Mathematical analysis and numerical methods for science and technology*, Vol. 1. *Physical origins and classical methods*, xviii+695 pp., Springer-Verlag, Berlin–New York, 1990.

[5] DAUTRAY Robert & LIONS Jacques-Louis, *Mathematical analysis and numerical methods for science and technology*, Vol. 2. *Functional and variational methods*, xvi+561 pp., Springer-Verlag, Berlin–New York, 1988.

[6] DAUTRAY Robert & LIONS Jacques-Louis, *Mathematical analysis and numerical methods for science and technology*, Vol. 3. *Spectral theory and applications*, x+515 pp., Springer-Verlag, Berlin, 1990.

[7] DAUTRAY Robert & LIONS Jacques-Louis, *Mathematical analysis and numerical methods for science and technology*, Vol. 4. *Integral equations and numerical methods*, x+465 pp., Springer-Verlag, Berlin, 1990.

[8] DAUTRAY Robert & LIONS Jacques-Louis, *Mathematical analysis and numerical methods for science and technology*, Vol. 5. *Evolution problems. I*, xiv+709 pp., Springer-Verlag, Berlin, 1992.

[9] DAUTRAY Robert & LIONS Jacques-Louis, *Mathematical analysis and numerical methods for science and technology*, Vol. 6. *Evolution problems. II*, xii+485 pp., Springer-Verlag, Berlin, 1993.

[10] DAUTRAY Robert & LIONS Jacques-Louis, *Mathematical analysis and numerical methods for science and technology*, Vol. 7. *Évolution: Fourier, Laplace*, xliv+344+xix pp., INSTN: Collection Enseignement. Masson, Paris, 1988 (reprint of the 1985 edition).

[11] DAUTRAY Robert & LIONS Jacques-Louis, *Mathematical analysis and numerical methods for science and technology*, Vol. 8. *Évolution: semigroupe, variationnel*, xliv+345–854+xix pp., INSTN: Collection Enseignement. Masson, Paris, 1988 (reprint of the 1985 edition).

[12]   DAUTRAY Robert & LIONS Jacques-Louis, *Mathematical analysis and numerical methods for science and technology*, Vol. 9. *Évolution: numérique, transport*, xliv+855–1303 pp., INSTN: Collection Enseignement. Masson, Paris, 1988 (reprint of the 1985 edition).

[13]   LIONS Jacques-Louis, *Problèmes aux limites dans les équations aux dérivées partielles*, 176 pp., Séminaire de Mathématiques Supérieures, No. 1 (Été, 1962) Les Presses de l'Université de Montréal, Montreal, Que., 1965 (deuxième édition).

[14]   LIONS Jacques-Louis & MAGENES Enrico, *Problèmes aux limites non homogènes et applications*, xx+372 pp., Vol. 1. Travaux et Recherches Mathématiques, No. 17 Dunod, Paris, 1968.

[15]   SCHWARTZ Laurent, *Théorie des distributions*, xiii+420 pp., Publications de l'Institut de Mathématique de l'Université de Strasbourg, No. IX-X. Nouvelle édition, entièrement corrigée, refondue et augmentée. Hermann, Paris, 1966.

[16]   SCHWARTZ Laurent, *Un mathématicien aux prises avec le siècle*, 528 pp., Éditions Odile Jacob, Paris, 1997. *A mathematician grappling with his century*, viii+490 pp., Birkhäuser Verlag, Basel, 2001.

[17]   SOBOLEV Sergei & VASKEVICH V. L., *The theory of cubature formulas*, xxi + 416 pp., Math. & Its Applic., V. 415. Kluwer Academic, 1997.

[18]   TARTAR Luc, *An introduction to Navier–Stokes equation and oceanography*, 271 pp., Lecture Notes of Unione Matematica Italiana, Springer, Berlin–Heidelberg–New York, 2006.

[19]   YOUNG, L. C., *Lectures on the calculus of variation and optimal control theory*, W. B. Saunders, Philadelphia, 1969.

Index

Editor in Chief: Franco Brezzi

Editorial Policy

1. The UMI Lecture Notes aim to report new developments in all areas of mathematics and their applications - quickly, informally and at a high level. Mathematical texts analysing new developments in modelling and numerical simulation are also welcome.

2. Manuscripts should be submitted (preferably in duplicate) to
Redazione Lecture Notes U.M.I.
Dipartimento di Matematica
Piazza Porta S. Donato 5
I – 40126 Bologna
and possibly to one of the editors of the Board informing, in this case, the Redazione about the submission. In general, manuscripts will be sent out to external referees for evaluation. If a decision cannot yet be reached on the basis of the first 2 reports, further referees may be contacted. The author will be informed of this. A final decision to publish can be made only on the basis of the complete manuscript, however a refereeing process leading to a preliminary decision can be based on a pre-final or incomplete manuscript. The strict minimum amount of material that will be considered should include a detailed outline describing the planned contents of each chapter, a bibliography and several sample chapters.

3. Manuscripts should in general be submitted in English. Final manuscripts should contain at least 100 pages of mathematical text and should always include
   - a table of contents;
   - an informative introduction, with adequate motivation and perhaps some historical remarks: it should be accessible to a reader not intimately familiar with the topic treated;
   - a subject index: as a rule this is genuinely helpful for the reader.

4. For evaluation purposes, manuscripts may be submitted in print or electronic form (print form is still preferred by most referees), in the latter case preferably as pdf- or zipped ps- files. Authors are asked, if their manuscript is accepted for publication, to use the LaTeX2e style files available from Springer's web-server at
ftp://ftp.springer.de/pub/tex/latex/mathegl/mono.zip

5. Authors receive a total of 50 free copies of their volume, but no royalties. They are entitled to a discount of 33.3% on the price of Springer books purchased for their personal use, if ordering directly from Springer.

6. Commitment to publish is made by letter of intent rather than by signing a formal contract. Springer-Verlag secures the copyright for each volume. Authors are free to reuse material contained in their LNM volumes in later publications. A brief written (or e-mail) request for formal permission is sufficient.

图书在版编目（CIP）数据

索伯列夫空间和插值空间导论 = An Introduction to Sobolev Spaces and Interpolation : 英文 /（美）塔塔（Tartar, L.）著 . — 影印本 . — 北京：世界图书出版公司北京公司 , 2012.8（2023 年 4 月重印）
ISBN 978-7-5100-5043-5

Ⅰ . ①索… Ⅱ . ①塔… Ⅲ . ①索伯列夫空间—英文②插值—英文 Ⅳ . ① O177.3 ② O174.42

中国版本图书馆 CIP 数据核字（2012）第 186189 号

中文书名	索伯列夫空间和插值空间导论
英文书名	An Introduction to Sobolev Spaces and Interpolation Spaces
著　　者	Luc Tartar
责任编辑	高　蓉　李　黎
出版发行	世界图书出版有限公司北京分公司
地　　址	北京市东城区朝内大街 137 号
邮　　编	100010
电　　话	010-64038355（发行）　　64033507（总编室）
网　　址	http://www.wpcbj.com.cn
邮　　箱	wpcbjst@vip.163.com
销　　售	新华书店
印　　刷	北京建宏印刷有限公司
开　　本	711mm × 1245mm　1/24
印　　张	10.25
字　　数	172 千字
版　　次	2012 年 8 月第 1 版
印　　次	2023 年 4 月第 4 次印刷
版权登记	01-2020-0333
国际书号	ISBN 978-7-5100-5043-5
定　　价	35.00 元